W9-ATP-650

Basic Ideas of Statistics

Basic Ideas of Statistics

Bernard W. Lindgren
University of Minnesota

Macmillan Publishing Co., Inc.
NEW YORK
Collier Macmillan Publishers
LONDON

Macmillan Publishing Co., Inc.
866 Third Avenue, New York, New York 10022

Collier-Macmillan, Canada, Ltd.

Library of Congress Cataloging in Publication Data

Lindgren, Bernard William (date)
Basic ideas of statistics.

Bibliography: p.
1. Mathematical statistics. 2. Statistics.
I. Title.
QA276.L536 519.5 74-6036
ISBN 0-02-370750-X

Printing: 1 2 3 4 5 6 7 8 Year: 5 6 7 8 9 0

Preface

This is a textbook intended for a freshman/sophomore course of about 40 to 45 hours. Only high school algebra is assumed as a prerequisite. I have tried to make the material as easy as possible to digest for those who are not mathematically inclined.

My intention is to show the student some of the things statistics can and cannot do, emphasizing ideas and concepts rather than methods. I do not believe a student with such a brief course in statistics should be led to think he is prepared to tackle significant statistical problems on his own. The most that can be expected is that he will acquire sophistication sufficient to challenge the statistical inferences he encounters on TV and through other media, to avoid being completely overwhelmed by reports whose conclusions are phrased in statistical terms, and to recognize the need for a competent statistician when he meets a statistical problem himself.

In regard to coverage of statistical topics, this text is rather traditional. Some nonparametric methods are included, but I have neglected the Bayesian approach. I do not think one can teach *all* statistical ideas in a brief course such as this; and although the Bayesian concept is not difficult to comprehend, it is nontrivial to implement in any problems that have a semblance of realism. Nor have I attempted a breakthrough along the lines of what is called "data analysis" (as opposed to classical testing). Its proponents are themselves steeped in classical experience and training, and have not yet shown us how to reduce their insights to the level needed for elementary instruction. One very eminent biostatistician told me that he had never, in his 40-odd years of experience, tested an hypothesis; but some people do, or say that they do, and it still seems as though an introduction to the idea of testing is in order.

I have departed from tradition in some minor points: I have used n rather than $n - 1$, as a divisor in defining a sample standard deviation, since I have never seen a valid reason to use the $n - 1$, and since n is surely more natural. I have not specifically or directly introduced the mathematician's delight, the binomial distribution, for it has been my experience that the students' biggest headache in a beginning course is deciding whether to use npq or pq/n. (If one uses the sample mean consistently, rather than the sample sum, there is no problem.) And I have not included the usual exercise in combinatorics, believing that the topic of "permutations and combinations," a traditional bugaboo for nonmathematicians, is not essential to the elementary notions of sampling and inference.

In preparing the text, I have used a computer (actually, a programmable desk calculator) in Monte Carlo studies. The idea of a sampling distribution seems to be hard to grasp, and it is not enough just to talk about many possible samples. Seeing many samples and seeing the variability of statistics from sample to sample are much better. Ideally, students should have access to a desk calculator for calculating statistics, and it would be an added bonus to have a computer terminal for setting up simulations of sampling.

Some problems and examples in the text use real data, and some, data that are invented but that might have resulted from real problems. It should be mentioned, on the other hand, that *real* problems are seldom as simple and straightforward as textbook examples need to be. (Indeed, the real problem, not treated at all here, is the defining of the appropriate problem.) If a reader has a set of real data that he thinks would be better than one of my artificial sets, and if he thinks the text is worth improving, I should be pleased to receive such sets for use in a second edition. I shall also welcome any criticisms that would be helpful in making a revision.

I am grateful to Professor E. S. Pearson for permission to use certain tables from *Biometrika Tables for Statisticians*, and to the Rand Corporation, for permission to use extracts from its book of random numbers. I acknowledge receiving help from all of my colleagues in statistics at the University of Minnesota, although they may not always have realized that they were giving it for this purpose. The expert typing of the manuscript by Gerald DuChaine, Linda Gardiner, Kay Halvorsen, and Claudia Tysdal is appreciated.

B.W.L.

Contents

Basic Ideas of Statistics

1

Data and Variability

What is statistics? What are statistics? Both questions are proper, since the word *statistics* is used in both the singular and the plural. In the singular sense statistics is the subject that deals with the collection, analysis, and interpretation of data. In the plural, statistics are the raw data themselves, the numerical facts that are the results of (or that describe the results of) measurement, or experimentation, or activity. The original meaning, implied in the root, "state," of the word "statistics," suggests that these data or facts have to do with a political or geographical state; but the meaning has become so broadened that the data can be related to the state of affairs in any field or endeavor.

Why are data gathered? The precise motivation is peculiar to each particular situation, but it may be said generally that data are gathered to provide knowledge or understanding of some aspect of the world around us, either as a contribution to the sum of man's knowledge, or as a means for making one's actions in specific instances more beneficial or profitable.

1.1 Statistics Do Not "Show"

Drawing conclusions from data is usually a personal matter. Sometimes, to be sure, different people will draw the same conclusion from a given set of data; but more often than not, different people draw somewhat different conclusions. Yet the language usually used in reporting investigations makes the process of inference sound objective: "Statistics show that . . . ." Let the reader beware of any conclusion that begins in this manner.

A syndicated newspaper column included the following item:

> Q. My wife and I are thinking of taking up bicycling as a hobby but, in view of the number of cars on the road, we want to know how risky it is?—M. L., Houston.
>
> A. Statistics show you're safer on two wheels than on four—some 800 bicyclists are killed in accidents yearly, while over 56,000 motorists die during the same time.

In fact, the statistics actually given in this answer do not in themselves show a thing about relative safety. Consider the following points.

1. How many cyclists were there? If there were only 800, then cycling is indeed a risky pastime. Surely the populations of cyclists and of motorists would be relevant; the size of each would have to be known to make sense out of the statistics given.
2. Even more significant than the numbers of people involved is the length of exposure to the environment of risk. Accident statistics are usually more properly quoted in terms of "passenger miles" or "motorist miles" or "bicyclist miles."
3. What is the age distribution among the cyclists? It may be that all the fatalities were cases of young children whose inexperience and resulting poor judgment were the main factors. If so, the man and his wife might be quite safe.

The reader may think of other reasons for dismay at the answer that was given. The reasons stated should make it clear that no conclusion can be drawn from the data given. Certainly, the statistics that were cited did not "show."

The popular advice column "Dear Abby" is not above misinterpreting statistics, albeit unintentionally and with the best of motives, the motive of correcting incorrect and misleading statistics. A column headed (in one paper) "Figures Show Sex Education Is Easing Sweden's Problems" was triggered by a deluge of letters from readers opposed to sex education in the schools, who asserted that since Sweden had instituted such education, it has become a "nation of degenerates—leading the world in suicides, alcoholism, divorce, and venereal disease." That this had happened, if indeed it has, *because* of the introduction of sex education is of course a wild claim, with no foundation in logic. Rather than attack the logic, Abby wanted the correct statistics, and so she consulted the World Health Organization in Geneva.

In reply to the assertion that Sweden leads the world in alcoholism, Abby says that "the people of France and the people of the United States both consume more alcohol per capita than do the people of Sweden." But alcoholism and the consumption of alcohol are not the same thing. For example, the people of France, who consume quantities of alcohol in wine with meals, may even have less alcoholism than Sweden; but no statistics about alcoholism are given by Abby, so no conclusion can be drawn.

To counter the claim about Sweden's divorce rate, Abby asserts that the true rates are one of every six marriages in Sweden, and one of every three in the United States. If her statistics are accurate, this is not incompatible with the theory of Swedish degeneracy; for in a country where families occur without the step of a legal marriage, it may be that only those unions are legalized that are more certain to be lasting. But it could also be noted that the statistic giving the proportion of marriages ending in divorce would actually be almost impossible to determine. Marriages begun in the last five years, say, have not yet run their courses; undoubtedly some have already ended in divorce, and just as surely some will end in divorce in the future. However, the latter number is unknown and, if changing social values and systems involve a changing divorce rate, quite risky to estimate.

Replying to a charge that the incidence of rape increased by 55 per cent in two years, Abby says:

> For 1965 and 1966 the "arrest" figures per 100,000 population were respectively 87 and 78, showing a decrease of 10 per cent—not an increase of 55 per cent. In a population of 7,847,395, this works out to approximately 1.1 per 100,000 in 1965 and 1.0 per 100,000 in 1966, one of the lowest rates in the world. By comparison, the United States had 10,734 rape arrests in 1965 (or 5.36 per 100,000 population) and increased 7 per cent to 11,609 in 1966—or 5.8 per 100,000 population.

Here two figures are given for the same thing: 87 per 100,000 and 1.1 per 100,000, the latter achieved by dividing 87 by 78.47395. Which is correct? One figure is above and one is below the U.S. figure of 5.36 per 100,000. Apart from this obvious error that crept into the writing or printing of the column, notice that the figures quoted are for arrests. These may or may not be related to incidence figures, for it may be that with relaxed attitudes, or under different statutes, not many who commit the act are arrested for it. And if it is the amount of the *increase* that is significant—in which case it is the 10 per cent decrease compared with the 7 per cent increase that is relevant, and not the 87 or the 1.1—the inference that Sweden is really improving is also invalid. For, perhaps as a nation "degenerates," the act in question becomes more tolerated, not likely to be reported nor the perpetrator to be pursued and arrested. Moreover, in a more degenerate climate, the physiological or psychological tensions that lead to the act may be present in much lesser amount. Thus it is possible that increasing degeneracy and a declining rape incidence might be compatible. So, if it is a matter of establishing the per cent increase of rape, the respective sources would have to be compared, and the precise definitions of what was being counted would have to be made clear (i.e., incidence, arrest, or conviction). But if it is a matter of establishing degeneracy, there is much more to be considered than the statistics presented.

Abby's own summary is that she has presented the facts in fairness to a much maligned nation. The column's headline, no doubt written locally, asserts that her facts show that sex education is easing Sweden's problems. When even such an honest attempt as this to present the truth encounters such problems without recognizing them, it is not too surprising that the phrase "lies, damn lies, and statistics" has become so overworked.

Another example of what statistics do not show is the following quotation from an AAA publication (*Minnesota Motorist*):

> The Minnesota Highway Department recently reported on the results of a coroners' testing program started July 1, 1967. Unfortunately, it shows that 54 per cent of the drivers killed in automobile accidents in 1968 had measurable traces of alcohol in their blood. Even worse . . . 43 per cent of these fatalities showed better than the .10 per cent presumptive level of intoxication. Perhaps some day people will learn that drinking affects their driving capability, and park their car rather than run the risk of killing themselves, or some innocent bystander.

The conclusion is clearly drawn that drinking causes highway deaths. This may be true, but the statistics given do not establish it. Quite apart from the fact that statistics can only suggest, and never establish any *causal* relationships, there is a gross omission in what is given. Suppose, for instance (and one must suppose, since the information is not given), that 80 per cent of *all* persons have measureable traces of alcohol in their blood. One might then conclude that it is safer to drink! In any case, the facts given do not prove that drinking caused any of the accidents. It is probably the case that, say, 60 per cent of the drivers killed carried collision insurance; would this mean that is is safer not to carry such insurance? And perhaps 80 per cent of those killed had just eaten within three hours before the accident; is it dangerous to drive after one eats?

It is often said that one can prove anything using statistics. In fact, one can prove *nothing* using statistics. Statistics can suggest and point, and guide, but only when intelligently used. Numerous examples of the misuse of statistics are to be found in the books by Huff and by Wallis and Roberts, listed at the end of the chapter.

1.2 Kinds of Data

Data may be classified as *numerical* or *categorical*, and somewhat different models and inferential techniques are employed in the two cases. *Numerical data*, as the term implies, are just numbers—results that are either inherently ordinary numbers or are numbers as the result of some scaling or coding. Measurements are usually expressed as numbers. The age of a person drawn from a population is a number. The number of customers entering a certain store on a given day is a number. The yield of corn from a certain plot can be given as a number of bushels. A patient's blood pressure is expressed as a number. And so on.

A further classification of numerical data into *discrete* and *continuous* types is useful. Data are discrete if, from the outset, it is known that only numbers of a certain finite* list of numbers will ever be encountered. Thus the number of children in a given family unit must be an integer, a number from the list 0, 1, 2, 3, . . . , 100. (If you do not want to limit the list to 100, use 200 or some other finite number. Or, as suggested in the footnote, do not put a limit on the integer that might be encountered.)

EXAMPLE 1.1

The students in a class were asked to state how many brothers and sisters each had. The results were as follows:

3, 4, 4, 2, 4, 4, 2, 2, 2, 2, 1, 1, 5, 2, 3, 4, 3, 5, 1, 4, 2, 7, 3, 4.

These numbers are integers, of course, and each one is included in the finite list 0, 1, 2, . . . , 100.

The age of a person is not restricted to a finite list of numbers; it can be any nonnegative number, rational or irrational, from 0 up to, say, 200. Such a

* To be more idealistic, one would include in this category a list that is *countably infinite*, i.e., infinite but not so infinite that the items cannot be listed in some order (i.e., "counted"). In practice there are always limitations that make it unnecessary to consider anything but finite lists, but it is frequently easier to construct a mathematical model if infinite lists are permitted.

variable is assumed to be *continuous*, since age is measured in time, and time is thought of as varying continuously, not jumping (like an old-fashioned schoolroom clock) from one value to another. Similarly, such measured quantities as weight, height, density, volume, distance, level of noise, etc., are conceived of as continuous—not restricted to taking on only isolated values.

Whereas the idea of discreteness is that there are spaces between the possible values, in the case of continuous data there are no gaps in the set of values that might be encountered. It is important to notice, however, that even in the case of a variable such as time, or some space dimension, which is ideally continuous, observed values are necessarily discrete and must be recorded as though discrete. We have only discrete systems for writing down numbers and for reading scales on measuring instruments. For example, in the decimal system, if one limits himself (as he must) to a fixed number of decimal places, the possible values can be listed discretely. It is common and necessary practice to read a measuring instrument to the nearest hundredth, say, and to write down the corresponding number as a terminating decimal.

EXAMPLE 1.2

In a study of 50 lung cancer patients, hemoglobin levels were measured and recorded as follows, in numbers of grams per milliliter:

13.5	15.6	16.3	12.3	13.1	14.2	12.4	11.3	14.0	14.6
13.6	14.8	12.7	10.9	11.0	11.4	15.0	10.1	15.4	11.3
10.7	14.6	13.5	15.1	12.1	12.0	14.2	11.4	15.0	13.3
13.2	9.1	16.9	14.2	15.0	13.6	14.8	11.4	14.8	15.7
13.5	13.5	12.9	13.8	13.8	13.7	16.3	11.6	14.2	10.7

The recorded values have been rounded off to the nearest tenth of a gram, even though, conceptually, any number of grams (in some range) is possible.

Categorical data are those resulting when each experimental unit is classified according to some scheme—put into one of several possible categories (usually a finite list). For each person in a population of people, one could observe and record sex (male or female), eye color (brown, blue, hazel), blood type, political preference, nationality, etc. In each instance there is a given set of categories into one of which each person may be classed. Similarly, each automobile may be classed according to its manufacturer or according to its body style. In the toss of an ordinary die, the outcome is taken to be the face that turns up, which is one of six categories.

Sometimes the categories in a given classification, unlike those mentioned so far, are *ordered*. For instance, a person may be classed as poorly dressed, moderately well dressed, or very well dressed. Themes written in a freshman composition class are placed by the instructor in one of the classes A, B, C, D, and F. In these and similar instances it is usually the case that one might really prefer to have a numerical rating but for practical purposes is content with crude categories. Other examples are states of the sea, the force of an earthquake, and a certain magazine columnist's classification of movies on an "embarrassment" scale; in each of these, the categories are numbered from 0 to 10. The dangers in such scaling are evident. For example, in the familiar assignment of numbers to grades (A = 4, B = 3, etc.), one is tempted to conclude that an A is twice as good as a C, or that the difference between A and B is exactly the same as between C and D.

Observe that numerical data that are discrete (with a finite set of possible values) can be thought of as categorical. Even continuous data become categorical when rounded off. However, when data are numerical, on meaningful scales, methods are available that are more powerful than those devised for the case of data that are truly categorical, with no order among the categories.

Sometimes data are artificially coded so as to be numerical when the experimental results are really categorical. An ordinary die, for example, is supplied with a coding; the number of dots on the upturned face is the result of quite an arbitrary scheme of identification, turning categories into numbers.

Data may be classified further according to the number of characteristics of interest for each experimental unit. In studying a relationship such as that between height and weight, for example, one would measure these two quantities for each person drawn in a sample from the population. These numbers, one for height and one for weight, would be kept in pairs, so that each "observation" or piece of data is a pair of numbers. The data are said to be *bivariate*, as opposed to data in which the observations are single numbers for each experimental unit, data termed *univariate*. Multivariate data would consist of observations each of which is a set of more than one number—bivariate if two, trivariate if three, and so on.

EXAMPLE 1.3

Information on sex, height, weight, age, hair color, and eye color was obtained from each of a class of 30 students, the results shown in Table 1. For student 1, the observation is

(male, 73, 175, 228, brown, green).

This is multivariate (if one permits the term "variate" for purely categorical data). The first component is categorical, with two

Table 1.1

SEX	HEIGHT	WEIGHT	AGE	HAIR	EYES
M	73	175	228	Brn	Grn
F	64	110	226	Brn	Hzl
F	66	128	246	Brn	Bl
M	72	165	288	Brn	Brn
F	64	135	240	Brn	Bl
F	63	115	232	Bld	Bl
F	70	150	256	Brn	Bl
F	63	100	265	Bld	Bl
M	70	180	245	Brn	Brn
M	72	160	281	Brn	Bl
M	73	145	216	Bld	Bl
M	70	165	262	Brn	Bl
M	69	137	281	Brn	Grn
M	68	145	290	Brn	Bl
M	71	145	230	Bld	Bl
M	73	180	378	Brn	Brn
M	68	135	232	Brn	Bl
F	62	95	254	Blk	Brn
M	65	140	224	Blk	Brn
M	71	175	240	Brn	Grn
F	63	145	218	Blk	Brn
M	71	200	276	Brn	Bl
F	66	113	219	Brn	Grn
F	66	115	224	Bld	Bl
M	71	139	258	Bld	Bl
M	70	125	221	Blk	Brn
M	70	170	236	Blk	Brn
F	64	134	249	Brn	Hzl
F	68	134	443	Brn	Hzl
M	71	140	217	Brn	Bl

categories. The next three components are numerical and continuous, although in recording them the results have been rounded—height to the nearest inch, weight to the nearest pound, and age to the nearest number of months. The fifth and sixth components are categorical, with four categories for hair color and four for eye color. (Incidentally, the number of categories is often set for the convenience of the data collector; the category "brown," for instance is a very broad catch-all category for all hair color that is not black, not blond, and not red. More realistically, hair color is much more continuous; but for some purposes the given scheme of classification is adequate.)

1.3 Sources of Variability

The word *datum* is seldom heard; it is always *data*. Why does one almost always encounter *many* observed values or numerical facts rather than just one? The answer to this involves the ubiquitous phenomenon of *variableness*. Results vary; and a single observation ordinarily provides only a very incomplete picture of the state under study. Variability seems to be a fact of life; and since we must live with it, we must study patterns of variation to know how best to live with it. Variability implies uncertainty, and the subject of statistics is sometimes said to deal with the drawing of conclusions and the making of decisions in the face of uncertainty.

One source of variability in data is what in some contexts would be called *product variation*, and in others, *response variation*. In the fabrication or manufacture of an article, variable factors, which cannot always be controlled, predicted, or even identified in some cases, give rise to variations in product characteristics. The lengths of bolts produced by a certain process will exhibit variations, perhaps slight, from bolt to bolt, even though ostensibly identical procedures are carried out in each case. In packaging goods for marketing, the machine that fills the box or cuts the material to a given specification will not be able to do so exactly the same twice in a row, partly because of variable factors in the machine and partly because of variation in the composition of the material to be packaged. Similarly, agricultural products, such as apples, corn, and milk, exhibit variations in their size or amount and quality, even when produced under supposedly identical conditions.

When an object is treated, and perhaps allowed to develop in some way, a subsequently observed characteristic or attribute is thought of as a *response*. Animal or vegetable growth, over a given period, is a response to the various environmental factors that exist, as well as to treatments such as those involving nutrients, artificial light, preventative inoculation or spraying, etc. Such growth will exhibit variableness from subject to subject. A student exposed to a certain course of instruction by a given method experiences an educational growth that is a response to the way he was treated, and this response varies from student to student, even under the same treatment process. And in the medical field, the response to a given medical treatment is far from consistent, variations from patient to patient occurring even if the treatment is the same. These variations in response exist partly because of the variations in the subjects that are treated, in the seed, or the animal, or the student, or the patient; partly because the treatment itself may vary, even when it is not supposed to; and perhaps for other, unaccountable reasons. For example, the response of a patient to a given amount of a drug will depend on that patient's individual makeup, on the actual potency of the drug (which may be variable), perhaps on the time of day, and even, conceivably, on the attitude of the nurse who administers it. It should be clear that there is not a clear distinction between product variation and response variation. The former may, indeed, be considered to be a special case of the latter.

Variability is encountered in *survey data*. One is often interested in a specific group or *population* of people (or of other creatures or things, such as the population of pheasants in North Dakota, or the population of television sets in Memphis). More precisely, the interest usually lies in one or more characteristics, numerical or otherwise, of the individuals in the given population.

Populations of people (or of birds or television sets) are composed of individuals that are different with respect to the characteristic of interest, and it is the patterns of variation in this characteristic across the population that are the object of investigation. In the case of the U.S. census, the constitutional mandate is to make a complete survey, primarily to determine the numbers of people according to geographical location, but secondarily to determine other of their characteristics, such as age and sex. It may be, however, that a complete survey of a population is too costly, or impossible, and the results for the whole population may be adequately approximated using results for only *part* of the population. Moreover, if one attempts to survey a whole population, he may find that various errors in such a huge task make the complete survey as inaccurate as the information contained in a subset, or *sample*, from the population. What is usually done in survey work, therefore, is to draw a sample from the population and determine the characteristic of interest for each member of the sample. These data will exhibit variability, in the sense that one could not predict the values that will turn up in a particular sample, even if he knows the pattern of variation in the population.

It might be supposed that the product variation of the lengths of bolts in a collection of 50 bolts taken from a production line can be thought of as a sample taken from a *population* of bolts, but the population is not something that is so obvious, as in the case of actual groups of people. The population would be, if anything, the collection of all bolts that could be manufactured if the process continued forever—an infinite population. And the yield of corn per stalk, for a collection of 50 stalks, could be thought of as a sample from a certain cornfield; but here, too, it could be imagined that the population consists of all conceivable cornstalks that could have been planted and grown (the same variety, under the same conditions). Similarly, the IQ's of senior students in a certain high school might be thought of as comprising a sample from the population of high school students in a certain city in a given year; or they might represent a sample taken from a population of all conceivable high school students. Surely, only the context of the problem, and sometimes not even that, would suggest which interpretation is correct. Indeed, the proper population of reference is often a controversial matter, although always pertinent. The term *sample survey* will refer to the case of an actual, finite population.

EXAMPLE 1.4

On May 18, 197–, a certain newspaper poll reported that on the basis of interviewing done May 5–9, Mayor S was leading challenger D in voter preference by 73 per cent to 23 per cent (4 per cent undecided). On June 1, on the basis of interviewing done May 19–24, Mayor S was leading by 71 per cent to 25 per cent (4 per cent undecided). In the election in June, Mayor S got 72.4 per cent of the votes. In the first polling, 600 voters were interviewed, and in the second the sample numbered 997.

Each voter was assumed to have the characteristic S (would vote for Mayor S) or the characteristic D (would vote for challenger D). The interviewers asked each voter in the sample whether he was an S or a D. Some of the problems in survey sampling are evident: The voter would not always know whether he was an S or a D, or at any rate was not willing to commit himself on it. And between the time of the interviewing and the time of publication of the poll people do change their minds (and even more so, between the interview and the election). Nevertheless, the sample proportion of decided voters, 74 per cent, is remarkably close to the final percentage who voted for Mayor S—even with a sample of only 997 voters.

The observing of phenomena in day-to-day operations gives rise to data that exhibit variability. The observations are unpredictable and involve what is thought of as random or chance variation. Such operational data usually describe happenings determined by a multitude of factors operating without any coordination, perhaps along with factors that constitute a more deterministic component and that appear in data as hints of underlying structure.

It is typical of operational data that they are used to provide a basis for making decisions and for helping to make operations more efficient and more profitable—for getting a job done optimally, whether the job be operating the production line for a refrigerator, fighting a war, transporting people, educating college students, marketing a new detergent, or planning reforestation. In planning such operations and in decisions as they proceed, data are needed that relate to and carry information about, respectively, the location and status of fabrication of components for the refrigerator; the disposition of enemy forces and materiel; the numbers of people needing transportation and the characteristics of available methods; the numbers of new births, from which one can infer future patterns of education and of dropouts or shifting objectives; consumer demand for and manufacturing and distributing capabilities for the detergent; and growth characteristics of trees and needs for lumber.

Further instances of data arising in operational situations include sales figures, the arrival data and length of the queues at a service station, consumption data for water or power, failure records of equipment, particle counts in pollution studies, the number of raisins per cookie in raisin cookies, and the records of winnings and losses in gambling games. In these and in the earlier examples, variability is the "name of the game." One does not expect to control or eliminate it but to understand and live with it as intelligently as possible. Thus rainfall data show variability, and the planning of water storage must take this into account.

It should perhaps be mentioned that there is a complicating element in most operational situations—the element of time. Data frequently are gathered from different points in time, a fact that makes the mathematical description of what is happening more difficult but does not make the situation any less random. A record of sales figures, for instance, will involve the unpredictable type of random variation as well as some kind of more systematic dependence on time. The best analyses of such situations would require a mathematical structure that incorporates a time variable, but such a degree of sophistication is beyond this elementary treatment.

The lines between operational and other types of variability are fuzzy. Product variations, for example, are a fact of life in industrial operations. The distinction between product variations and sales variations may not be worth dwelling on, but it is simply that in making products one aims at certain standards, variations about which are an undesirable, if to some extent tolerable, annoyance; whereas sales figures are determined by the actions of many independent buyers, the resulting variations being necessarily tolerated and not readily controlled. In the case of a gambling game, variability is its life's blood; without it, no one would play.

EXAMPLE 1.5

Given in Table 1.2 are "gallons pumped" figures for a certain gasoline station in a small town for three consecutive years, month by month. Some seasonal effects can be detected; for instance, there is a local peak in July of each year. And there is a year-to-year increase. But such systematic cycles and trends are corrupted and almost hidden by random variations in sales—variations that have no simple, deter-

Table 1.2

1967	1968	1969
10,569	13,260	15,911
10,952	12,978	15,515
11,825	14,488	18,048
11,307	13,294	17,333
11,068	13,294	17,986
12,031	14,812	18,248
12,454	16,082	19,187
11,498	15,576	18,442
13,395	14,114	17,051
12,641	17,277	18,516
15,828	17,582	17,096
13,135	17,218	16,507

ministic explanation and can only be taken into account by a model for random phenomena.

Another source of variability is that of the process of *measurement* itself. It might be naïvely assumed that an object has a length and that this length can be precisely determined by making a measurement with, say, a ruler or tape measure. Experience shows, however, that repeated measurings of the same physical object result, not always in the same measured value, but in different values. In making measurements of solubility, concentration, acidity, speed, force, mass, frequency, etc.—whether in the chemical, physical, or medical laboratory, or in other walks of life—it is found that repeated measurements of a given quantity will exhibit variation. Less clearly defined notions such as intelligence, achievement, assets, and quality, present even greater problems in their measurement. Measurement processes generally, even those using the most refined technique available, are subject to what are referred to as *random errors*.

Calling an error "random" does not really explain anything; the term is used to describe the fact that uncontrollable and unpredictable factors enter into the making of measurements and the reading of measuring instruments. Even such a relatively precise device as a micrometer has a certain amount of "play" and must be fitted onto the object whose dimension is being measured. Environmental conditions, such as temperature and humidity, also may affect measurements; these conditions can sometimes be adequately controlled but usually only at considerable expense. In the fields of educational and psychological testing, the measurements of a quantity are affected by the time of day, the phase of the moon, the degree of hunger, the state of personal relationships, and perhaps a host of other factors that one might not even think of, let alone control or eliminate.

Variations in measurements are often the result of an imperfect definition of the thing that is to be measured. In determining the length of a table, it will be found that variations from the perfect rectangularity of shape, which is usually taken for granted, introduce variations in the measurement of length. The concentration of a chemical in a solution will usually be slightly different in different portions of the liquid, even though the mixture was stirred vigorously to minimize such differences. The dimension of educational achievement, even in a particular subject, is even harder to pin down, and the measuring devices are very crude indeed.

Because responses, product characteristics, and other recorded quantities constituting the data are known only through processes of measurement, they are usually confounded with measurement errors. They will exhibit variability that includes random errors of measurement as well as the actual variation of the characteristic being measured from subject to subject. Now, it may be that the order of magnitude of the errors of measurement are such that they are insignificant when compared to the response variation, say, but it can also happen that measurement errors almost completely mask other kinds of variation.

The preceding discussion and examples of sources and types of variability in data is presented not to provide a tidy classification for the sake of classification, but to point out the variety of ways that variability is encountered and the fact that it does exist in almost every practical, quantitative situation of science, business and industry, medicine, agriculture, government, and politics—in nearly every phase of life.

Different experimenters do get different data with the same basic experiment. Yet it is soon noticed by anyone who gathers data that partially hidden in them is the notion of an underlying structure. In scientific experimentation this structure is a combination of deterministic, scientific laws and a law describing the random component. In political or geographical censuses, the underlying structure is just the population being sampled or the corresponding collection of values of characteristics of individuals in the population, together with the technique employed to obtain the sample. In operational data, the structure may be a combination of certain laws of human or economic interaction, say, and a law characterizing the nature of the randomness; or (as in the case of gambling games) it may be that the *only* law coming into play is one that has to do with randomness. In any case, one must seek to understand the nature of the variability, the element of randomness, in order to be able to draw inferences or make decisions on the basis of data.

Projects

Since this chapter did not introduce any material whose comprehension would be aided by "problems" of the usual sort, there is no problem set at this point. However, it is suggested that students do one or more projects of the following type.

Project 1

Watch the media (television, newspapers, etc.) for examples of the use or misuse of statistics—reports of data gathered and presented for information or for "proving" a point. Clip and prepare a brief discussion.

Project 2

Gather some actual data that might be used in statistical inference and save for use in Chapters 2 and 3. To prod the imagination, here are some examples of what one might do:

(a) Measure the widths of 50 oak leaves selected from those fallen under a given tree. (Appropriate for certain areas and seasons.)
(b) Count the numbers of words in the lines on a page of a book.
(c) Record the time intervals between passing cars during one hour on a street with little traffic.
(d) Weigh 25 candy bars of a certain kind—on a scale that reads at least as accurately as to the nearest tenth of an ounce.

Project 3 (in class)

With the students in class considered as a sample from the population of students on campus, data can be collected by the instructor by having each student give such information as his height, weight, age, eye color, hair color, sex, political preference, and number of brothers and sisters. Collected data can then serve as the basis for exercises in data presentation and the calculation of statistics that will be presented in Chapters 2 and 3.

Readings

Outside readings in certain other books will prove a useful supplement and can be started at once. In particular, the following books, all available in paperback editions, are very good:

Huff, Darrell, and Irving Geis. *How to Lie with Statistics* (New York: W. W. Norton & Company, Inc., 1954).

Tanur, J. M., F. Mosteller, W. H. Kruskal, R. F. Link, R. S. Pieters, and G. R. Rising. *Statistics: A Guide to the Unknown* (San Francisco: Holden-Day, Inc., 1972).

Wallis, W. A., and H. V. Roberts. *Nature of Statistics* (New York: The Free Press, 1965).

Frequency Distributions

In order to comprehend a mass of data—to appreciate its significance and to extract the information it may contain relative to the state of nature—various techniques are used to *reduce* the data to simpler forms, to graphical representations, and to a small number of descriptive or characteristic measures that relate directly to the problem being studied. Despite the well-developed mathematical theory of statistics that is directed at making the reduction of data more systematic and objective, good data reduction still involves a large measure of art and intuition. Thus the tools and devices to be considered here, without the mathematical preparation required for a more sophisticated approach, will be introduced as appealing to the intuition. Their justification lies in a history of their successful adaptation to practical problems.

The simplest case, and the one to be considered in this chapter, is that of *univariate* data. These are observations that arise in making a single measurement on each population element drawn—whether that measurement be numerical or categorical.

2.1 Tabular Presentation of Data

A first step in the organization of *categorical data* is the grouping of observations that fall in the same category. This can be accomplished as the data are being recorded, by prior preparation of a sheet with a listing of the various categories that will be encountered. The order in this list is arbitrary, unless the categories happen to have an intrinsic order. As the process of gathering data proceeds, each observation is recorded by making a mark opposite the label of the category into which it falls. When all the results are so recorded, the marks for each category may be counted, to obtain a *frequency* for that category. This frequency is simply the number of times that particular category is encountered in the process of observation. In this way, the data are "reduced" to a set of categories and corresponding frequencies.

EXAMPLE 2.1

To record the hair color of students in a sample from a population of students, the categories red, blond, brown, and black were established, with the assumption that any student's hair color could be classed in one of these four groups. Data were collected as follows (as taken from Table 1.1):

Red	
Blond	𝍸 |
Brown	𝍸 𝍸 𝍸 ||||
Black	𝍸

The frequencies are therefore as given in Table 2.1.

Table 2.1

COLOR	FREQUENCY
Red	0
Blond	6
Brown	19
Black	5
Total	30

Discrete, numerical data can be thought of as categorical, involving categories that are naturally ordered; the recording of such data proceeds just as described for categorical data. From this a table of frequencies is again prepared based on the preliminary reduction.

EXAMPLE 2.2

The numbers of siblings of each student in a certain class were listed in Example 1.1 as follows:

3, 4, 4, 2, 4, 4, 2, 2, 2, 2, 1, 1, 5, 2, 3, 4, 3, 5, 1, 4, 2, 7, 3, 4.

The frequency tabulation of these results is as shown in Table 2.2.

Table 2.2

NUMBER OF SIBLINGS	FREQUENCY
1	3
2	7
3	4
4	7
5	2
6	0
7	1
Total	24

Because the observed values of a *continuous* variable can fall anywhere in an interval, it is not possible to prepare ahead of time a list of all values that might be encountered. However, this impossibility refers to an idealization of what is done in actual observation.

When a continuous variable such as weight or height is measured by use of the imperfect measuring devices available to the experimenter, the result is recorded as a *rounded-off value*—a number that is given to the nearest number of years, or number of minutes, or number of milligrams, etc. The *recorded* data are then discrete. Moreover, actual data cover only a finite portion of the set of all numbers. For example, in making repeated measurements of the length of a rod that looks to be about 10 ft long, with a rule that can be read to the nearest tenth of an inch, the results would undoubtedly all be included in this finite list:

100.0, 100.1, 100.2, 100.3, . . . , 139.8, 139.9, 140.0.

In practice, such a finite list can be provided so that when a number (say, 119.6) is observed, the experimenter can simply make a mark opposite that value.

EXAMPLE 2.3

Sixty "1-lb" packages of bacon were weighed to the nearest $\frac{1}{8}$ ounce. The results were as follows:

$$\begin{array}{lllll}
16\tfrac{1}{4} & 16 & 15\tfrac{7}{8} & 16\tfrac{1}{4} & 15\tfrac{7}{8} \\
15\tfrac{3}{4} & 16 & 16 & 17 & 16\tfrac{1}{8} \\
16\tfrac{1}{8} & 16\tfrac{1}{8} & 15\tfrac{7}{8} & 16\tfrac{1}{4} & 16\tfrac{1}{4} \\
15\tfrac{7}{8} & 16 & 16 & 16\tfrac{1}{8} & 16\tfrac{1}{8} \\
15\tfrac{7}{8} & 16 & 15\tfrac{7}{8} & 16\tfrac{1}{2} & 16 \\
16 & 15\tfrac{7}{8} & 16\tfrac{1}{4} & 16\tfrac{1}{8} & 16\tfrac{1}{8} \\
16\tfrac{1}{8} & 16\tfrac{1}{4} & 16\tfrac{1}{4} & 16\tfrac{1}{4} & 16\tfrac{1}{4} \\
15\tfrac{7}{8} & 16\tfrac{1}{4} & 16 & 15\tfrac{7}{8} & 16 \\
16\tfrac{1}{8} & 16\tfrac{7}{8} & 15\tfrac{7}{8} & 15\tfrac{3}{4} & 15\tfrac{7}{8} \\
16\tfrac{1}{4} & 16 & 16\tfrac{1}{2} & 16\tfrac{1}{8} & 16 \\
16\tfrac{1}{8} & 16\tfrac{1}{8} & 15\tfrac{7}{8} & 16\tfrac{1}{8} & 16\tfrac{1}{4} \\
16\tfrac{1}{4} & 16\tfrac{1}{8} & 16 & 16\tfrac{1}{8} & 16
\end{array}$$

The corresponding frequency tabulation is given in Table 2.3.

Table 2.3

WEIGHT, LB	FREQUENCY
$15\frac{3}{4}$	2
$15\frac{7}{8}$	12
16	14
$16\frac{1}{8}$	15
$16\frac{1}{4}$	13
$16\frac{3}{8}$	0
$16\frac{1}{2}$	2
$16\frac{5}{8}$	0
$16\frac{3}{4}$	0
$16\frac{7}{8}$	1
17	1
Total	60

It should be pointed out that the very process of recording observations by making marks opposite categories, and so obtaining a list of frequencies, is itself a reduction of the data from what would be obtained if one simply wrote down the values in a sequence as observed. What is lost is the *order of observation*. From the list of values, written one by one in order as they are observed, it is possible to construct a table of frequencies; but from a frequency table one could not reverse the process and reconstruct the list of observations in the order obtained. The next example provides an illustration of what can happen in the reduction of data to a frequency table.

EXAMPLE 2.4

In Table 2.4 are given 200 observations of viscosity, obtained by measurement with an Ostwald viscosimeter.

It is possible to make a table of frequencies of the various possible observed values, but such a table would not show what the record in Table 2.4 shows, namely, that something is happening as the sampling proceeds. (This assumes, as was the case, that the data were written in groups of five across the page in the order in which they were obtained.) That is, the state of affairs seems to be different at the beginning of the sampling from what it is at the end of the sampling. Something obviously shifted, from the top to the bottom of the table, but this shift would be lost in a reduction to a frequency tabulation. Moreover, what is lost is something that would invalidate what would be presented in a frequency table, inasmuch as such a table is intended to convey information about a state of affairs—which in this case is not a fixed thing. An analysis of the clearly changing state of affairs can only be conducted from the original data, in their order of collection.

When, as in Example 2.4, the state of affairs is in the process of change, the manner in which it is changing would itself be an object for investigation. Thus data that give sales figures or stock market averages are data from a changing population, not from a fixed one. The analysis of such data, however important, is beyond the scope of this text. Rather, it will be assumed, whenever frequency tabulations are encountered, that the population or state of affairs did not change as the data were gathered and that nothing has been lost in this initial reduction to a frequency table.

The result of reducing data to values (or categories) and their corresponding frequencies is called a *frequency distribution*. It shows how the sample values or observations are distributed among the various possible values, with so many at this value, so many at that, and so on. A frequency distribution is defined by frequency tables such as those encountered in preceding examples, but it can also be characterized and represented in many other ways.

A variant of the frequency table is a table in which the *relative* frequencies are entered; each frequency is indicated as a fraction of the total number of observations in the sample. This relative frequency is particularly useful in dealing with samples of different sizes, among which comparisons of actual frequencies would not be meaningful.

Table 2.4 Viscosity as measured by an Ostwald viscosimeter.

37.0	36.2	34.8	34.5	32.0
31.4	31.0	30.8	30.0	33.9
34.4	33.5	32.9	32.8	31.3
33.3	33.7	34.3	35.9	31.0
34.9	33.4	33.3	32.4	32.0
31.7	32.8	27.6	31.6	35.0
30.0	30.7	32.2	35.3	35.5
34.4	30.2	32.0	31.3	31.6
32.0	33.1	31.2	31.9	31.8
29.8	29.6	31.7	32.0	29.9
31.6	34.4	33.3	35.9	35.4
31.6	32.1	34.2	33.8	33.4
31.3	33.1	32.2	33.3	31.2
34.6	32.7	32.1	32.0	30.5
32.6	31.5	30.8	31.2	31.0
31.5	30.7	32.5	33.0	32.9
34.5	33.7	32.0	31.3	30.9
32.3	32.0	31.9	30.0	33.9
33.8	34.8	34.2	33.8	33.8
31.1	31.8	26.8	32.3	32.8
31.7	33.5	34.5	33.6	32.5
32.5	32.7	32.4	33.6	34.1
31.8	30.5	30.8	30.6	30.7
31.9	31.4	30.9	32.0	30.8
29.9	31.6	29.5	31.7	30.4
30.7	31.3	28.7	28.8	31.0
27.6	31.9	31.3	29.9	32.1
31.2	30.5	30.9	29.8	28.7
30.0	29.1	29.7	31.0	29.0
30.9	28.9	29.5	28.8	29.3
28.4	30.0	29.3	30.2	31.5
31.3	30.3	32.5	31.8	30.8
28.7	30.7	32.3	33.5	32.9
32.1	29.8	29.9	30.8	30.7
31.9	32.0	29.5	30.6	30.1
32.8	31.0	30.0	29.2	32.5
30.2	31.3	29.5	29.0	30.3
30.2	30.7	30.8	31.6	32.7
32.4	32.8	31.3	31.4	30.7
30.7	30.6	31.3	30.6	31.7

EXAMPLE 2.5

Table 2.5 gives the results of netting fish, each observation having been obtained by counting the number of fish in a given setting of a net. In

Table 2.5

Number of fish	0	1	2	3	4	5	6	7	8	9	10 or more	Total
Frequency, lake A	6	9	4	4	0	1	0	1	0	1	3	29
Frequency, lake B	26	22	14	7	2	4	5	3	1	1	3	88

lake A, 6 settings of the net came up with no fish, and in lake B, 26 settings yielded no fish. But 6 is not properly compared with 26, for there were many more settings in lake B. The relative frequences are $\frac{6}{29}$ and $\frac{26}{88}$, or, in decimal form, about .207 and .295. The complete table of relative frequencies is as given in Table 2.6. (These relative frequencies should add up to 1 in each case, but they happen not to do so because of round-off errors in expressing fractions as decimals.)

Table 2.6

Number of fish	0	1	2	3	4	5	6	7	8	9	10 or more
Relative frequency, lake A	.207	.310	.138	.138	0	.034	0	.034	0	.034	.103
Relative frequency, lake B	.295	.250	.159	.080	.023	.045	.057	.034	.011	.011	.034

2.2 Graphical Devices

A graphical representation of the frequency distribution is often helpful in comprehending a collection of data. One obvious graph would plot the possible observed values along a first dimension and the frequencies along a second dimension. This is particularly natural when the "values" are actual numerical quantities but somewhat artificial when they are just unordered categories. As in making up a frequency table, so in making a corresponding graph, *some* order must be used in listing categories; and if there is not a natural order of numerical categories, an arbitrary order must be used—one that will then be evident in the graph but that will have no significance.

Sometimes, in such graphs, the order used is that defined by the frequencies themselves, the most frequent categories given first (or farthest to the left), say. This may serve erroneously to suggest that the order has some meaning, whereas another sample from the same population is likely to produce a different ordering of frequency magnitudes.

EXAMPLE 2.6

A survey was made of the eye color of 39 students in a statistics class. The results are given in Table 2.7. Figure 2.1 gives three graphical representations of this table, each using a rod whose length is proportional to frequency. The difference is in the order in which the categories are listed and in whether the frequencies are plotted horizontally or

Table 2.7

EYE COLOR	FREQUENCY
Blue	21
Hazel	9
Brown	7
Green	2
Total	39

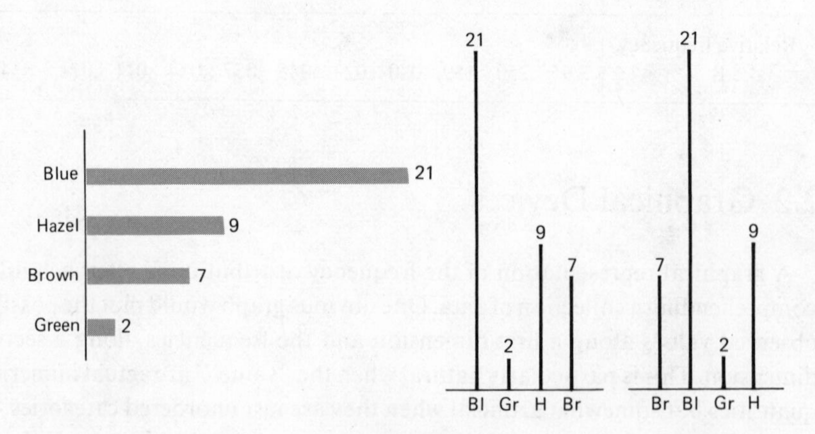

Figure 2.1
Eye color from a sample of 39 students.

vertically. Another variant shown is the use of a bar instead of a rod. Bars are often used simply for visibility, with no significance implied in the width of the bar.

EXAMPLE 2.7

The frequency table in Example 2.2, giving frequencies of the various numbers of siblings of students in a class of size 24, is repeated here:

Number of siblings	0	1	2	3	4	5	6	7
Frequency	0	3	7	4	7	2	0	1

Figure 2.2 is a graphical plot of this distribution.

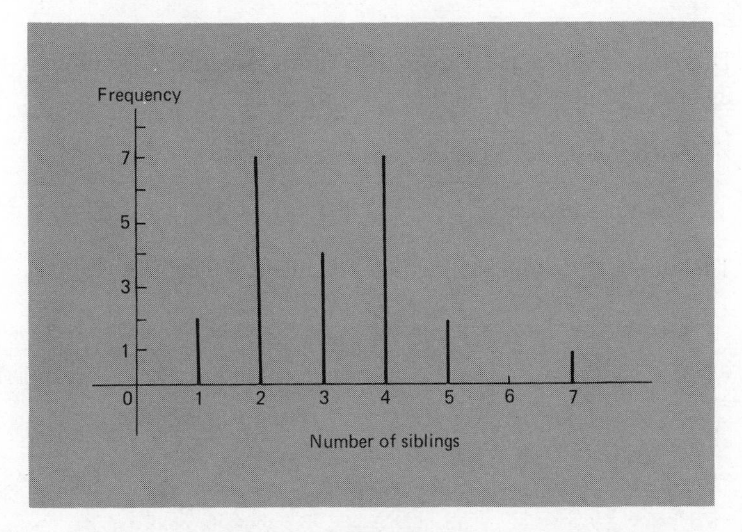

Figure 2.2
Frequency distribution of the number of siblings.

Another graphical device, applicable in the case of *numerical* data, is the *sample distribution function*. This function finds use in modern statistical inference. It is essentially a graph of the total of the relative frequencies accumulated in moving from left to right along the axis of possible values. Suppose, after a mark has been made on the scale of values corresponding to each observation, that you start at the low end of the scale and walk along the axis with a counter in your hand; as you encounter an observation mark you push the counter button, and if there happens to be several observations at one point you push the button once for each observation. The counter will keep track of how many observations you have passed by, and when you get to the right-hand end (or high end) of the scale, you will have a number on your counter equal to the number of observations in the sample. The record of the counter reading, divided by the number of observations in the sample to give *relative* frequencies, as a function of how far you have walked at any given point along the axis, is precisely the sample distribution function. At any point its value, and therefore the height of the graph, is the proportion of observations you have passed by in reaching that point in your walk.

The sample distribution function does not exhibit anything that is not exhibited (in a different way) in the frequency table or in graphs showing frequencies. On the other hand, it exhibits just as much, embodying exactly the same information. The sample distribution function can be constructed from a frequency table, and conversely.

EXAMPLE 2.8
The *cumulative* frequencies of the various numbers of siblings in Example 2.7 are shown in the following table:

Number of siblings	0	1	2	3	4	5	6	7
Cumulative frequency	0	3	10	14	21	23	23	24

The sample distribution function is shown in Figure 2.3.

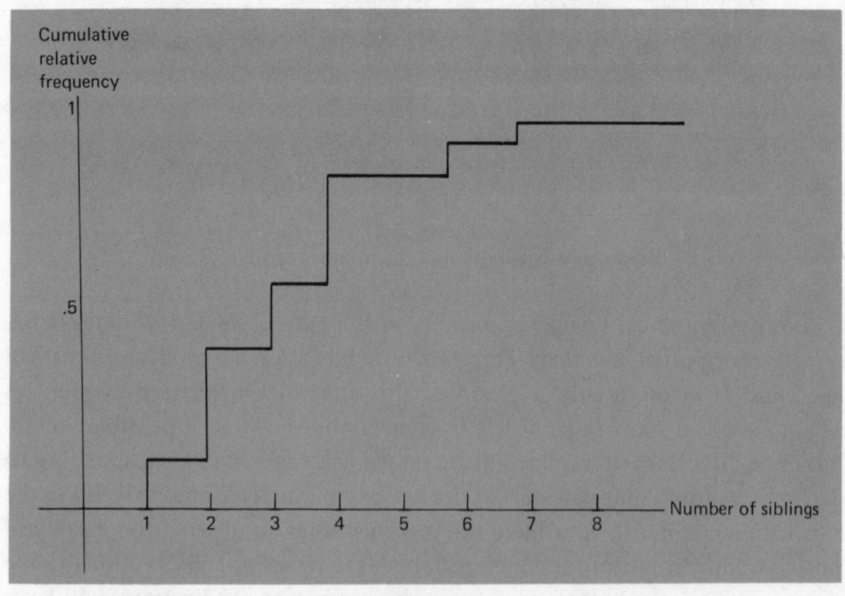

Figure 2.3
Sample distribution function.

EXAMPLE 2.9

The 12 girls included in the sample with characteristics given in Table 1.1 gave their weights as follows:

110, 128, 135, 115, 150, 100, 95, 145, 113, 115, 134, 134.

Rearranged into an increasing sequence, these numbers are

95, 100, 110, 113, 115, 115, 128, 134, 134, 135, 145, 150.

Figure 2.4 shows a scale or axis on which these weights are each marked with an x, and over which the sample distribution function corresponding to the observed weights is drawn. Observe that it jumps an amount $\frac{1}{12}$ at each observation (and $\frac{2}{12}$ at the points where there are two observations). One can imagine that the graph is drawn like this: As the

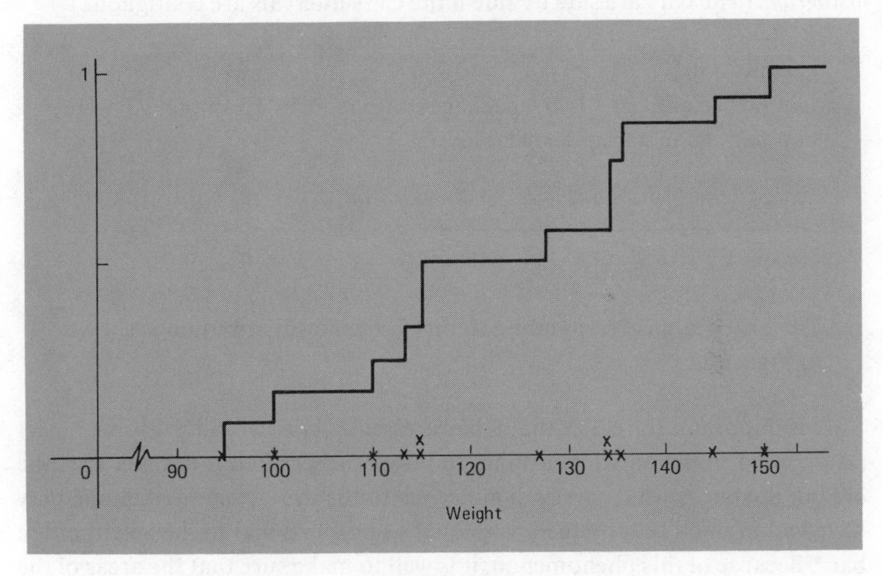

Figure 2.4
Distribution function of a sample of 12 weights.

hand, starting at the extreme left on the horizontal axis, moves a pencil slowly across the page to the right, it is bumped up one notch whenever an x is encountered on the axis—continuing to move to the right at the new level after being bumped. (If there are two marks at one x-value, the hand is bumped twice as hard and moves up two notches, etc.)

In the recording of continuous data the process of rounding off, which necessarily goes with the reading of actual dials and scales, amounts to what is termed a *grouping* of possible values into class intervals. Each such interval is represented by a typical value, namely, the value to which numbers in that interval are rounded. Such data are therefore referred to on occasion as *grouped data*. As discussed earlier, data in this form are ordinarily presented in a frequency table, with the representative values in one column and the corresponding frequencies in another. In making graphical representations of such distributions, cognizance should be taken of the fact that the f_i observations in the group of observations that are rounded to the value x_i were actually (as they occurred, before being rounded off) numbers that could have fallen anywhere in the class interval represented by the value x_i. Thus, instead of a rod or height f_i at the point x_i, a *bar* of height f_i and width *extending over the class interval* represented by x_i is used to display the results. When this is done for each class interval, the resulting collection of side-by-side bars is called a *histogram*. (The bars are side by side if the class intervals are contiguous.)

EXAMPLE 2.10

The 60 weights of "1-lb" packages of bacon in Example 2.3 were summarized in a frequency table:

Weight, lb	$15\frac{3}{4}$	$15\frac{7}{8}$	16	$16\frac{1}{8}$	$16\frac{1}{4}$	$16\frac{3}{8}$	$16\frac{1}{2}$	$16\frac{5}{8}$	$16\frac{3}{4}$	$16\frac{7}{8}$	17
Frequency	2	12	14	15	13	0	2	0	0	1	1

The histogram corresponding to this frequency distribution is shown in Figure 2.5.

It is important to realize that a bar is essentially a two-dimensional object (as opposed to the one-dimensional rod used earlier), and it is a fact that the *area* of the bar is what carries significance to the eye—even though one may have had in mind that frequency is equal or proportional to the height of the bar.* Because of this phenomenon, it is well to make sure that the areas of the bars do represent frequency, either by making all the bars (i.e., the class intervals) the same width, or, if there are bars of varying widths, by adjusting the heights so that the product of height and width is proportional to frequency. Ordinarily, the round-off process will involve rounded values that are equally spaced, implying class intervals of equal width; and then there is no problem.

* The book *How to Lie with Statistics*, by Darrell Huff and Irving Geis (New York: Norton, 1954) devotes an entire chapter called "The One-dimensional Figure," to the pitfalls encountered in using two-dimensional figures to convey one-dimensional information.

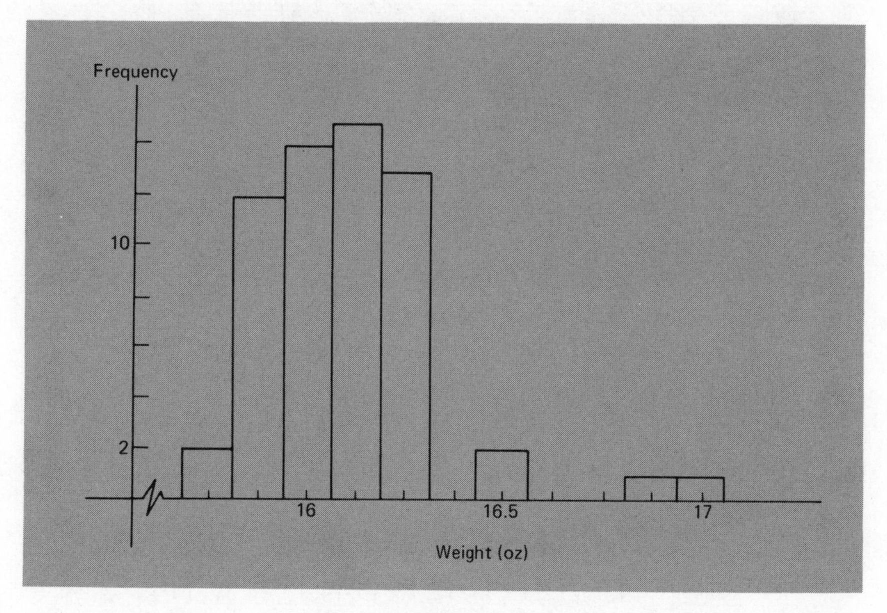

Figure 2.5
Histogram of bacon weights.

But when observations are sparse, it is tempting to group several intervals together, which is legitimate as long as the graph takes this into account.

EXAMPLE 2.11
The frequency table of bacon weights in Example 2.9 can be made more compact by lumping together some of the intervals on the higher end of the scale, as shown in Table 2.8.

Table 2.8

INTERVAL	FREQUENCY
$15\frac{11}{16}$ to $15\frac{13}{16}$	2
$15\frac{13}{16}$ to $15\frac{15}{16}$	12
$15\frac{15}{16}$ to $16\frac{1}{16}$	14
$16\frac{1}{16}$ to $16\frac{3}{16}$	15
$16\frac{3}{16}$ to $16\frac{5}{16}$	13
$16\frac{5}{16}$ to $17\frac{1}{16}$	4

A correct and an incorrect histogram are shown in Figure 2.6. The incorrect one clearly suggests many more observations in the largest group than are there.

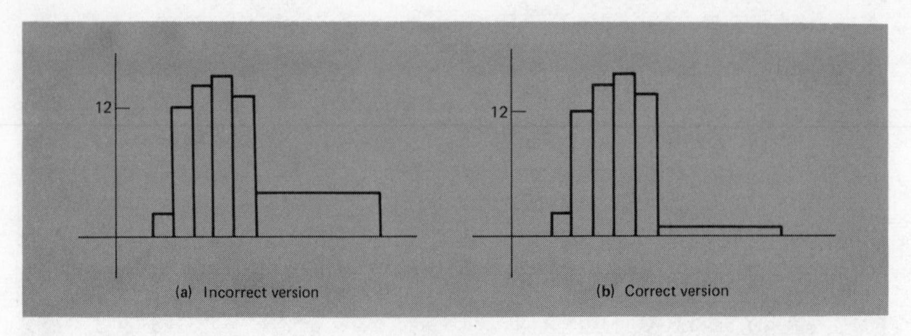

Figure 2.6
Adjustment of height in combined intervals (Example 2.11).

Problems

2.1. The number of licorice "snaps" in each of 25 boxes was counted, with these results:

14, 13, 13, 13, 14, 14, 11, 13, 13, 14, 12, 11

14, 13, 13, 13, 11, 12, 13, 15, 14, 13, 14, 12, 14.

Make a frequency table for these data and represent the results graphically.

2.2. The credit loads of 40 students in a certain fraternity in a given quarter were found to be as follows:

13, 18, 15, 16, 19, 18, 16, 14, 18, 18, 18, 20, 18, 16

16, 17, 20, 17, 19, 15, 17, 18, 22, 19, 14, 13, 18, 14

18, 18, 8, 19, 17, 18, 21, 13, 18, 9, 18, 13.

Construct a frequency distribution and a graphical representation.

2.3. A student observed the occupancy of cars passing a certain point near campus. He tabulated the results as follows:

Number of occupants in a car	1	2	3	4	5
Frequency	7	8	5	3	2

(a) How many cars did he observe?
(b) Represent the data graphically.

2.4. In a dormitory sample survey of 50 students, ages of the respondents were tabulated as follows:

Age	17	18	19	20	21	22	23
Frequency	4	16	14	10	4	1	1

(a) Why is a histogram (i.e., using bars) especially appropriate here? Construct the histogram for the given frequency distribution.

(b) If (as is sometimes the case) the question on age had been stated in this way: "Give your age at your nearest birthday," how would the histogram bars best be located? [If you already so located them in part (a), rethink your reasoning there. How does a person usually give his age if you have not given him specific roundoff instructions?]

2.5. A lightly traveled street is being checked for traffic density. The times between cars (passing a certain point) are recorded, with these results (in seconds):

8, 26, 12, 4, 43, 28, 71, 3, 33, 2.

Plot these as marks on a horizontal axis and then construct a sample distribution function for these data. (Would a histogram be helpful here?)

2.6. Twenty-five students were asked their opinion of a certain course, their answers being one of the following choices:

E: excellent—recommend without hesitation.
G: good—only minor reservations.
F: fair—could take it or leave it; uninspired.
P: poor—waste of time and effort.

Their answers were tabulated in a frequency distribution:

Opinion	E	G	F	P
Frequency	3	14	7	1

Represent this graphically. (Is there an obvious ordering among the categories of answers? Would a bar be appropriate? If so, would the bars be of equal width? Contiguous?)

2.7. A Department of Commerce "Fact Book" gives the following frequency table of illegitimate births in a certain year according to the age of the mother:

AGE	NUMBER (1000)
Under 15	7
15–19	144
20–24	102
25–29	35
30–34	17
35–39	10
40 and over	3

Construct a histogram, with special thought as to the proper treatment of the first and last class intervals.

2.3 Regrouping Data

Experience with histograms shows that they often will exhibit erratic fluctuations in frequency from class interval to class interval. (See Example 2.11, for example, particularly Figure 2.7.) One might well wonder whether these fluctuations are telling him something about the phenomenon being sampled or are simply a manifestation of the randomness encountered in processes of sampling. In practice one has only a single such sample, but it is quite instructive to see what *might* have happened, by looking at the histograms of many samples. This is not usually practical, but it can be done through the device of constructing artificial samples using tables or generators of what are called *random numbers*. The actual construction is beyond our scope; suffice it to say that modern digital computers can be used, and they were used in the examples that follow.

EXAMPLE 2.12
The viscosimeter readings in Example 2.4 were reduced to a frequency table and plotted in a histogram, the result being shown in Figure 2.7. Observe the large variation in adjacent frequencies, a factor of five or more in some cases.

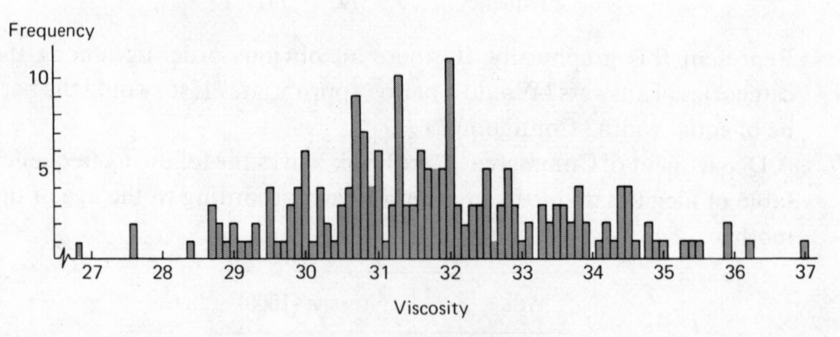

Figure 2.7
Histogram of viscosity data.

A population closely resembling that from which the viscosity data in Figure 2.7 might have come was constructed on a computer, and four samples of size 1600 drawn. A plotter linked to the computer produced the histograms in Figure 2.8. Notice that even though the

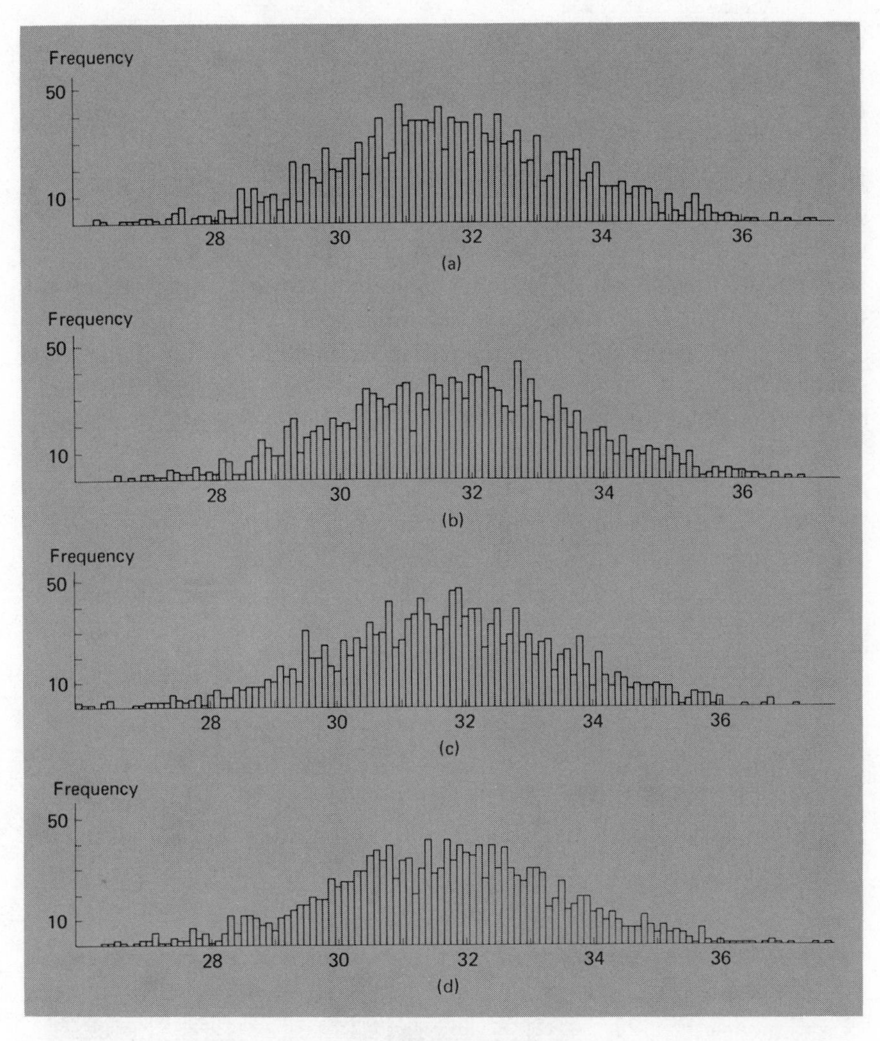

Figure 2.8
Artificial samples of size 1600.

samples were much larger than that of Figure 2.7, there is still considerable fluctuation from sample to sample as well as among the adjacent frequencies.

Inspecting the histograms of Example 2.12, it is evident that there is considerable similarity in the crude features of the histograms, and it is not unreasonable to conclude that there is information about the population being sampled in the common aspects of the samples. The roughness, the erratic variation from sample to sample, is thought of as "noise," uninformative and unwanted, not as a facet of the population of interest. It can be eliminated in a single sample, to some extent, by a process of *regrouping*.

Data that are rounded off and presented in a frequency tabulation are referred to as *grouped data*, and when a different roundoff scheme is used, the data are said to be *regrouped*. Using wider class intervals is equivalent to a coarser rounding off of the observations and will result in a new frequency table and new histogram that exhibit somewhat less variability from interval to interval. Some of the sampling variability is gone.

Some sampling variability was reduced in Example 2.12 by obtaining larger samples. But this is not always possible, and even when it is, there will usually be some undesired roughness that a suitable regrouping can largely eliminate.

EXAMPLE 2.13

The viscosity data shown graphically in Figure 2.7 were regrouped into class intervals of equal width centered at 27.3, 28.4, 29.5, etc. The resulting frequency distribution was used to construct the histogram in Figure 2.9. Observe that the erratic appearance of Figure 2.7 has been smoothed out by the regrouping. To show that regrouping does not take care of everything, six samples of 200 were obtained artificially from a population approximately the same as that from which the viscosity measurements might have been obtained. The results are shown in Figure 2.10. Although each histogram is indeed smoother, there is variation from sample to sample, indicating that not all the sampling fluctuations have been removed. If the histograms for the samples of 1600 observations in each of the samples represented in

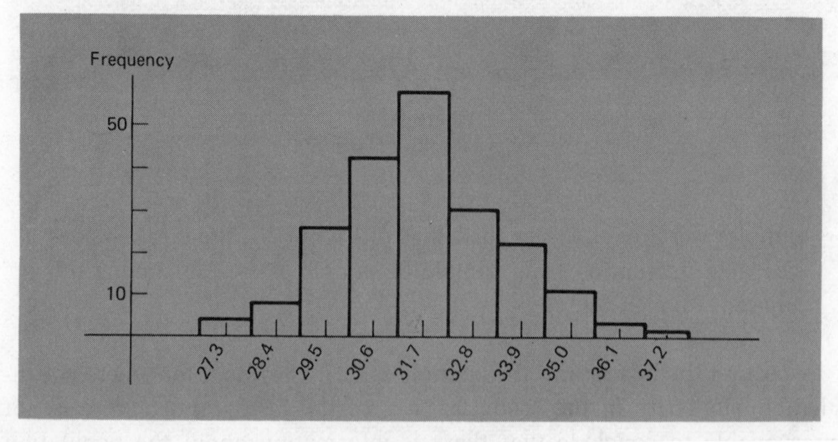

Figure 2.9
Regrouped viscosity data.

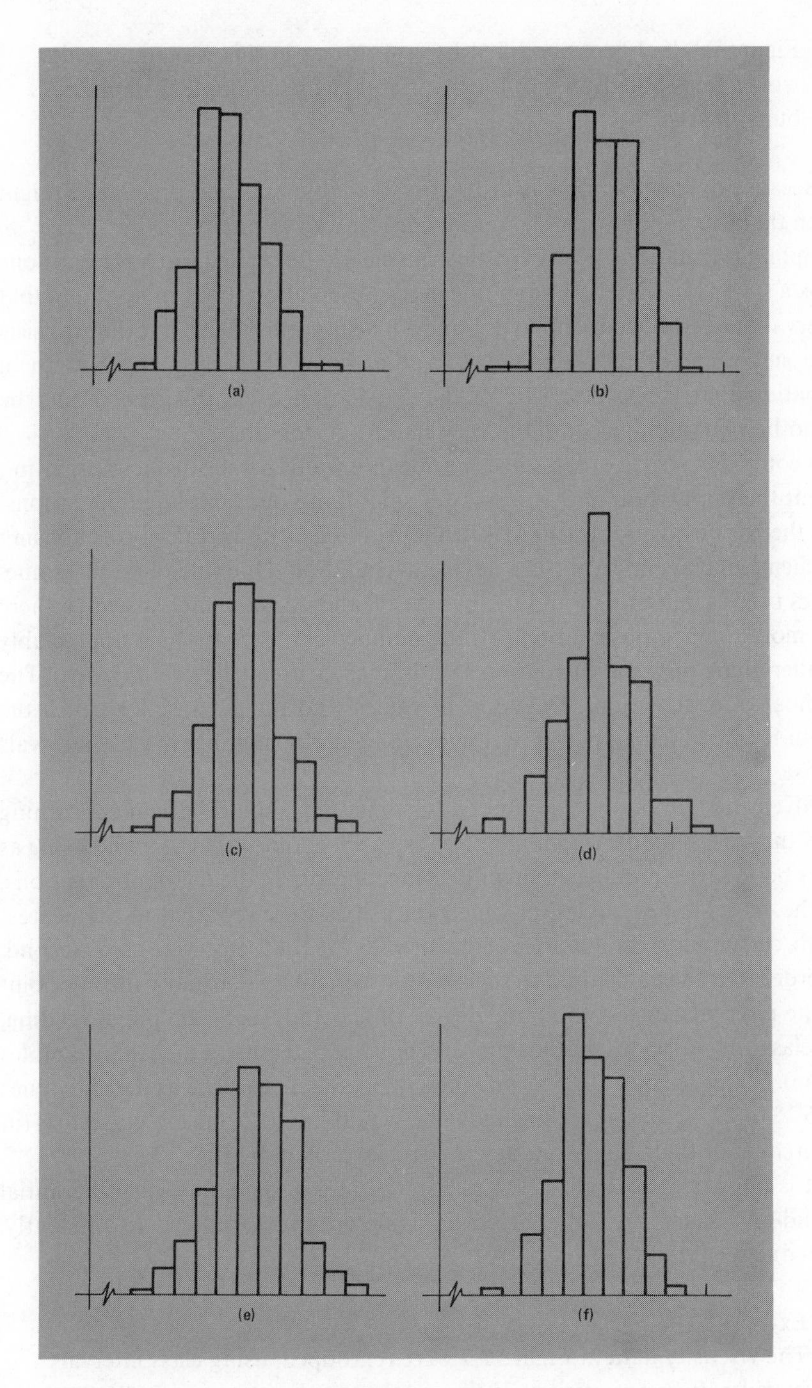

Figure 2.10
Artificial samples of size 200, grouped into 11 class intervals.

Figure 2.8 had been modified by regrouping in this way, the results would undoubtedly exhibit less variability from sample to sample—but still *some*.

Because only *one* sample is ordinarily available in actual practice, it might seem that the use of wider class intervals is a way to minimize the "noise" of sampling fluctuations. However, this can clearly be carried too far. For, if one took a *single* class interval, rounding every observation to the midpoint of that interval, the result would be a very smooth histogram indeed, one that consists of a single bar or rectangle. On the other hand, if there is any pattern of variation that is characteristic of the population itself, this, too, would be smoothed out and be lost in the very smooth histogram.

A compromise is in order, since one does not want to smooth out information about the population in the process of smoothing out sampling fluctuations. But there is no invariable rule in this compromise, no universal way of choosing a scheme of class intervals or a degree of round-off. One rule of thumb sometimes used is that one should use between 8 and 20 class intervals when there are more than 50 observations. (If the number of observations is appreciably smaller than this, the histogram is not such a useful device anyway.) The number of observations, the range of values, and the purpose for which the result is to be used all play a part in deciding on the number of class intervals to use.

Given the number of class intervals, something should be said concerning how the class boundaries should be chosen. First, one ought to avoid using as class boundaries numbers that will be encountered in the list of observations. In the preceding examples, in which viscosities were reported to the nearest tenth, class interval boundaries of the form 26.75, 27.85, etc., were used. Second, in order that the value used to represent a class interval, usually the midpoint of the interval, suggest the same degree of accuracy as in the initial reading, the class interval width should be an odd number of tenths (in the same examples again). Using 26.75 to 27.85 as one class interval, one has 27.3 as the midvalue; if 26.75 to 27.75 were used, the midvalue would be 27.25, falsely suggesting (in the frequency table, say) accuracy to the nearest hundredth.

It is important to realize that the regrouping of data, and even the initial round-off scheme, are rather arbitrary. Different people will group differently, with different results.

EXAMPLE 2.14

The viscosity data in Figure 2.7 were regrouped, using class intervals of width .9, first starting at 25.65, and then again starting at 25.95. The resulting histograms are shown in Figure 2.11. Both histograms are

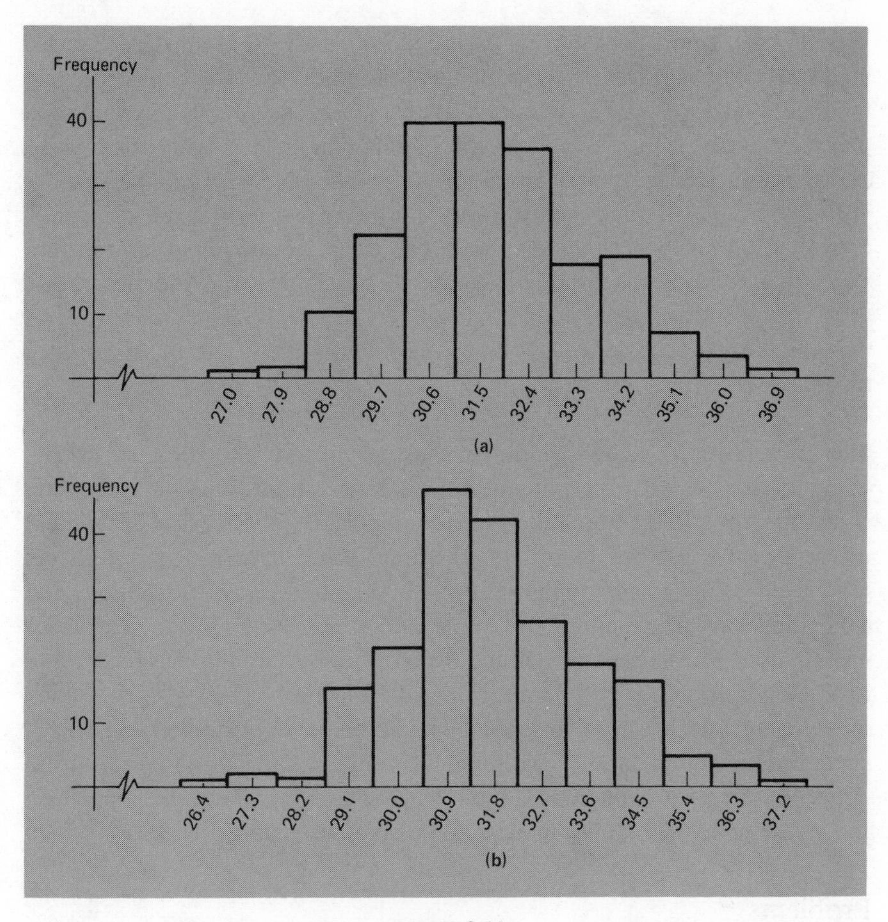

Figure 2.11
Viscosity data, regrouped with interval width .9.

relatively free of the irregularities obviously attributed to sampling fluctuations; yet, owing to the different location of the class intervals, they are quite different. Even though they were made up from the *same* 200 observations, these graphical representations are quite different in appearance.

The point of this example is to show that care must be used in drawing conclusions about a population from a histogram, when just the arbitrary starting point of the class interval scheme can make such a difference in the appearance. And, of course, a different choice of class interval width can also make no little difference in the appearance. [Compare the histograms in Figure 2.11 (width of .9) with that of Figure 2.9 (width of 1.1).]

It has been pointed out that even continuous observations must be recorded in discrete fashion, and the use of the histogram bar was explained as a reminder that observed values were rounded off. On the other hand, some data that seem at first glance to be from continuous experiments really are discrete, as in the case of the number of boxes per carton in Problem 2.7 and the earnings in Problem 2.8. Although it may be out of order to use bars for the given data, it is often useful to regroup such data—as in the continuous case—so that a class interval contains several of the discrete possible values. And then, rather than a single rod at the center of the class interval, a bar is again used to suggest that the observations in a class interval actually were scattered throughout the interval.

In reducing data to a frequency distribution, and in the representation of such a distribution by a bar diagram or a histogram, the motivation is to extract and to present visually the content of the data—such information as the data contain relating to the population being sampled, i.e., to the basic experiment whose repeated performance yields the data. It has been seen that different samples from the same population *vary*; and so frequency distributions and histograms vary from sample to sample, and also according to the particular scheme used for grouping or rounding off. These sampling variations, and variations introduced by different round-off schemes necessary in the actual recording of data, are not of interest, being unrelated to the underlying experiment. But the ingredient of variation in the samples that is seen to persist from sample to sample *is* of interest. It is this that, in a sense, defines the population and gives rise to probability models, to be taken up in Chapter 3.

Problems

2.8. The birth weights of 30 babies were recorded as follows (in pounds):

8.2	7.5	7.0	5.5	7.1
8.0	8.3	6.5	6.75	5.9
8.3	7.75	5.25	8.25	6.7
9.0	5.8	7.1	7.25	7.5
8.0	8.25	6.75	6.7	5.4
7.7	7.25	5.75	6.6	6.75

Construct histograms with about five to seven class intervals, using at least two different class interval schemes.

2.9. Hamburger boxes furnished to a large chain of hamburger shops come in cartons containing, nominally, 800 boxes. Actual counts in 30 cartons showed that there are seldom exactly 800 in a carton:

768, 786, 787, 788, 790, 790, 793, 793, 794, 799

799, 799, 800, 800, 800, 800, 801, 806, 809, 815

815, 815, 816, 817, 826, 826, 831, 845, 847, 862.

Construct a histogram with about six class intervals.

2.10. A waitress kept a record of her average hourly earnings during each of 25 weeks (wages plus tips):

$2.61, 2.81, 2.83, 2.37, 2.78, 3.25, 3.19, 3.30, 3.02,

3.60, 2.79, 2.79, 2.64, 2.32, 3.42, 2.73, 2.47,

2.75, 2.20, 2.59, 2.52, 2.36, 3.05, 2.90, 2.18.

Construct a histogram.

2.11. Devise a scheme of class intervals and make a frequency tabulation and histogram for the following data, which give hemoglobin levels of 155 cancer patients at a veterans' hospital (data in g/100 ml):

13.5	13.1	12.7	13.8	14.2	12.8	14.6	16.2
13.6	11.0	13.7	11.5	14.2	18.2	14.4	13.8
10.7	12.1	12.1	15.3	14.8	14.8	12.9	15.3
13.2	15.0	5.2	12.0	13.5	15.6	11.4	12.1
13.5	13.8	13.8	15.5	11.7	13.3	14.0	13.5
15.6	14.2	13.8	12.9	13.5	10.8	13.8	16.7
14.8	11.4	15.4	12.7	15.1	13.0	16.0	16.4
14.6	12.0	12.4	14.0	10.9	12.6	14.2	15.4
9.1	13.6	16.3	11.4	12.7	12.7	11.7	13.0
13.5	13.7	13.6	13.7	13.7	11.0	14.0	14.8
16.3	12.4	11.9	14.6	11.5	14.0	15.4	12.0
12.7	15.0	13.4	11.9	12.7	11.0	15.0	10.1
13.5	14.2	12.4	15.2	12.7	12.7	14.8	18.6
16.9	14.8	12.4	13.8	15.0	11.8	14.2	16.0
12.9	16.3	13.2	14.0	10.0	13.8	14.6	14.0
12.3	11.3	12.7	13.7	15.0	16.2	11.3	
10.9	10.1	15.7	15.7	13.5	15.4	13.3	
15.1	11.4	14.8	13.8	14.0	15.4	15.7	
14.2	11.4	12.0	11.1	11.9	15.8	10.7	
13.8	11.6	14.8	16.3	13.2	10.8	15.5	

3 Probability Models

Because of the variability and consequent unpredictability involved in obtaining a piece of data, one speaks of the process as an *experiment of chance*. Performing the experiment will result in one of various possible outcomes, the list of which may have been prepared for the purpose of making a frequency distribution.

The list or set of possible outcomes of an experiment of chance is called the *sample space* of the experiment, since it includes all the outcomes that will be encountered in sampling, or making observations. In a given set of data resulting from a finite number of observations on the experiment, each outcome in the sample space will occur a certain number of times, a number called earlier the frequency of the outcome. (This frequency can be zero for a particular outcome.) The collection of sample space outcomes and corresponding frequencies in a particular sample has been termed the *frequency distribution* of the observations in that sample.

3.1 Events

A useful concept is that of an *event*, which is the term used to mean any set of outcomes in a sample space. Such a set is usually defined by some condition or description, which is satisfied by certain outcomes and not by others. Alternatively, the set may be defined by simply making a list of the outcomes in it.

The condition that the outcome of an experiment is a particular *one* of the possible outcomes is a special case of an event; i.e., the individual outcomes are themselves events, sometimes called *elementary events*, or *elementary outcomes*, since they are the elements from which other events are constructed.

EXAMPLE 3.1

When a die is tossed, the result is one of the six faces of the die—that is, the face that turns up when the die comes to rest. The *sample space* of this experiment is the list of possible outcomes, the faces—identified by the number of dots that have been marked on the faces by the manufacturer: $\{1, 2, 3, 4, 5, 6\}$. Each sample observation is one of these numbers.

Some examples of an *event* in this sample space are these, referred to by arbitrary capital-letter names:

E_1: The outcome is an even number.
E_2: The outcome exceeds 3.
E_3: The outcome is a 1.

The first condition, E_1, is satisfied by the outcomes 2, 4, and 6, and giving this list of outcomes is another way of specifying the event:

$$E_1 = \{2, 4, 6\}.$$

Similarly, event E_2 is defined by the list

$$E_2 = \{4, 5, 6\},$$

including precisely those outcomes that exceed 3. Event E_3 is an example of an *elementary event*. It is said that event E_1, for example, is *composed* or is the *union* of the elementary events $\{2\}$, $\{4\}$, and $\{6\}$.

Just as the number of times a given *outcome* occurs in a sequence of observations can be counted, so can the number of times that any given *event* occurs be counted. This number is the *frequency* of the event. Every event has a frequency in a given set of observations, and this frequency can be computed from the frequency distribution of the sample. For the number of times an event occurs is just the *sum* of the numbers of times its various component elementary outcomes occur—the sums of the frequencies of the outcomes that make up the event. An event is said to have *happened* if, when an observation is obtained, that observation satisfies the condition defining the event—i.e., if it is one of the elementary outcomes in the event.

EXAMPLE 3.2

A die was tossed 120 times, with results given in the following frequency table:

Number of points	1	2	3	4	5	6
Frequency	12	24	21	16	19	28

Consider again the events $E_1 = \{2, 4, 6\}$ and $E_2 = \{4, 5, 6\}$, as defined in Example 3.1. Each of these events has a frequency in the sample which is the sum of frequencies of elementary outcomes:

$$\text{freq}(E_1) = 24 + 16 + 28 = 68,$$

$$\text{freq}(E_2) = 16 + 19 + 28 = 63.$$

In the 120 tosses, event E_1 was encountered 68 times, the sum of the numbers of times the elementary events 2 and 4 and 6 were encountered. The *relative frequency* of E_1, of course, is $\frac{68}{120}$.

In the case of a numerical variable that can attain any of a continuum of values (such as the *time* elapsed during which a radioactive material decays to half the intensity of radiation), the sample space is a set of numbers—the set of all real numbers, or the set of all positive numbers, or the set of numbers in a given interval, to mention some commonly occurring examples.

As discussed in Chapter 2, in practice one will never encounter numbers that are specified more precisely than to a certain number of significant digits, owing to the limitations of available measuring instruments. And there may be practical limits as to how big or small the measured value may be. Nevertheless, it is often the case that a model whose sample space consists of *all* positive real numbers, say, may be simpler to work with (believe it or not!) than one that matches the practical limitations—one involving a discrete approximation mandated by the inadequacies of available or "real" instrumentation.

The sample space for a continuous numerical variable will be taken to be an *interval* of numbers, perhaps infinite on one end or the other (or both). An *event* in such a sample space could be any set of these numbers. However, the only events of practical significance are *intervals* (i.e., *sub*intervals of the sample space) or events made up of two or more intervals.

An individual elementary outcome or an elementary event is indeed an event; but it is not the sort of event one finds useful in studying the model for a continuous variable. Thus in the idealization in which *all* numbers on an interval are possible, the number $\frac{7}{3}$ would almost never be encountered, so why bother to watch for it? (Of course, the number $\frac{7}{3}$ would likely be encountered if one were *rounding off* to the nearest $\frac{1}{3}$; but really, then, the event $\frac{7}{3}$ in the rounded values would consist of the *interval* of numbers from $\frac{13}{6}$ to $\frac{15}{6}$, all of which are rounded off to $\frac{14}{6}$ or $\frac{7}{3}$.)

After the experiment is actually performed, it will be observed that a particular number has indeed occurred (and this might have been $\frac{7}{3}$). But the probability model here is to represent what *can* happen, not what *has* happened.

An interval is defined by its end points. The interval from a to b is sometimes denoted by (a, b), or by $[a, b]$ if it is intended that the end points be included. The same interval can be expressed in fancier mathematical notation as

$$x\colon a < x < b,$$

which is read "the set of all numbers x such that x is between a and b." More simply, one can write

$$a < x < b,$$

where x is thought of as a dummy variable, in the sense more explicitly defined in the more complicated notation and its meaning.

EXAMPLE 3.3

Consider the time elapsed from one arrival to the next, as one keeps track of arrivals at a service facility (such as a post office window, a gasoline pump, a barber shop). The elapsed time is theoretically unrestricted and can be any number of time units that is not negative. If T denotes the elapsed time, the interval $0 \leq T < \infty$ constitutes the sample space. An event in this sample space can be any set of non-negative T-values, but the simplest events are intervals, such as

$$E_1\colon 10 < T < 15$$

(defined by the condition that the elapsed time is between 10 and 15 time units), or perhaps

$$E_2\colon T > 18.$$

A somewhat more complicated event is one of the form

$$E_3: |T - 10| > 2,$$

or the set of T-values farther than 2 units from 10. The inequality is satisfied by T's that are greater than 12 or less than 8. Thus E_3 is the set consisting of the two intervals $T > 12$ and $0 \leq T < 8$. The event E_3 *occurs* if the observed time T is in one or the other of these two intervals.

The condition that the outcome of an experiment is not in an event E is itself an event—that is, the negation of any condition is a condition. The outcomes not in E constitute the event called the *complement of* E, relative to the given sample space. In a sample of n observations this complement has the frequency $n - \text{freq}(E)$.

EXAMPLE 3.4
The sample space for observing eye color in Example 2.6 consisted of the four "points" blue, hazel, brown, and green. If E denotes the event that the eye color of an individual is blue (which happens to be an elementary event, incidentally), then the complement of E is the event consisting of hazel, brown, and green. If an individual's eye color is not blue, then of course it is one of the other three colors. In the sample described in Example 2.6, the elementary events or outcomes had these frequencies: blue, 21; hazel, 9; brown, 7; green, 2. The frequency of the complement of E (blue eyes) is the sum of 9, 7, and 2, or 18. It is also the sample size, 39, minus the frequency of blue: $39 - 21 = 18$.

3.2 Long-run Tendency of Relative Frequencies

It was mentioned in Chapter 2 that in dealing with samples of various sizes, it is awkward to deal with frequencies of outcomes, since these are significant only when related to the sample size. So in the case of events generally, it is more meaningful to use *relative* frequencies when considering samples of varying size. If $\text{freq}(E) = f$ in a sample of n observations, the relative frequency of E is the ratio f/n.

Relative frequencies of the elementary events of an experiment (i.e., of the distinct possible outcomes) must add up to 1, being proportions of the whole. Thus, if the elementary outcomes are $o_1, o_2, \ldots, o_k$, and if $\text{freq}(o_i) = f_i$, then

$$f_1 + f_2 + \cdots + f_k = n.$$

As is evident upon division by n,

$$\frac{f_1}{n} + \frac{f_2}{n} + \cdots + \frac{f_k}{n} = 1.$$

This equation, stating that the proportions of the whole add up to 1, will often be written, using *sigma notation*, as

$$\sum_{i=1}^{n} \frac{f_i}{n} = 1.$$

(The Greek capital sigma is an instruction to sum terms of the type that follow it, as the subscript i courses through the integers $1, 2, \ldots, n$, this particular range of values of i being indicated by the $i = 1$ and n that grace the $\sum$.)

EXAMPLE 3.5

The table of frequencies for 120 tosses of a die given in Example 3.2 is reproduced here as Table 3.1 but extended with a list of corresponding

Table 3.1

NUMBER OF POINTS	FREQUENCY	RELATIVE FREQUENCY
1	12	$\frac{12}{120}$
2	24	$\frac{24}{120}$
3	21	$\frac{21}{120}$
4	16	$\frac{16}{120}$
5	19	$\frac{19}{120}$
6	28	$\frac{28}{120}$
Total	120 (n)	1

relative frequencies. Notice that *events* in this experiment also have relative frequencies. For instance, if, as before, $E_1 = \{2, 4, 6\}$ and $E_2 = \{4, 5, 6\}$, then

$$\text{rel. freq}(E_2) = \tfrac{68}{120} = \tfrac{24}{120} + \tfrac{16}{120} + \tfrac{28}{120},$$

being the sum of relative frequencies of the elementary events that make up the event E_1. Similarly,

$$\text{rel. freq}(E_2) = \tfrac{63}{120} = \tfrac{16}{120} + \tfrac{19}{120} + \tfrac{28}{120}.$$

In following up the speculation that larger samples give a more accurate picture of the basic experiment or underlying population, it is interesting to study the results of some "actual" sampling from a known population, as more and more observations are taken, to note any long-run tendencies. The results presented in the examples of this section are indeed actual, in the sense that they were not an invention of the author's imagination; but the word "actual" is in quotes because the results were obtained by computer simulation of a sampling process. This process of artificial or simulated sampling makes it feasible to study long-run tendencies to an extent that would be difficult to achieve by means of real sampling in actual problems.

EXAMPLE 3.6

The artificial population sampled in Example 2.12 was sampled 1000 times, each of the 1000 observations being classified as to whether or not it fell in the event $E: 31.05 < x < 31.95$. The first several results were as follows, recorded S for success (the observation *did* fall in E) or F for failure (the observation did *not* fall in E):

S F F F F F F S F F F F F F F S F F S F S

F F F F F F F F S F F S S S F F F F F F S F

F F F F F S F F F F S F S F S F F F S F F F.

After each observation, the relative frequency of E among the observations obtained up to that point can be calculated. By referring to the above sequence, the following sequence of relative frequencies can be calculated:

$$1, \tfrac{1}{2}, \tfrac{1}{3}, \tfrac{1}{4}, \tfrac{1}{5}, \tfrac{1}{6}, \tfrac{1}{7}, \tfrac{2}{8}, \tfrac{2}{9}, \tfrac{2}{10}, \tfrac{2}{11}, \tfrac{2}{12}, \tfrac{2}{13}, \tfrac{2}{14},$$

$$\tfrac{2}{15}, \tfrac{2}{16}, \tfrac{3}{17}, \tfrac{3}{18}, \tfrac{3}{19}, \tfrac{4}{20}, \tfrac{4}{21}, \tfrac{5}{22}, \tfrac{5}{23}, \text{etc.}$$

A plotter was used to obtain the graphical record of this sequence and its continuation to 1000 observations; the result is given in Figure 3.1. Although for small numbers of observations the relative frequency of occurrence of E is somewhat erratic, it tends to settle down or stabilize at a limiting value. The horizontal line in the graph of Figure 3.1 is at the height .1963, which is the limiting value theoretically anticipated in setting up the computer simulation. The actual graph of the relative frequency does get rather close to this and appears to be heading for it (or something very close to it) as a limiting value.

EXAMPLE 3.7

Three sequences of tosses of a computer-simulated die were carried out, the results recorded in terms of frequency of a 6. The plots of relative

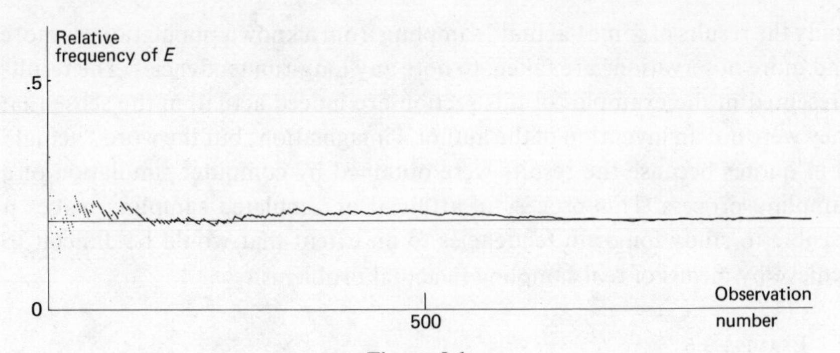

Figure 3.1
Tendency, after 1000 trials, of the relative frequency of an event (Example 3.6).

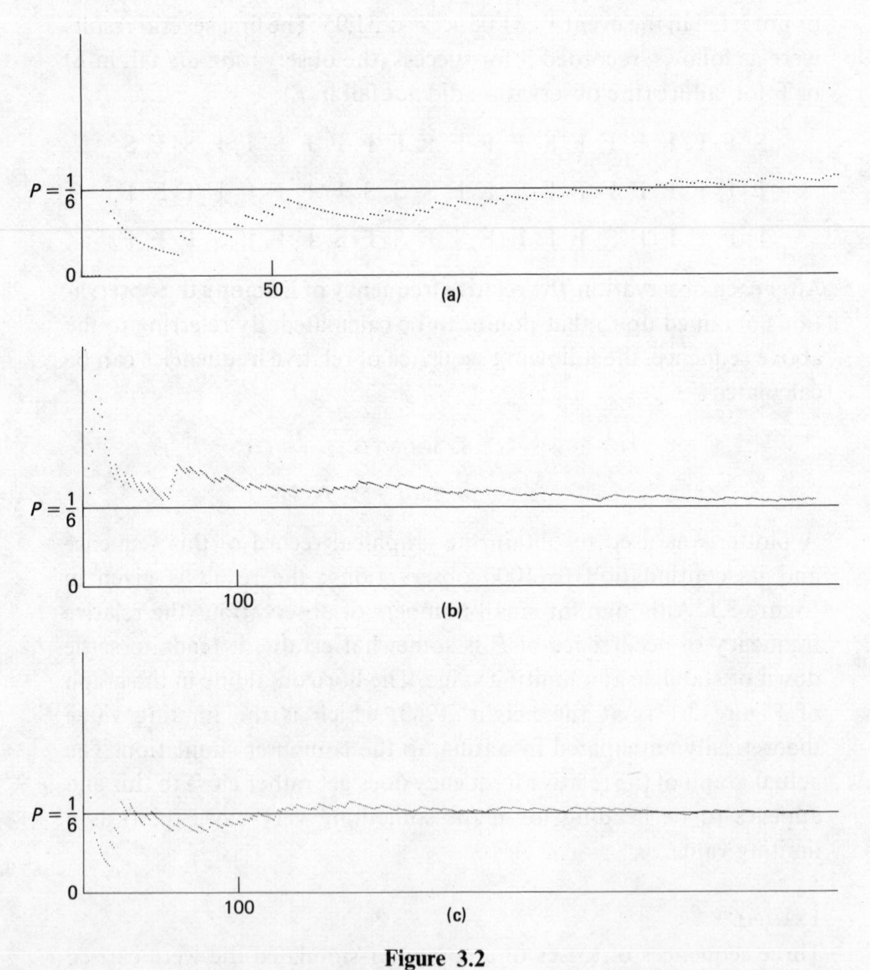

Figure 3.2
Relative frequency of 6 in simulated die tossing: (a) 200 tosses; (b) and (c) 500 tosses.

frequency for the one run of 200 tosses and two additional runs of 500 tosses, as a function of observation number, are shown in Figure 3.2. The horizontal lines are drawn at the level $\frac{1}{6}$, which experience shows to be the long-run limit for an ordinary die—after a *large* number of tosses. Notice that after 200 tosses in the one experiment, the relative frequency is not quite showing signs of stability. In the second run, even after 500 tosses, the apparent limiting value is not $\frac{1}{6}$, although in the third run it seems to be heading for $\frac{1}{6}$ by the time 500 tosses are recorded. It might be observed that in Figure 3.1 the limiting value that is emerging after 1000 trials is not yet apparent after 500 trials.

Although the event counted in this experiment was the turning up of a "6" in a (simulated) die, which is an elementary event of the experiment, the relative frequency of *any event* would clearly have to exhibit the same stability as do the relative frequencies of the *elementary* events; for the relative frequency of an event is the sum of the relative frequencies of the elementary events that make it up. For instance, the sum of the relative frequencies of $\{2\}$, $\{4\}$, and $\{6\}$ is the relative frequency of the event $\{2, 4, 6\}$. If the relative frequencies of $\{2\}$, $\{4\}$, and $\{6\}$ have long-run limits, then the relative frequency of $\{2, 4, 6\}$ will also have a long-run limit, which is the sum of the long-run limits of the component outcomes or elementary events.

Problems

3.1. What would be an appropriate sample space for each of the following experiments?

(a) Counting the number of telephone calls coming into a switchboard in a 1-minute period.

(b) Drawing a card from an ordinary deck of playing cards.

(c) Drawing a person's name blindly from a voter list for jury duty.

3.2. A gambling game consists of randomly cutting a deck of cards. A reward is paid for a face card and a higher reward for an ace. Would you need the sample space of 52 elements in Problem 3.1(b)? What sample space *would* be appropriate?

3.3. A die is tossed 20 times with these results:

$$4, 6, 2, 3, 3, 6, 1, 4, 5, 1, 5, 6, 1, 3, 1, 2, 6, 4, 4, 6.$$

Determine the relative frequencies, in this sample, of these events:

(a) The result is even.
(b) The result exceeds 4.
(c) The result is a 1.
(d) The result is even or is a 1.

3.4. A swimmer keeps track of his times in a series of 400-meter races. What is an appropriate sample space for his time? If he kept track only of his finishing spot, what is the sample space?

3.5. Drivers are categorized according to eye color and hair color. What is the sample space? Suppose that 100 drivers were classified, with the following results:

	BLUE	BROWN	HAZEL
Brown	20	10	9
Black	4	8	0
Blond	17	5	6
Gray	8	4	2
Red	4	2	1

Determine the relative frequency of each event:
(a) Blue eyes. (b) Blond hair.
(c) Blue eyes or blond hair. (d) Blue eyes and blond hair.
[Are there any relationships among (a)–(d)?]
(e) Gray hair.
(f) Gray hair and blue eyes, or gray hair and brown eyes, or gray hair and hazel eyes. (Any comment?)

3.6. In the experiment yielding the data on hemoglobin level in Problem 2.9, what is the sample space? Determine the relative frequency of the event $H > 15$ (where H denotes the hemoglobin measurement) and also the relative frequency of $|H - 15| > 4$.

3.7. Toss a coin (or a die—or a thumbtack, if you have one) 100 or more times and keep track of the relative frequency of some event as you toss. Is there a limiting tendency? [The interesting thing about the thumbtack, which (like the coin) can fall in one of two ways, is that you probably cannot guess the limiting relative frequency ahead of time.]

3.3 Probability

It is a phenomenon observed in real life, as it is in these examples, that the long-run relative frequencies of events tend to stabilize, to approach long-run or limiting values. Because the approach takes place as n, the sample size, increases *without limit*—something one can only think about, and not carry out, the limiting values are said to be *ideal*. They are thought of as intrinsic characteristics or properties of the experiment being performed, characteristics that may not be evident in one or a small number of performances of the experiment, but that begin to emerge after many performances.

The ideal or long-run limiting relative frequencies of events are called their *probabilities*. The probability of an event is its limiting relative frequency in any sequence of observations of the basic experiment that continues without limit, and it is axiomatic that this limit is the same in essentially all such sequences of observations. Intuitively, the probability of an event is related to what one thinks of as the *chance* of the event's occurrence in a given, single trial of the experiment. Having observed that approximately one-half of the tosses in repeated tosses of a coin result in *heads*, one has the feeling that in any single toss the two outcomes *heads* and *tails* have the same chance of turning up—are equally likely to occur. He would bet on one outcome or the other with even odds, in a single toss of the coin.

Mathematically, rather than base a theory of probability on a definition of probability as a long-run relative frequency, it is easier to take the existence of long-run trends as evidence on which to assume the existence of a number called probability for each event, a number that satisfies (as a set of axioms) certain properties of a relative frequency. The properties are these:

(i) The probability of the event that consists of all outcomes in the sample space is 1.
(ii) Probability is a nonnegative number.
(iii) Probabilities are additive over nonoverlapping events.

Relative frequencies, in any given sample, constitute a measure of events that satisfies these properties; and with these as axioms for probabilities of events, a consistent and useful mathematical theory of probability can be developed.

The notation commonly used for the probability of an event E is $P(E)$. In terms of this notation, and denoting by Ω the whole sample space, the above properties can be written as follows:

(i) $P(\Omega) = 1$.
(ii) $P(E) \geq 0$, for every event E.
(iii) $P(E_1 \text{ or } E_2) = P(E_1) + P(E_2)$, for any two nonoverlapping events E_1 and E_2.

Here the event written "E_1 or E_2" means the event consisting of those outcomes that satisfy condition E_1 or condition E_2 (or both). The term "nonoverlapping" means that no outcomes satisfy both conditions E_1 and E_2; i.e., the conditions defining them are mutually exclusive.

Probabilities of events in a sample space of an experiment, being idealizations that are only *assumed* to exist, constitute, together with their properties and rules of combination, what is called a *mathematical model* for the experiment. This model is a structure of mathematical symbols, definitions, axioms, and rules of logic, together with consequences of those axioms and rules.

EXAMPLE 3.8

Axioms (i) and (iii) can be used to derive a useful relation that gives the probability of the complement of an event in terms of the probability of the event. For, if E_2 is the complement of E_1, then

$$\{E_1 \text{ or } E_2\} = \Omega,$$

and, by axiom (iii), since an event and its complement are nonoverlapping,

$$P(\Omega) = P(E_1) + P(E_2).$$

Since $P(\Omega)$ is 1 by axiom (i), it follows that

$$P(\text{complement of } E_1) = 1 - P(E_1).$$

In a probability model, the mathematical quantities and their relationships are intended to represent a given phenomenon of chance so that one can use the mathematics to describe and predict various aspects of the phenomenon. As in the case of the mathematical model for any real-life phenomenon, one will never really *know* whether the model is correct, i.e., perfectly accurate and reliable in its representation. Nevertheless, it is assumed that there does exist a correct model; and it is often the aim of a scientific or empirical investigation to learn as much as possible about the true or correct model, or to construct a model that, in its representation of nature, is adequate for the investigator's purposes.

3.4 Discrete Models

A *finite probability model* is the model for an experiment with a finite number of possible outcomes. The data for such an experiment are given in terms of frequencies—the numbers of times (in the sample) each possible outcome occurs. If the outcomes are called, using general, mathematical names, $o_1, o_2, \ldots, o_k$, and the corresponding frequencies $f_1, f_2, \ldots, f_k$, the frequency distribution of a sample of n observations can be given in a table of relative frequencies, Table 3.2.

Table 3.2 Data from n performances of a discrete experiment.

OUTCOME	RELATIVE FREQUENCY
o_1	f_1/n
o_2	f_2/n
$\vdots$	$\vdots$
o_k	f_k/n
Sum	1

The probability model for this experiment, based on the assumption that relative frequencies have long-run limits as observations accumulate without limit, is given in a table similar to that for the frequency distribution of a sample, in terms of the idealizations called "probabilities." If the outcome or elementary event o_i has probability p_i, the table is as given in Table 3.3. The

Table 3.3 Model for a single performance of a discrete experiment.

OUTCOME	PROBABILITY
o_1	p_1
o_2	p_2
$\vdots$	$\vdots$
o_k	p_k
Sum	1

probabilities of the elementary events add up to 1, because the relative frequencies of these outcomes in any sample add up to 1.

An *event* E, being made up of elementary outcomes, has a probability that can be expressed in terms of the probabilities of elementary outcomes, in terms of the numbers $p_1, p_2, \ldots, p_k$, just as it has a relative frequency that can be expressed in terms of the relative frequencies $f_1/n, \ldots, f_k/n$ (see Example 3.5). Probabilities of events, defined by such computations, will satisfy the conditions given earlier as axioms for a probability model.

EXAMPLE 3.9

The result of selecting a card from an ordinary deck of cards can be classified as one of the following five outcomes:

o_1: ace,
o_2: king,
o_3: queen,
o_4: jack,
o_5: 10 or below.

In a sampling experiment the results shown in Table 3.4 were obtained from 1000 trials, each trial being the blind selection of a card from a shuffled deck.

Table 3.4

OUTCOME	RELATIVE FREQUENCY
o_1	$\frac{81}{1000} = .081$
o_2	$\frac{70}{1000} = .070$
o_3	$\frac{75}{1000} = .075$
o_4	$\frac{84}{1000} = .084$
o_5	$\frac{690}{1000} = .690$
Sum	1

One model proposed for the basic experiment of drawing a card from the deck is that given in the table of outcomes and probabilities, Table 3.5. The model is based on the observation that in a large number of trials, the face cards appear to occur with about the same frequency, and the other cards about nine times as often as any one face card.

Table 3.5

OUTCOME	PROBABILITY
o_1	$\frac{1}{13} \doteq .077$
o_2	$\frac{1}{13} \doteq .077$
o_3	$\frac{1}{13} \doteq .077$
o_4	$\frac{1}{13} \doteq .077$
o_5	$\frac{9}{13} \doteq .692$
Sum	1

The probability of any *event* can be computed in this model from these probabilities of elementary events. For example, the relative frequency of the event "the card is a face card" can be computed from the relative frequencies of its constituent elementary outcomes:

$$\begin{aligned} rf(\text{face card}) &= rf(\text{king}) + rf(\text{queen}) + rf(\text{jack}) \\ &= .070 + .075 + .084 = .229, \end{aligned}$$

where *rf* is a very temporary notation for relative frequency. In the same manner, the probability of a face card can be computed from the probabilities of the individual outcomes:

$$\begin{aligned} P(\text{face card}) &= P(\text{king}) + P(\text{queen}) + P(\text{jack}) \\ &= \tfrac{1}{13} + \tfrac{1}{13} + \tfrac{1}{13} = \tfrac{3}{13} \\ &\doteq .231. \end{aligned}$$

Similarly, every event in the sample space has a probability that can be computed from the probabilities of the elementary outcomes.

The approach described above for finite models can also be used if the list of outcomes is not finite, as long as it is indeed a *list*. In order to handle these discrete but nonfinite models it is necessary to know how to add infinite sequences of numbers (i.e., of probabilities). This matter will not be pursued further here.

A discrete probability model, finite or not, is often referred to as a *discrete probability distribution*, a term analogous to the earlier "frequency distribution." A *frequency* distribution tells how the *observations in a sample* are distributed among the various possible values; a *probability* distribution tells how probability is distributed among the various possible values. In both cases, various devices are used to describe the distribution, the frequency table and the probability table being commonly used in the discrete case. Another descriptive device will be taken up in Section 3.6.

How does one determine the true probability model for an experiment of chance? It cannot be done, really, but two different approaches are used in correspondingly different contexts. One approach is to postulate a certain simple model based on a study of the nature of the experiment, and the other is, by repeated performances of the experiment, to seek to learn from the data as much as possible about the true model. Both approaches are used in statistical inference; the first, in which a priori models are postulated, will be discussed next.

In many games of chance an object or device is used that has a certain symmetry which serves to suggest a particular model, even before any experimentation (hence the term *a priori*). Games involving coins, dice, and roulette wheels fall in this category.

EXAMPLE 3.10

An experiment consists of tossing a die and noting the number of points showing when it comes to rest. To the eye, the die is symmetrical—a perfect cube—and in betting on the toss without prior experiment one would usually give even odds on all the faces. That is, he would bet in a manner consistent with the conviction that all six faces have the same chance of occurring in a particular toss. This conviction may involve the belief that if the die were tossed a great many times, the relative frequencies of the six faces would be about equal. It may involve past experience with similar dice, but ordinarily that experience is not sufficiently well organized nor extensive as to permit basing a model on it. (Example 3.1 gave the results of 120 tosses, with resulting relative frequencies that were *not* all equal. But 120 is not a very large number.)

The mathematical or *ideal die* is then defined by the following table of equal probabilities:

Number of points	1	2	3	4	5	6
Probability	$\frac{1}{6}$	$\frac{1}{6}$	$\frac{1}{6}$	$\frac{1}{6}$	$\frac{1}{6}$	$\frac{1}{6}$

Because dice are used in games that assume, for fairness, the equal likelihood of the six faces, a real die should be manufactured out of a homogeneous material, and so that it is as nearly a perfect cube as possible. In tossing this die, one should pick it up carelessly and give it a healthy toss and spin, again so that no single side of the cube can be favored. To verify that one is successful in achieving conditions under which the ideal model truly represents the experiment, one would have to repeat the toss infinitely often and verify that the relative frequencies of the faces all approach $\frac{1}{6}$, which is clearly a most time-consuming task.

EXAMPLE 3.11

The author has a record, whose source has been forgotten, of a sequence of 200,000 tosses of a drilled die, as well as one of 200,000 tosses of a precision die. In particular, the number of even numbers showing in each sequence of tosses is given as follows:

drilled die: 100,916 even numbers in 200,000 tosses.
precision die: 100,196 even numbers in 200,000 tosses.

The corresponding relative frequencies are .50458 and .50098, respectively. Even with as prodigious a number of tosses as this, the agreement with the ideal value of .50000 is not overwhelming. The problem of deciding whether the agreement is good enough to accept the ideal model—to proceed as though the die is fair—will be discussed in Chapter 8.

Another type of experiment, including certain games of chance, for which one can construct an a priori model is that in which an object is selected from a group of objects in a way that the man on the street thinks of as "at random." Picking a card from a shuffled deck, picking a draft number from a rotating drum containing 366 capsules with numbered days of the year, and picking a name from a list of voters to send a questionnaire—these are examples of the type of experiment called *random selection*, provided they are conducted so that the ideal model is applicable. To be considered fair, one would want all cards or capsules or names to have the same chance of being chosen. That is, one would strive to emulate a mathematical or ideal selection in which the probabilities postulated for the possible outcomes are all equal. Thorough mixing or shuffling and blind selection are both essential.

To select an object *at random* will mean, by definition of the phrase, that the mathematical model with equal probabilities for all available objects is the one to be used, with the implication that in carrying out the experiment there will be an attempt to achieve experimental conditions such that this particular model is the true model. The random sampling model is useful not only in lotteries and card games but in sampling from actual finite populations of people for the purpose of learning about the characteristics of the population. The application to this situation will be discussed later in more detail.

The third type of a priori model to be considered is that for a spinning pointer or wheel. The roulette wheel is one that is constructed with 38 (or 37) compartments of equal size, into one of which a small ball falls when it slows down after being spun around the wheel. It would be fair if these compartments are equally likely outcomes, each with probability $\frac{1}{38}$ (or $\frac{1}{37}$). The wheel of fortune is fitted with equally spaced pegs, so that when it is spun and gradually slows down, a pawl pivoted near the edge falls down between two pegs and causes the wheel to stop in one of the finitely many positions determined by the pegs. In Russian roulette the chamber of a revolver, with six positions, is spun and comes to rest with one of the chambers in firing position; these are presumed to be equally likely. The spinner that is furnished with certain childrens' games is a pivoted pointer that stops, after being spun in a horizontal plane, in one of several sectors; these sectors are constructed with equal central angles, so as to be equally likely as regions in which the pointer might stop.

In each of these examples, the number of possible outcomes is a finite number, say k, and the a priori model is that in which each outcome is assigned probability $1/k$. A device is fair if it can be operated so that this a priori model represents its operation, in terms of long-run relative frequencies. A behavioristic judgment of fairness would be indicated by a willingness to bet with equal odds on all outcomes, in a single spin.

Problems

3.8. A ball in a roulette wheel has 38 equally likely stopping positions (if it is a "fair" wheel), numbered 0, 00, 1, 2, 3, . . . , 36. Of the numbers from 1 to 36, half are red and half are black; the 0 and 00 are green.

(a) What is the probability that the outcome is not red or black?

(b) What is the probability that the outcome exceeds 18?

(c) What is the probability of red?

3.9. When two cubes are tossed, there are 36 ways a side of one cube can be paired with a side of the other. Under the usual a priori model, these are taken to be equally likely.

(a) For ordinary dice with faces identified as 1, 2, 3, 4, 5, 6, list the 36 outcomes of the toss of two dice. What is the probability of a pair? Of a total of seven points on the two dice?

(b) A pair of dice is custom made, one having five dots on each of its six sides, the other having four sides with two dots each and two sides with six dots each. What is the probability of a sum of 7?

3.10. Three families, who are going to spend a vacation together, have children as follows:

Smith: 2 boys, 1 girl,
Jones: 1 boy, 3 girls,
Doe: 5 boys.

In assigning a chore, each child's name is put on a slip of paper, one being drawn at random. What is the sample space, and what are the probabilities of elements of the sample space?

Let events be defined as follows:

S: the surname on the slip is Smith,
B: the name on the slip is that of a boy,
J: the surname on the slip is Jones.

Determine

(a) $P(B)$ and $P(S)$.

(b) $P(J \text{ or } B)$.

(c) $P(\text{name is not Doe})$.

3.11. Each card has the same chance of being drawn as any other, in the a priori model for the random selection of a card from a standard deck of playing cards.

(a) What is the probability assigned to each card?

(b) What is the probability that the card drawn is

(i) a heart? (ii) a king?

(iii) a face card? (iv) a heart or a face card?

3.12. It is not hard to see that the number of distinct pairs of cards (from a deck of 52 playing cards) is

$$51 + 50 + 49 + \cdots + 1 = 1326.$$

The a priori model for the random deal or selection of two cards from the deck assigns probability 1/1326 to each distinct pair. Determine the probability of each event:

(a) The two cards form a pair (i.e., are of the same denomination—such as a pair of tens or a pair of aces).

(b) The two cards add up to 21 in a blackjack game. (Ace counts 1 or 11, a face card counts 10, and a numbered card counts the number on the card.)

(c) The pair of cards has a total of 17 points, under the counting scheme of (b).

3.13. As in Problem 3.5, a driver is classified according to color of hair and color of eyes. A driver is picked at random; the sample space for his hair–eye combination consists of the 15 combinations corresponding to the cells in the tabulation of Problem 3.5. A similar table is given here, but with entries that represent *probabilities* for the various outcomes:

	BLUE	BROWN	HAZEL
Brown	a	p	u
Black	b	q	v
Blond	c	r	w
Gray	d	s	x
Red	e	t	y

(Thus the s in the table is the probability that the driver has gray hair and brown eyes.) These probabilities are assumed to exist but are not known—which is the reason for the letter "names" rather than specific values. In terms of these names, give formulas for

(a) P(brown eyes). (b) P(blond hair).

(c) P(blue eyes but not blond). (d) P(not a redhead).

3.14. Consider the following game: Player A wins over B if a coin shows heads at least once when it is tossed twice. The sample space can be taken to consist of these elements: TT, TH, H. (The single H recognizes that if the first toss is heads, the game is won.) Mathematicians of the seventeenth century debated whether the correct model should assign *equal* probabilities or the probabilities $\frac{1}{4}$, $\frac{1}{4}$, and $\frac{1}{2}$ (respectively). Toss a coin often enough for you to convince yourself one way or the other, and in the model you choose calculate the probability that A wins.

3.5 Continuous Models

The mathematical model for the case of a variable of the *continuous* type is somewhat different from the discrete models considered so far. Recall that data from a continuous phenomenon are recorded by a process of *discretizing* by rounding off each observation to one of a list of values, a discrete list such as might be encountered in a discrete experiment. The summary of such data in a frequency table will look exactly like the summary of a set of observations on a discrete phenomenon. A difference is that in the continuous case the round-off procedure used is arbitrary, with the result that a variety of different frequency distributions can represent the same set of data.

At least two lines of reasoning can be used to go from the idea of a sample from a continuous population to the notion of a continuous probability model. One to be taken up in Section 3.8, idealizes the notion of sample distribution function. The approach here will utilize the histogram and its behaviors as the sample size increases to infinity and the class interval width shrinks to zero.

In constructing a histogram for a particular sample from a continuous population, having specified certain class intervals (perhaps by specification of the round-off procedure), one represents the fact that the observations in a given class interval could have come from anywhere on that interval by erecting a bar over the whole interval. When the class intervals, and hence the bars, are of equal widths, the frequency of observations in an interval consisting of a series of adjacent class intervals is proportional to the area over that interval and under the outline of the tops of the bars.

Since it is the height of the bars that represents frequency, the sides of the bars have no particular significance except in outling the bars; when considering two adjacent bars, the boundary of their common side is irrelevant in computations of area. Two histograms from Figure 2.8 are reproduced in Figure 3.3, one as it was earlier and another with the sides of the bars removed where two bars touch each other. The result is a *curve* (mathematicians use this term even when describing something with corners).

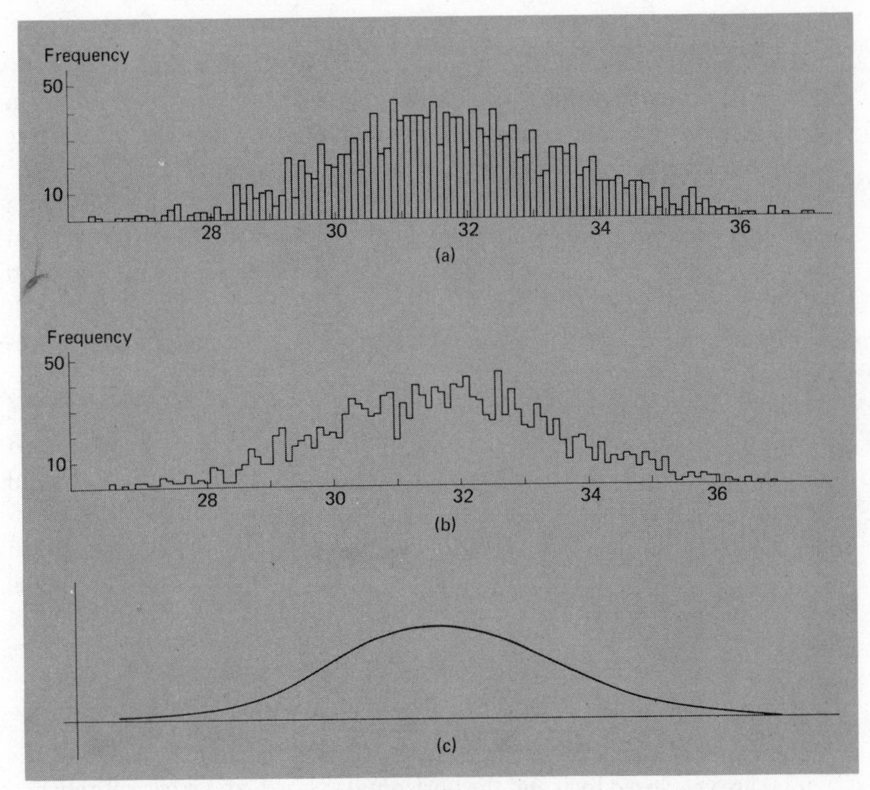

Figure 3.3
(a) and (b) histograms; (c) idealization to a continuous model.

The curve defined by the top of a histogram will naturally be different for each sample, since the histograms are different. As the sample size is increased, however, the erratic nature of the curve will be smoothed out to some extent—not completely as long as the system of class intervals is fixed. If one takes correspondingly shorter class intervals along with the increasing sample size, it is not difficult to imagine that a limiting curve would emerge, one that is more like what one thinks of as a curve, without so many square corners. Such a limiting curve, drawn by sheer speculation from the histograms in Figure 2.8, is also shown in Figure 3.3. This curve, or rather the *true* or correct one, something like what is drawn, is used to define a continuous probability model or continuous distribution of probability over the sample space.

To interpret such an ideal curve, it suffices to recall how to interpret a histogram. Values or regions of values where the tops of the bars in a histogram are high above the horizontal axis indicate regions that occur with high frequency. Even so, regions of values where the smooth, ideal curve is high above the horizontal axis indicate regions of high probability. The relative frequency of occurrence of an interval of values, in a given sample, is proportional to the area within the histogram, above the axis, and within the interval. Even so, the probability that a value will fall in a certain interval of values, in making one observation, is proportional to the area under the ideal curve defining the distribution, above the axis, and within the interval.

The curve used to define a continuous model is called a *density curve*, or a *density function*, or a *probability density function*, provided that the vertical scale is chosen so that the total area under the curve represents the total available probability, 1. With this convention, it follows that probability is not just proportional to area; it is *equal* to area, since the whole to which proportions relate is then 1.

Actually, *any* function whose graph lies above (or on, but not below) the horizontal axis and is such that the total of the enclosed area is 1 can be used to define a probability distribution of continuous type. Whether it is a useful distribution, in the sense of representing a real phenomenon (even approximately), is a separate question. Useful or not, the model exists as a mathematical entity, and one can compute probabilities in this model as areas under the density function's graph above the horizontal axis and over the events of interest.

EXAMPLE 3.12

The graph of Figure 3.4 can be used as a density function, because the total area enclosed by it and the horizontal axis is 1, and it lies entirely above or on the x-axis. It is possible to express the function in terms of formulas, but the graph serves perfectly well, and areas over intervals are composed of trapezoids and are easily calculated by elementary methods. For instance, the probability that an observation would fall between $-\frac{1}{2}$ and $+\frac{1}{2}$ is the area shaded in Figure 3.4. It is easily calcu-

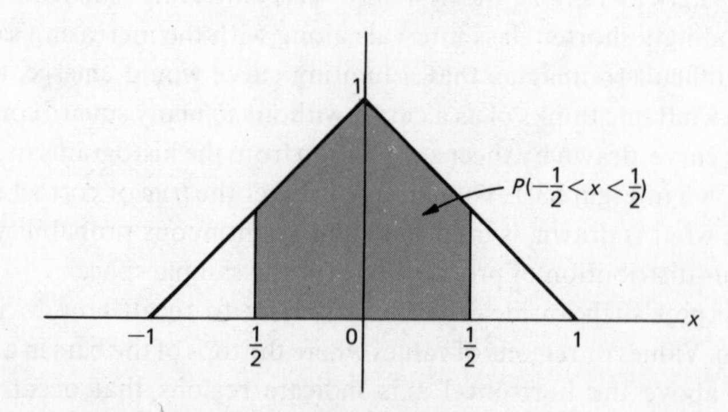

Figure 3.4
Density function for Example 3.12.

lated upon noticing that the *unshaded* area consists of two triangles, each of area $\frac{1}{8}$:

$$P(-\tfrac{1}{2} < x < \tfrac{1}{2}) = 1 - \tfrac{1}{8} - \tfrac{1}{8} = \tfrac{3}{4}.$$

Is there a phenomenon for which this is the model? The question might well be raised, in view of the sharp point, but it is not hard to construct an experiment whose results are distributed this way. (See Problem 3.18.) Be that as it may, the significance of the shape of the given density is that values near zero are more likely than values near 1 or -1. Probability is thought to be more highly concentrated where the density function is large.

The idea of a spinning pointer can be extended to one whose sample space is continuous. Imagine that the number of ratchet positions of the wheel of fortune is increased. The ratchet serves the purpose of automatically rounding off and discretizing what would otherwise be a continuous sample space, and increasing the number of stopping positions corresponds to increased accuracy in the round-off process. With 1000 positions, the stopping position on a scale from 0 to 1 around the circumference would be a number on that range rounded off to the nearest thousandth. In the a priori model for a fair wheel, the probability of each stopping position—each possible outcome—would be $\frac{1}{1000}$. And in the sense that the ratcheting provides an automatic round-off, each stopping position can be thought of as representing a range of values; for instance, .137 would represent the range from .1365 to .1375. But, then, just as the bar is used in a histogram to suggest that the values in a class interval could have come from anywhere in an interval, so one might represent probabilities by bars instead of rods and use a histogram type of graphical representation of the probability distribution. This would consist of 1000 side-by-side bars of equal height; the tops would constitute a horizontal straight line, and this would in effect be the idealization of the sample histogram, for this specified degree of round-off, as the number of observations increases without limit.

EXAMPLE 3.13

A wheel of fortune with 10 stopping positions was simulated on a computer. It was "spun" 1500 times, and the results are shown as histograms in Figure 3.5, which summarizes the results of the first 50, the first 150, the first 500, and then all 1500 spins. These illustrate the claim that as the number of spins increases, the curve defined by the tops of the bars in the histogram for the accumulated results becomes smoother and tends toward a horizontal line—if the wheel is fair,

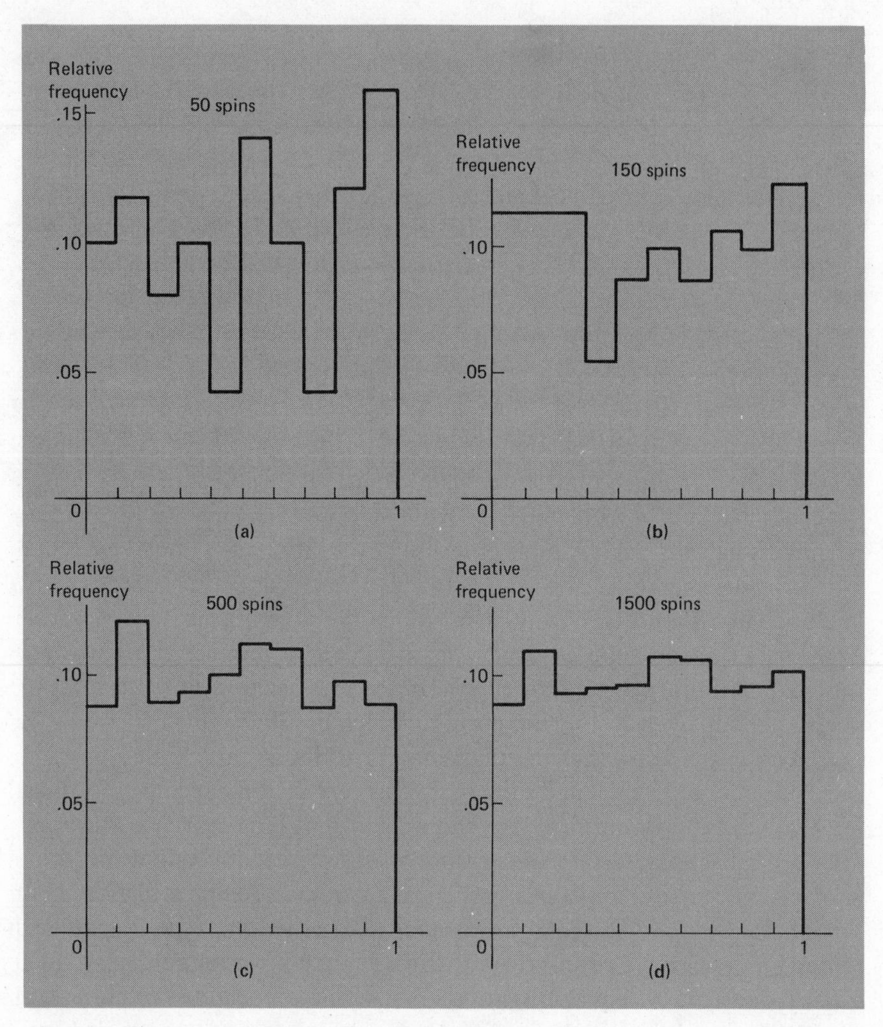

Figure 3.5
Histograms for spins of a wheel of fortune with 10 positions (scale from 0 to 1).

which in this case means that the simulation would have to be such that the 10 positions of the wheel are equally likely or equally represented in a large number of trials.

For *any* given number of ratchet positions, the tendency for the histogram to approach one with tops of equal height would be observed, as it was in the case of 10 positions in Example 3.13. As the number of positions increases without limit, the mechanism becomes indistinguishable from the freely spinning, or unratcheted, wheel. The model for this limiting device is thus taken to be defined by a density function that is of constant height over the interval of values marked on the perimeter of the wheel. This density is shown in Figure 3.6. It is an a priori model for an ideal or mathematical wheel.

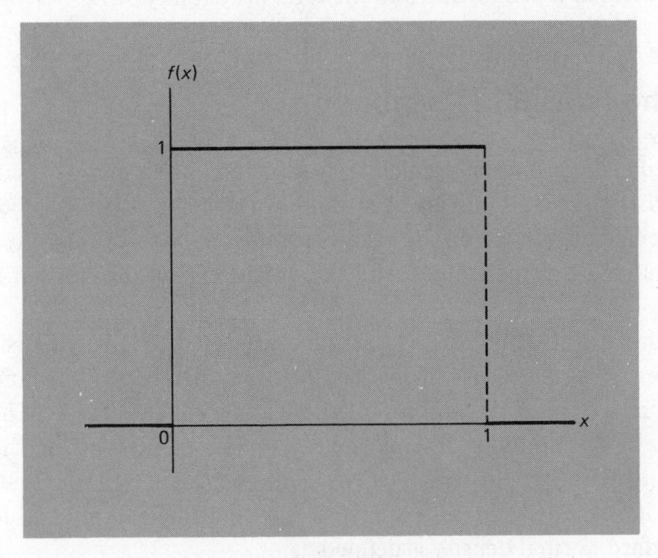

Figure 3.6
Density function for an ideal continuous wheel (scale 0 to 1).

The probability of an interval (i.e., a subinterval of the range of values from 0 to 1 marked on the circumference of the wheel) is proportional to the length of the interval:

$$P(a < x < b) = k(b - a).$$

Since

$$P(0 < x < 1) = k(1 - 0) = 1,$$

it follows that $k = 1$. This means that the height of the density curve at each point is 1, as marked in Figure 3.6. The probability of any interval is the area under the density function over that interval, and the probability of the whole interval from 0 to 1, i.e., of the whole sample space, is 1.

The above density function is called the *uniform* density, because of its constancy and because of the analogy with the idea of mass density of a uniform rod. It is the ideal model, representing long-run tendencies, but also what can happen in a single spin of the wheel—in terms of possible values and how one should bet.

A density function is a device for describing a probability distribution, and the uniform density just defined describes a distribution of probability that can be thought of as a uniform coat of paint, say, along a wire representing the x-axis. Any interval of length b contains a proportion b of the total amount of paint used, or a proportion b of the total probability of 1.

3.6 The Normal Distribution

A family of continuous models of particular importance is the family of *normal distributions*. A normal distribution is not an a priori model in the sense of being postulated on the basis of symmetry, but there are often reasons (somewhat more complicated) why one might expect the normal model to apply.

The family of distributions is basically generated by a single distribution, called the *standard normal distribution*, the other members of the family being obtained by a change of center (or location) and a change of scale. The various members of the family are therefore very similar to each other; the density functions are either congruent or can be made congruent by a simple stretching. The densities are *bell-shaped*, symmetrical, with a single hump and long tails. The standard normal density is defined as

$$f(x) = Ke^{-x^2/2}$$

(K being an unimportant constant of such size* as makes the area under the curve 1), and the graph is shown in Figure 3.7. Areas under the curve are not simple to compute, but tables are available in which can be found areas over

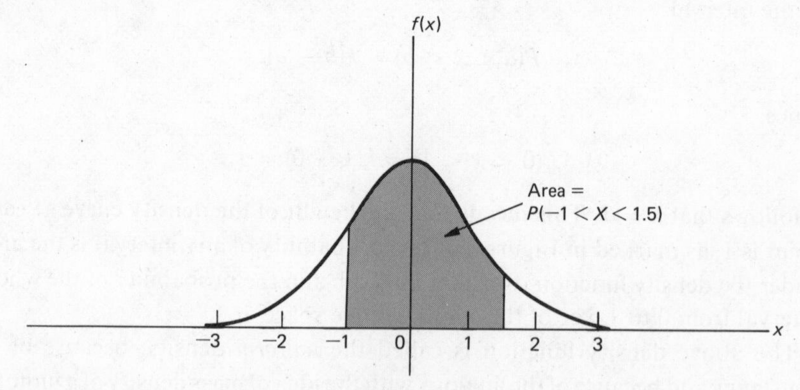

Figure 3.7
Standard normal density function.

* It can be shown that $K = 1/\sqrt{2K}$.

the intervals of the form $x < a$. Table I of the Appendix gives the area over $x < a$ as a function of a called $\Phi(a)$; i.e., to determine the probability of obtaining a value less than a one enters the table at a and reads out the probability $\Phi(a)$ in the body of the table. A brief extract from the table is the following:

a	−2.5	−2.0	−1.5	−1.0	−.5	0	.5	1.0	1.5	2.0	2.5
$\Phi(a)$	.006	.023	.067	.159	.308	.500	.692	.841	.933	.977	.994

EXAMPLE 3.14
The probability of observing a value less than 1.5, when an experiment is performed whose model is the standard normal distribution, is .933 (to three decimal places). The probability of observing a value in some interval, say between $x = -1$ and $x = 1.5$, can be computed by subtraction:

$$.933 - .159 = .774.$$

The .933 is the area under the normal curve to the left of $x = 1.5$, and .159 is the area to the left of $x = -1$. The difference, then, is the area between (see Figure 3.7).

The basic shape of a normal distribution is embodied in that of the standard normal distribution. Probabilities for the general normal distribution can be obtained from the standard normal curve, but this matter will be deferred until Section 4.6.

3.7 Random Variables

It is high time to introduce a language and notation that has become rather universal in probability and statistics. The result of a phenomenon that is numerical, or that has a numerical value assigned to it and is recorded in terms of that number, is called a *random variable*. The language refers not to the specific outcome, such as the number 5, which is one specific outcome when a die is tossed, but to the collection of values and the distribution of probability in the model assumed for describing the phenomenon. The term *probability distribution* has already been used to denote this concept, but this term refers to both the collection of values and the probability distribution, whereas the term *random variable* refers to the numerical outcome itself, the probability distribution being implicit rather than explicit. Thus the result of a toss of a fair die is a random variable; the list of possible values is implicit, as is the distribution of probability among those values.

The number at which a freely spinning pointer stops, on a scale marked on the circle traced by the pointer's tip, is a random variable. The uniform distribution for an ideal pointer is usually understood but is not referred to explicitly. The yield in bushels from an acre of corn is a random variable, and there is an implicit distribution of probability on the range of possible values of this yield. The term *random variable* refers to this distribution of possible yields rather than an actual observation such as 45.3.

It is convenient to use mathematical symbols for random variables, and the customary symbol is a capital letter. One might denote by X the outcome of the toss of a die, by Y the stopping point of a spinning pointer, and by Z the yield of corn in bushels from a 1-acre plot.

The convenience of this notation is that it is easy to denote events in terms of it. For example, the statement $X = 3$ defines the event that the die (if X denotes the outcome of the toss of a fair die) lands with the number 3 showing. The statement $.5 < Y < .8$ defines the event that the spinning pointer (if Y denotes the stopping point) stops at a number on the range from .5 to .8. The inequality $Z > 60$ (where Z is the number of bushels of corn per acre) defines the event that the yield is a number of bushels in excess of 60. And so on. Previously, the notation $\{x : a < x < b\}$ or, more simply, $a < x < b$, with x thought of as a "dummy" variable, was used to describe an interval of numbers from a to b; such an interval is an event in the sample space of real numbers. The new notation for this event is $a < X < b$, which is read "The random variable X falls in the interval from a to b."

The advantage of the random variable terminology and notation is perhaps not clear at this point, but it will emerge with use. Basically it permits one to say simply Z, in repeated references to yield, for example, instead of saying, each time, "the number of bushels of corn per acre."

Associated with a given random variable, say, X, there is a corresponding probability distribution in the space of its possible values. To describe this distribution, one might use a probability table or probability function in the discrete case, or a probability density function in the continuous case. One speaks of the "density function of the random variable X," meaning, of course, the density function of the probability distribution among the values of X.

Perhaps a more enlightening example of a random variable is that of determining some numerical characteristic for each person in a population of people, in connection with the chance experiment of picking a person at random from the population. The value of the numerical characteristic is a random variable; it has a certain range of possible values and, as will now be explained, an associated distribution. The basic distribution in this situation is defined by the method of sampling; it is a distribution in the population of people. With this distribution, each person in the population is equally likely to be drawn. But from this distribution one can calculate, from the value associated with each person, the probability of each particular value of the measured characteristic that might occur. (Or, in the idealization to an infinite population, with a continuum of possible values, one can calculate, in principle, the density of the distribution of values.) Thus there are two distributions in the picture—one in the population of *people*, and one in the population of *values* assigned to people. The notion of random variable refers to the assigned numerical value (such as the weight of a person) and to the distribution in the space of possible values (i.e., of possible weights of people).

The notation for a random variable is especially useful in writing probability statements. Thus, if W denotes the weight of a person chosen at random from a certain population of people, then $P(W > 150)$ denotes the probability that the weight of a person so drawn will exceed 150 lb. The inequality $W > 150$ is an event, and $P(W > 150)$ is the probability of that event.

Problems

3.15. A spinning pointer has a scale marked (uniformly) from -5 to 5. Determine the probability that the number at which the pointer stops
(a) is between -1 and 1.
(b) is *not* between -1 and 1.
(c) exceeds 3 in magnitude. (The *magnitude* of a number ignores its sign.)

3.16. Suppose that the model for a hemoglobin measurement H in a certain population of patients is given by the density function in Figure 3.8. Let q, r, s, t, and u shown on the figure be the areas indicated under the curve, above the axis, and between the vertical lines.

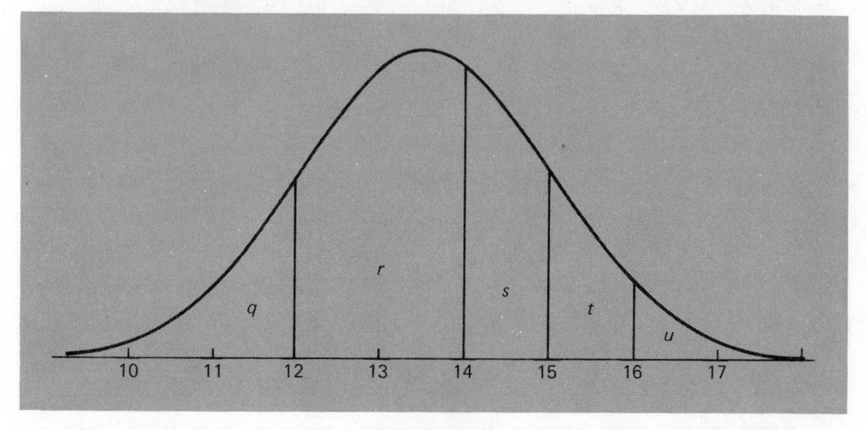

Figure 3.8
Density for Problem 3.16.

(a) In terms of these areas (which are probabilities) give formulas for
   (i) $P(H > 12)$.
   (ii) $P(H < 15)$.
   (iii) $P(|H - 14| > 2)$.

(b) Is there a relationship among the probabilities q, r, s, t, and u?

3.17. A random variable Z has a standard normal distribution. Use the brief table in Section 3.6 to determine the following:

(a) $P(Z < 2.0)$. (b) $P(Z > 2.0)$.
(c) $P(Z < -2.0)$. (d) $P(|Z| > 2.0)$.
(e) $P(-2.0 < Z < 1.5)$.

3.18. When a spinning pointer with scale 0 to 1 is spun *twice*, the a priori model for the difference of the outcomes (first outcome minus second) is given by the density in Figure 3.9. Determine the probability that

(a) the two outcomes are within .1 of each other.
(b) the outcomes differ by more than .5.

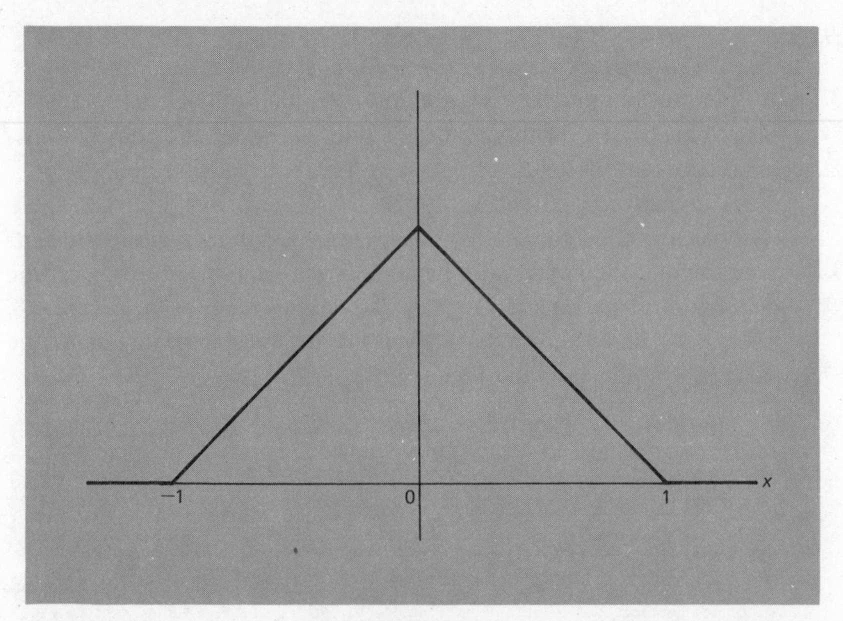

Figure 3.9
Density for Problem 3.18.

3.19. Go back to the histogram you drew for the hemoglobin data of Problem 2.9. Transfer it to (if it is not already on) a piece of graph paper marked with squares large enough so that you can easily, but crudely, estimate areas by counting squares. Sketch a smooth curve over the histogram, a curve that you think might be the population density from which the data are taken. Determine roughly (by counting squares) the probability in your sketched model that a measurement falls between 14 and 16. From the original data, determine the relative frequency of the event $14 < H < 16$ and compare with your "probability."

3.20. Let Z have a standard normal distribution; use Table I to determine the following:

(a) $P(Z < 1.96)$. (b) $P(Z > 1.96)$.

(c) $P(|Z| > 1.96)$. (d) $P(-1.645 < Z < 1.645)$.

3.8 The Distribution Function

Chapter 2 introduced a descriptive device for samples of numerical observations, called the *sample distribution function*. This function describes in a cumulative fashion the same information as is given in a frequency table or in the list of ordered observations. It has a counterpart in describing probability distributions of random variables, called the *probability distribution function*, or just the *distribution function*, or sometimes the *cumulative distribution function*—popularly abbreviated *c.d.f.*

The sample distribution function gives, as a function of a dummy variable x, the relative frequency of the sample observations that do not exceed x. Similarly, the distribution function (or c.d.f.) gives, as a function of x, the probability that one will observe a value that does not exceed x.

In repeated performances of the experiment yielding a random variable X one can watch for any particular event, say, $X \leq 10$, and record its frequency of occurrence as shown in Table 3.6. The relative frequency f/n will tend to a

Table 3.6

OUTCOME	FREQUENCY	RELATIVE FREQUENCY
$X \leq 10$	f	f/n
$X > 10$	$n - f$	$1 - f/n$
Sums	n	1

limiting value, as n increases, called the *probability of the event* $X \leq 10$. Similarly, the event $X \leq 2$ has a relative frequency and limit thereof called its *probability*; as does the event $X \leq -37$, etc. Indeed, for *any number* u, the event $X \leq u$ has a relative frequency (in a given sample) and a long-run relative frequency, as n increases, that gives the odds applicable to a single trial. This probability of $X \leq u$ depends on the number u, so it is a *function* of u. This function is the distribution function.

For a finite probability distribution—a distribution of probability among a finite set of values—the distribution function looks exactly like a sample distribution with the same possible values; and it is computed in the same way, but using probabilities instead of relative frequencies. Thus a sample distribution function for possible values $x_1, \ldots, x_k$ can be written

$$F_n(u) = \sum_{\substack{i \text{ for which} \\ x_i \leq u}} \frac{f_i}{n},$$

which means, simply, just the sum of all relative frequencies of possible values that do not exceed u. In exactly the same fashion, the population distribution function for a random variable with possible values $x_1, \ldots, x_k$ can be written

$$F(u) = \sum_{\substack{i \text{ for which} \\ x_i \leq u}} p_i,$$

which is just the sum of the probabilities of all possible values that do not exceed u. The next example illustrates each of these concepts and the convergence of the sample distribution function to the population distribution function as the sample size increases.

EXAMPLE 3.15

The data used for Example 3.13 could have been generated by tossing a symmetrical 10-sided die (rather, a long cylinder with a cross section that is a regular 10-sided polygon), whose a priori model assigns equal probabilities to the 10 possible outcomes. Let the outcomes (or sides of the cylinder) be numbered 0, 1, 2, ..., 9, a process that defines a random variable, namely, the number on the side that turns up when the cylinder is rolled and comes to rest.

Tables 3.7 and 3.8 give the relative frequencies for the samples of sizes 50, 150, 500, and 1500 that were obtained for Example 3.13, along with a column corresponding to the ideal as $n \to \infty$. Table 3.7 gives the relative frequencies and probabilities, and Table 3.8 gives cumulative frequencies and cumulative probabilities that define the sample and population distribution functions.

The graphs in Figure 3.10 show the cumulative relative frequency for $n = 50$ and the cumulative distribution function (for the a priori model). Even for $n = 50$ the graphs are not far apart, and if the graph for $n = 1500$ had been shown, it would have been almost indistinguishable, to the eye, from the c.d.f. of the population model.

Table 3.7

VALUE	RELATIVE FREQUENCY				PROBABILITY
	$n = 50$	$n = 150$	$n = 500$	$n = 1500$	
0	.10	.113	.088	.090	.1
1	.12	.113	.122	.112	.1
2	.08	.113	.090	.093	.1
3	.10	.053	.094	.096	.1
4	.04	.087	.102	.097	.1
5	.14	.100	.114	.109	.1
6	.10	.087	.112	.108	.1
7	.04	.107	.088	.095	.1
8	.12	.100	.100	.097	.1
9	.16	.090	.090	.103	.1

Table 3.8

VALUE	CUMULATIVE RELATIVE FREQUENCY				CUMULATIVE PROBABILITY
	$n = 50$	$n = 150$	$n = 500$	$n = 1500$	
0	.10	.113	.088	.090	.1
1	.22	.226	.210	.202	.2
2	.30	.339	.300	.295	.3
3	.40	.392	.394	.391	.4
4	.44	.479	.496	.488	.5
5	.58	.579	.610	.597	.6
6	.68	.666	.722	.705	.7
7	.72	.773	.810	.800	.8
8	.84	.873	.910	.897	.9
9	1.00	1.000	1.000	1.000	1.0

In the graphs of Figure 3.10, as sometimes in later graphs, vertical lines are drawn in to connect the horizontal sections for the purpose of guiding the eye; only the horizontal sections actually give values of the function. It is to be noted that the value *at* one of the integers 0 through 9 is the height of the graph as measured to the upper edge of the stair riser. (Such functions are called *staircase functions*.)

In the case of a continuous population, the probability of the event $X \leq u$, although again a function of u and called the *distribution function*, is not so easily calculated as it is in the discrete case—where it is simply a matter of adding probabilities of values no larger than u. Here, in the continuous case,

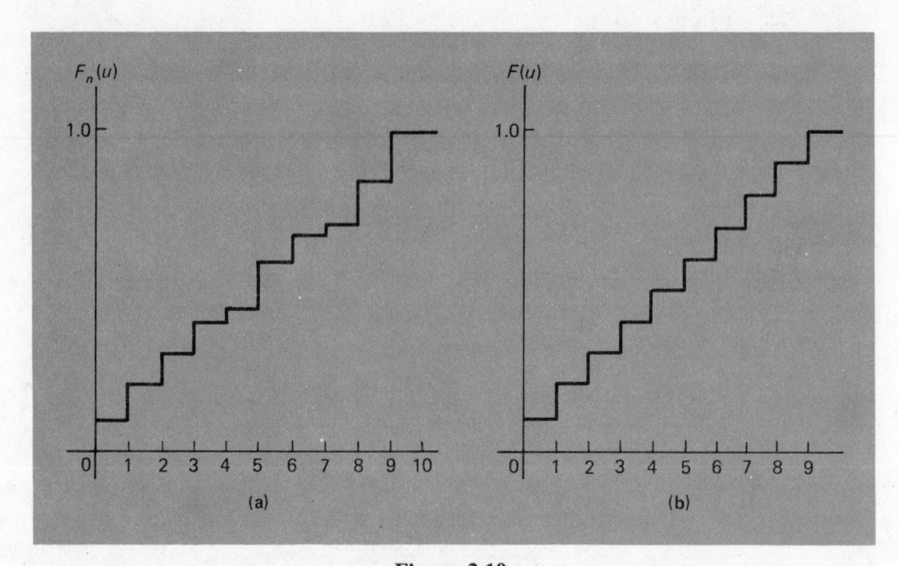

Figure 3.10
Distribution functions for Example 3.14. (a) $n = 50$; (b) ideal 10-sided die ($n = \infty$).

probabilities are *areas* under the density curve that define the model, and unless the curve is especially simple these areas are not easily calculated. The probability $P(X \leq u)$ that defines the c.d.f. is the area under the density, above the horizontal axis, and to the left of u. (See Figure 3.11, which gives a typical density function along with a graph of the corresponding c.d.f.)

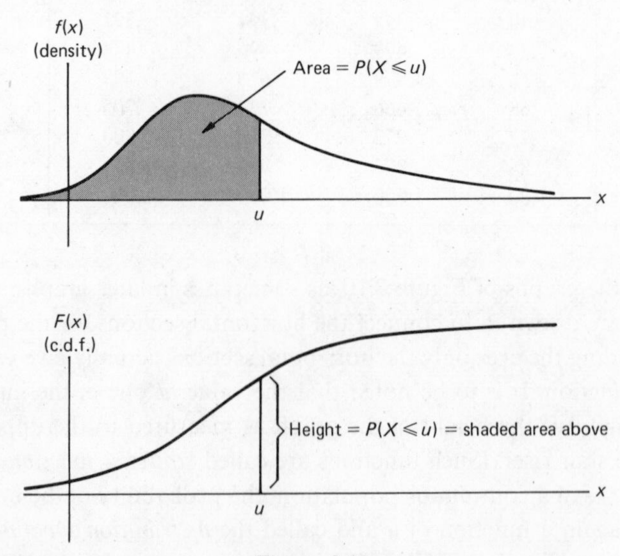

Figure 3.11
Relation between density and distribution functions.

The distribution function of a continuous distribution is a record of how much area is swept out to the left of a vertical line as this line is moved from left to right. Since area is added continuously and smoothly as the line moves to the right, the amount of area swept out grows continuously. This means that $P(X \leq u)$ is a continuous function of u and rises continuously to a final value 1. (That is, a pencil can draw the curve without being lifted off the paper.) The height of the distribution function at each point is just equal to the number of area units to the left of a vertical line through that point under the density curve. (See again Figure 3.11.)

EXAMPLE 3.16

The a priori model for the ideal spinning pointer, which stops at some point on the interval from 0 to 1, is defined by a density curve of constant height over that interval. The corresponding distribution function is easy to compute, because the areas involved are simple, geometric figures. The area to the left of u and under the horizontal density function is just the area of a rectangle of base u and height 1:

$$F(u) = P(\text{pointer stops between 0 and } u)$$

$$= \begin{cases} 0 & \text{if } u \leq 0, \\ u & \text{if } 0 < u \leq 1, \\ 1 & \text{if } u > 1. \end{cases}$$

The graph of the density and of this distribution function $F(u)$ are shown in Figure 3.12. As a vertical line moves from left to right across the density function graph, area is added on at a uniform rate; this results in a c.d.f. that increases linearly over the interval of possible values from 0 to the final height 1. Again at any particular point, the height of the c.d.f. is equal (as a number of linear units) to the area (as a number of areal units) to the left of the corresponding point on the density curve. And both fractions represent the probability that a value will be observed that is to the left of that point.

The fact that the sample distribution function $F_n(x)$, which records the relative frequency of observations in a sample of size n that do not exceed x, tends toward the population distribution function $F(x)$, which gives the probability that an observation will not exceed x, is illustrated in Figure 3.13. There are shown sample distribution functions corresponding to samples from a uniform spinning pointer of sizes 20, 50, 100, and 200.

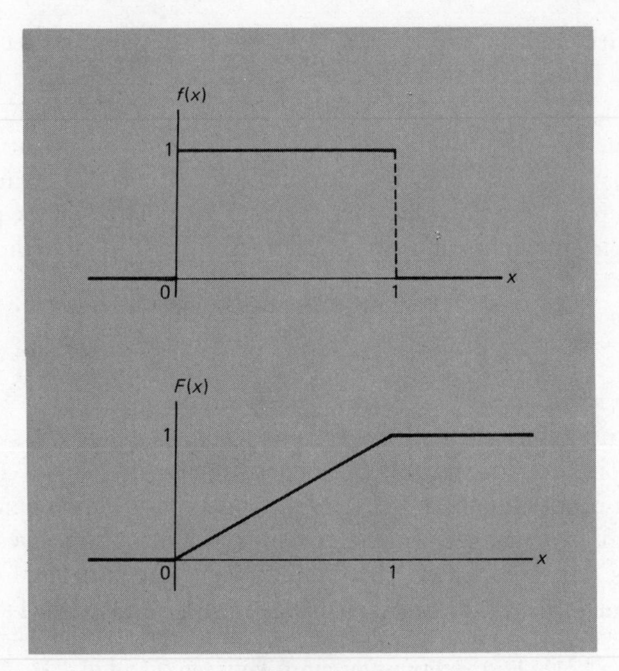

Figure 3.12
Uniform distribution.

EXAMPLE 3.17
The distribution function of the standard normal distribution (see Section 3.8) is tabulated in Table I of the Appendix. This table gives the area to the left of any point z as a function of z, or the probability that a value less than z is observed. And this is precisely what is called the distribution function. A sketch of the standard normal distribution function $\Phi(z)$ is given in Figure 3.14. Sample distribution functions for samples of sizes 20, 50, 100, and 200 from a standard normal population are shown in Figure 3.15, illustrating again that the sample distribution function approaches the population distribution function as the sample size increases.

A distribution function does not contain information that is not already contained in the probability table of a discrete distribution or in the density graph for a continuous distribution. It simply provides another way of describing a distribution or defining a population. Its usefulness lies primarily in the fact that it is a unifying concept, for with it, both discrete and continuous populations can be described. In the discrete case the c.d.f. is a staircase function, whereas in the continuous case it is a continuously increasing function.

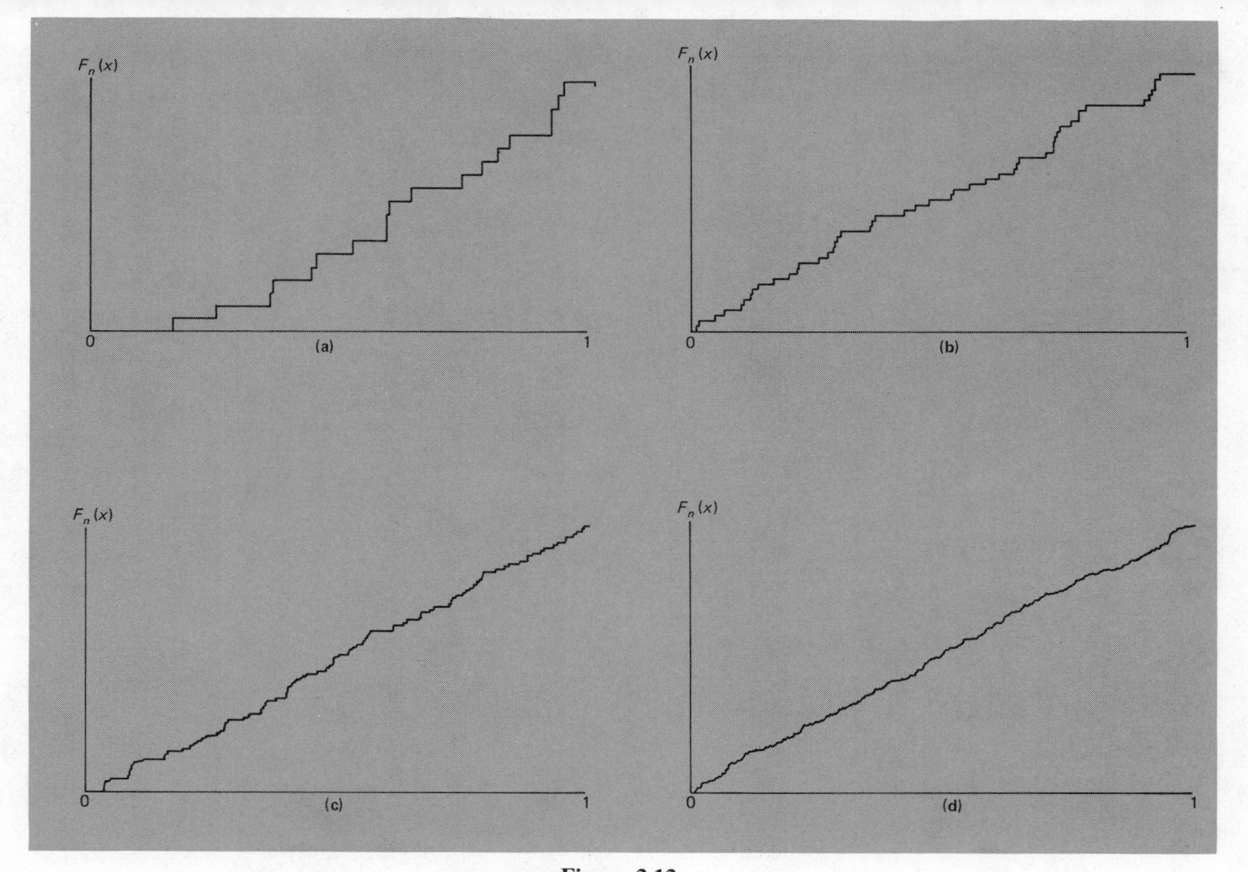

Figure 3.13

Sample distribution functions from uniform population. (a) $n = 20$; (b) $n = 50$; (c) $n = 100$; and (d) $n = 200$.

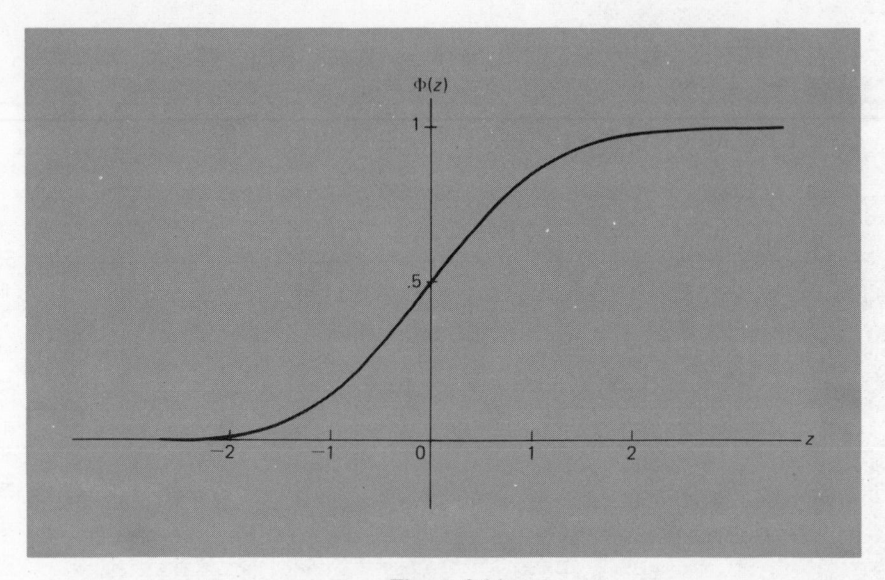

Figure 3.14
Standard normal distribution function.

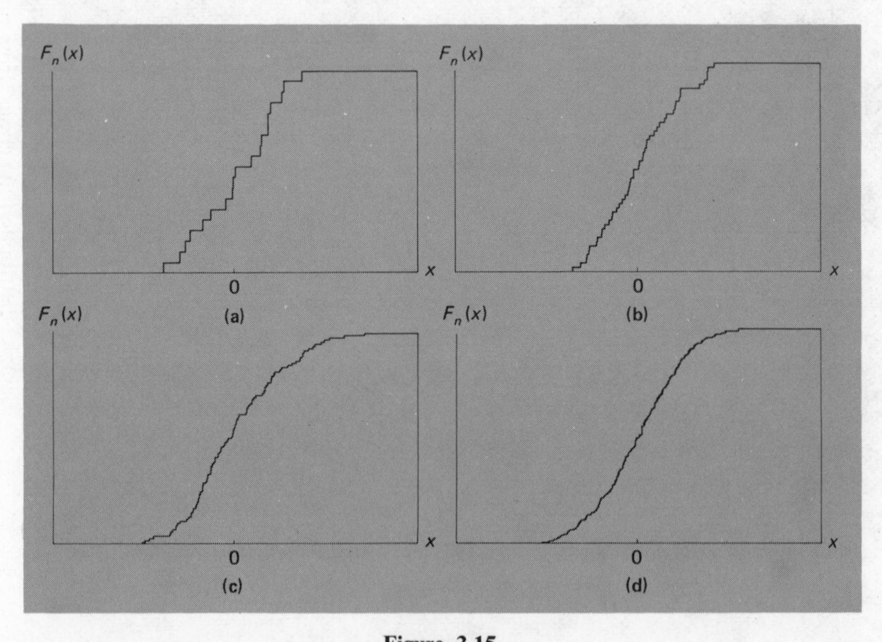

Figure 3.15
Sample distribution functions from normal population. (a) $n = 20$; (b) $n = 50$; (c) $n = 100$; and (d) $n = 200$.

Given a probability function or a density function, one can construct a corresponding distribution function; conversely, given a distribution function, it is possible to construct a probability table (in the discrete case) or a density function (in the continuous case). The probability table entries are simply the magnitudes of the stair-risers in the staircase distribution function. The density function, in the continuous case, happens to be the slope of the distribution function, but this concept from calculus will not be developed here.

A discrete population or distribution is most commonly given by means of a table of probabilities of individual values but sometimes by tables of cumulative probabilities (i.e., by the distribution function). A continuous population or distribution is almost always given by tables of distribution function values. Several such tables are included in the Appendix, although some are given in the inverse form, in which one specifies a value for $F(x)$ and looks in the table for the corresponding x.

3.9 Random Number Tables

A sequence of numbers obtained by random sampling is said to be a sequence of *random numbers*. Although the population from which these numbers come could be any population, sometimes the term "random number" is used to refer in particular to a sequence of observations taken from a uniform population. The Rand Corporation has published a book of random numbers.* These were obtained by repeating (in independent trials) an experiment whose model has 10 equally likely outcomes:

Model for random digits

Outcome	0	1	2	3	4	5	6	7	8	9
Probability	.1	.1	.1	.1	.1	.1	.1	.1	.1	.1

A table of random numbers has at least two uses: (1) as a tool for obtaining a random sample from an actual, finite population, and (2) as a source of manufactured data to use in studying sampling phenomena. For either use it should be noted, first, that although the table contains random "digits," successive pairs of these digits would be observations from the 100 equally likely two-digit numbers: 00, 01, 02, ..., 97, 98, 99. By putting a decimal point in front of these, one has observations from $0 < x < 1$ that are rounded off to two decimal places. Similarly, successive triples give random numbers from 000 to 999 (or from .000 to .999), etc. Groups of five digits can be treated

* Rand Corporation. *A Million Random Digits* (New York: Free Press, 1955).

as observations from the continuous spinning pointer that have been rounded to five decimal places.

To use the table for sampling from a real population, one proceeds as follows. Suppose, for example, that a population of 850 persons is to be sampled. First each person is assigned a number from 1 to 850, and then, using successive triples from the random number table, one obtains successive three-digit numbers to use in picking the sample—ignoring those that fall outside the range 1–850. Thus, if the successive triples are

$$192, 721, 867, 023, 471, 222,$$

one picks persons numbered 192, 721, 023 (ignoring the 867 that exceeds 850), 471, and 222. This could be done for as many persons as it is desired to have in the sample. In "simple random sampling" (random sampling without replacement) one would ignore a number if it had already occurred—having picked a person once, one would not pick him again. However, for "random sampling" (with replacement and mixing between selections) one would use a number even if it had been encountered before—corresponding to the continuing availability of everyone in the population when it is understood that those picked are put back and shuffled in with the rest, to restore the population to its original state. (If the population is very large with respect to the sample size, repetitions are uncommon, and it is practically immaterial whether one replaces those drawn.)

For artificial data from a population of uniform one-digit or two-digit (or more) numbers, or (approximately) from a spinning pointer population on $0 < x < 1$, the random number table is used as it is. Data from other populations can be simulated by applying a suitable transformation to the uniform numbers—but this will not be exploited (nor explained further) here.

It should be pointed out that in sampling one should not always start at the beginning of the table of random numbers. So doing would destroy the element of randomness. (Indeed, some would claim that random numbers are no longer random when they are published.) One should start at a "random" point in the table, which can be interpreted practically as just putting one's finger down blindly in the table to get a starting point.

A digital computer can be programmed to create, in effect, a random number table. Although the sequence of numbers is necessarily generated deterministically, with some ingenuity it is possible to obtain sequences that for all practical purposes are equivalent to such as would be generated by a truly random process (such as actually spinning a pointer). Here, for example, are 30 numbers generated on the computer used for the artificial sampling in this text:

.622690485	.894449791
.058024051	.939043925
.682697465	.232273811
.798226471	.735940505
.148567645	.342274631
.308461691	.925964285
.945389025	.852964251
.416281711	.735963265
.072169605	.342934671
.092918531	.945105445
.694637385	.408057891
.144484151	.833678825
.190040356	.176685911
.511170571	.123891405
.823946545	.592850731

Randomness, or lack of it, is not apparent; but on the surface these certainly might have come from spinning a pointer 30 times, except that real spinning pointers are not usually so precisely read.

Problems

3.21. Figure 3.16 gives a c.d.f. for a discrete random variable Y. From this graph reconstruct a probability table for Y (giving possible values and corresponding probabilities).

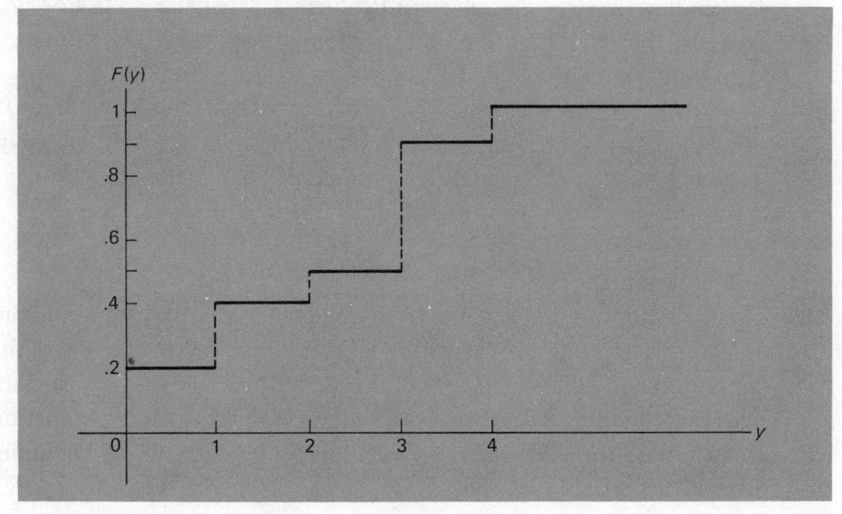

Figure 3.16
Distribution function for Problem 3.21.

3.22. Let X denote the IQ measurement of an individual. For a certain population the distribution function of X is that shown in Figure 3.17. From the graph determine approximately the probability of each event:
(a) $X > 90$. (b) $X < 110$.
(c) $90 < X < 110$. (d) $|X - 100| > 20$.

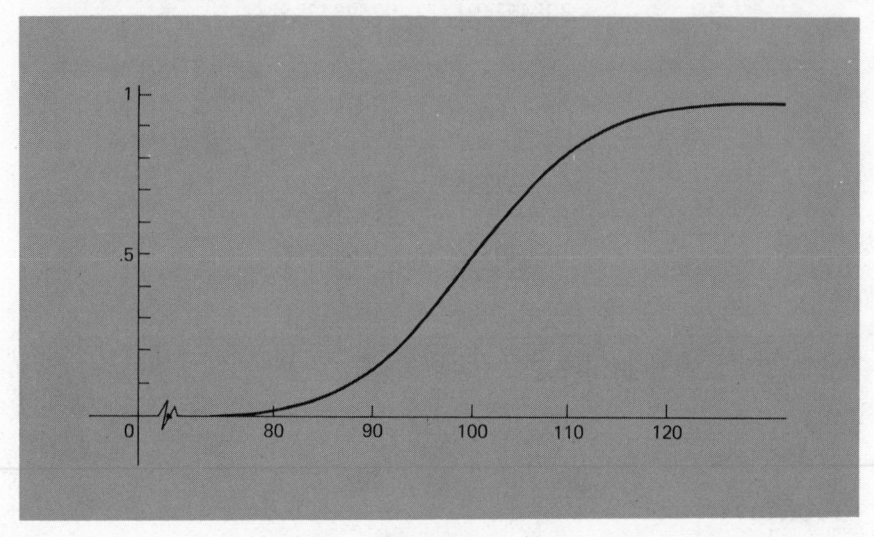

Figure 3.17
Distribution function for Problem 3.22.

3.23. Figure 3.18 shows the density function for the random variable X whose distribution function was given in Problem 3.22 (Figure 3.17). For each event in that problem, give the probabilities in terms of the letters identifying areas in Figure 3.18.

3.24. Use the random number table (Table VIII of the Appendix) to obtain 100 random digits. (Start anywhere in the table and use 100 successive integers.) Make a frequency tabulation of these digits, and sketch a corresponding cumulative distribution function. Superimpose on your sketch the distribution function of the a priori model (equal probabilities for the 10 integers).

3.25. Use the random number table to obtain 10 observations on a spinning pointer. (Start at any block in the table and use successive sets of five digits as five-digit decimals.) Plot the observations as x's on a horizontal scale, and then construct the corresponding sample distribution function. Superimpose the ideal a priori distribution function for the spinning pointer.

3.26. A student directory at a large university has 148 pages of names, about 265 names per page. How might you use the random number table to

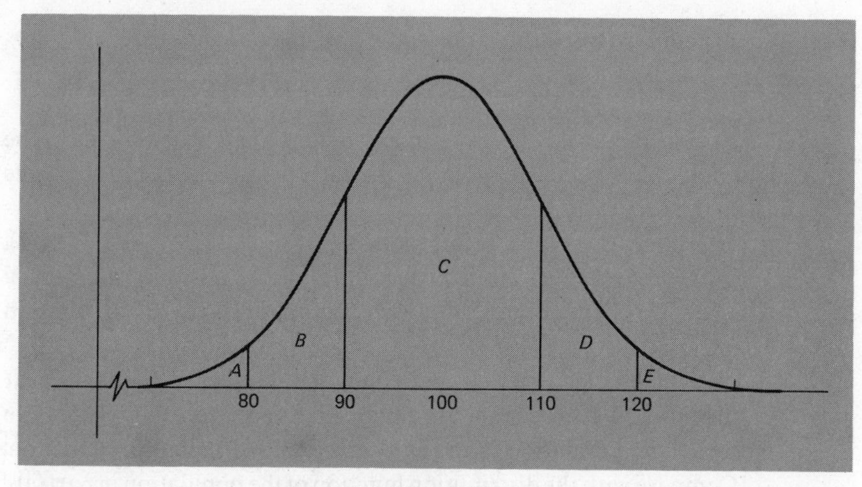

Figure 3.18
Density function for Problem 3.23.

obtain a random sample of 100 students? If your school has a student directory, use such a procedure to obtain a random sample of, say, 25 students. For each, note what class he is in (assuming that this information is given), and compare the results with what you know or can find out about the distribution of all students among the classes.

3.27. Given here are the names of 24 kindergarten students and their scores (assigned on a scale of 0 to 10 by a judge) on a drawing of a barn.

(a) Think of these data as constituting a sample of 24 from a large population of kindergarten students and construct a frequency distribution for the sample. Make a graphical representation.

(b) Think of the 24 students as comprising a population and use the random number table to obtain a sample of 10 students from the 24. Construct a sample distribution function of the 10 scores and compare with the population distribution function [which, of course, would be the same as what was called in part (a) the sample d.f.].

NAME	SCORE	NAME	SCORE	NAME	SCORE
Terri	4	Jon	3	Chris	4
Scott	1	Shari	6	Jennifer	4
Ruth	2	Cheryl	7	Paul	4
Sherry	2	Mark H.	3	Wayne	2
Melissa	6	Theresa	5	Lori	3
Lynn	3	Lance	4	Mark J.	2
Corrine	6	Larry	2	Todd	0
Gregory	5	Joseph	2	Ann	3

3.28. When a coin is tossed three times, the possible outcomes are

HHH, HHT, HTH, THH, TTH, THT, HTT, TTT.

The a priori model for this experiment assigns *equal* probabilities to the eight outcomes. Now let X denote, for each possible result of the three tosses, the *number of heads*.

(a) Determine (from the a priori model) $P(X = 3)$, $P(X = 2)$, $P(X = 1)$, and $P(X = 0)$. These four probabilities of the possible values of X define its distribution; represent it graphically by a distribution function.

(b) Toss three coins together and count the number of heads. Repeat this until 24 such counts are obtained. Summarize the results in a frequency tabulation and plot a sample distribution function. [Compare with the distribution function of the population in part (a).]

3.10 Subjective Probability

It has been intimated that probability models serve to represent experiments that can be repeated infinitely often—that probability is a long-run limit of relative frequencies in an infinite sequence of trials. Yet there are numerous instances in which people do treat nonrepeatable experiments as experiments of chance and bet (or otherwise take some action) according to the odds they feel apply, just as though they were betting on dice or card games.

Even if an experiment is repeatable, it sometimes is the case that a person will have an incorrect belief about the odds, in the sense of not corresponding to the long-run limit of relative frequencies. These personal odds define probabilities that he uses to base his actions on, even though they may not agree with the commonly accepted model or with a "true" model.

EXAMPLE 3.18

An advertisement* for the telephone company shows a man who is about to place a call by direct distance dialing rather than having an operator place the call. He "plays the percentages," he says, and dials direct even though "the odds are only 50–50" that the party he is calling will be there. The quoted even odds refer to the fact that the man placing the call would view the situation in the same way that he would view the toss of a fair coin, in which the two outcomes are considered equally likely. But the evenness of the odds simply expresses the caller's

* *Time* magazine, May 31, 1971.

> *belief* about his party's availability, a belief on the basis of which he is willing to act.

In the case of the toss of a coin there seems to be a number representing the probability of heads that is *objective*, a property of the coin and the tossing mechanism without regard to any person or persons who might be contemplating the experiment. In a sense the number is not debatable, and to some extent it is verifiable, by performing the experiment a great many times; and it is agreed on by everyone. On the other hand, the probability that a Mr. A will be elected president in 1984 does not seem to be detachable from the persons who are contemplating the experiment. Certainly, different people will give different odds, and it would be hard to insist that one person or another is right.

The term *personal probability* or, alternatively, *subjective probability* is used to refer to the probability that represents a person's degree of belief concerning the occurrence of an event. Although it can be assumed that the subjective probability of a given event exists, it is seldom easy to determine that probability. In principle, the following procedure can be used: Suppose that someone says to you: "I am going to offer you one of two prospects, and it is up to you to choose the one you prefer. Either I shall give you \$10 in 1984 if a coin I toss now turns up heads (and nothing at all if it falls tails), or I shall give you \$10 in 1984 if Mr. A is elected president then (and nothing at all if he is not elected)." Presumably the toss of the coin has an objective probability model, in which the probability of heads is $\frac{1}{2}$; if you choose the prospect of the \$10 when Mr. A is elected president, it would be said that your subjective probability of his being elected is greater than $\frac{1}{2}$. If you are indifferent between the two prospects, your subjective probability of Mr. A's election is precisely equal to $\frac{1}{2}$. Thus it ought to be possible, by confronting you with various such choices involving objective experiments with various odds, to find an objective experiment such that you are indifferent to (1) receiving \$10 as the result of a certain event E in that objective experiment, and (2) receiving \$10 as the result of Mr. A's election. The objective probability of E is then taken to be your subjective probability of Mr. A's election in 1984.

With such an approach, one could set up (in principle) a probability model for any given experiment in which subjective probabilities are to be defined. A consistent person would define a model in which the axioms for probability models are fulfilled, so that one treats subjective and objective probabilities in the same way mathematically.

4 Statistics and Population Parameters

The construction of a frequency distribution from a mass of data is an instance of reducing the data to a more comprehensible form. The purpose of such reduction is to make more apparent and usable such information about the ideal model, or population, as may be contained in the particular sample at hand. One seeks to distill this information, discarding what is irrelevant, without throwing out something useful in the process.

A manipulation of data is a *reduction* when it is not reversible. Thus, from a mass of data one can construct a frequency distribution and a corresponding histogram; but given the frequency distribution one could not re-create the raw data that led to it. The frequency distribution is a reduction of the data.

Similarly, in regrouping data according to fewer class intervals, one may succeed in eliminating sampling fluctuations while preserving some of the essential information about the population that may be in the sample; but the original frequency distribution is not then recoverable from the distribution that results from the process of regrouping.

Further reductions are often desirable, since a given scientific or operational problem calls for knowledge only of certain simple characteristics of that model, not of *every* aspect of the true mathematical model. For instance, it may be necessary to make some kind of welfare decision based on the average or typical income of a certain population. Or, some type of measuring instrument may have to be discarded or repaired if it has too much variablility. In such cases one might hope to reduce a set of observations to measures that relate to average or variability, measures that would be easier to comprehend and work with than the frequency distribution or the original data.

It is often sufficient in statistical inference to work with a *single* number computed from a given set of data, i.e., to reduce the data to such a number, chosen to embody the particular information in the sample that is relevant to the problem at hand. This leads to the following definition:

> **DEFINITION 4.1** *A* statistic *is a number computed from the observations in a sample.*

The force of this definition is that nothing but the sample data is used in computing a statistic.

A statistic is a number that describes some aspect of a set of observations; but if it is truly a reduction of the data, that description is not a complete characterization of the data. Even so, a person's body is a complex thing that can be described in terms of its height, weight, hat size, foot width, distribution of weight, etc. (It will be recalled that such dimensions as 38-28-38 are often referred to as "statistics.") Any one of these quantities describes some aspect of the body, and it may be that a single one is relevant to a particular problem. For instance, in buying shoes, only the foot size is pertinent. Whether a suit will fit a man is determined by his complete outline, but the "statistic" usually extracted from his body is his chest size. This seems to be the most relevant, simple characteristic of the body for the purpose of fitting a suit coat that is "ready made." (Actually, the additional classification as to short, regular, or long is also used, so the pertinent statictic has two dimensions, as in "42 long.")

Populations also need to be described; but again the complete description is often unnecessarily cumbersome, and some one or more simple characteristics will suffice. (A complete description is given by the probability distribution that defines the ideal model for an experiment of chance or by the probability table or the density function that defines the distribution.) There is a name for a quantity that describes a probability model or a population in part:

DEFINITION 4.2 *A* population parameter *is a number computed from the probability distribution that defines the population.*

Thus a population parameter is to a probability distribution what a statistic is to a sample distribution (or frequency distribution). A population parameter, being a quantity that describes an ideal model, is a number associated with the given experiment of chance. (A statistic, on the other hand, is a quantity associated with a particular set of observations on an experiment, determined by the particular observations that constitute that sample.)

A *problem* of statistical inference will often be phrased in terms of a population *parameter* (e.g., the "average" income of a population mentioned earlier). A *solution* to such a problem is expressed in terms of a *statistic* computed from a sample of observations on the population.

It is seldom completely clear *which* statistic ought to be computed and used, which one of the countless possible reductions of a set of data is best for the particular inference that has to be made. The mathematical theory of statistics gives some clues and offers some suggestions in this regard, but in many cases one's intuition alone will suggest a statistic that turns out to be both practical and theoretically sound. The approach here will be to introduce some simple descriptive sample measures that seem intuitively as though they ought to be good for something, and that are in fact commonly used in inference, and then to see how to idealize these to the case of a population.

The more obvious characteristics of either a sample or a population, characteristics that want describing, are perhaps

1. Location, or centering.
2. Variability, or dispersion.

The first relates to such questions as "What is the middle of the data?" and "What is the typical value?"; the second relates to the questions "How reliable is the measurement?" and "How are the observations spread or dispersed about their middle?" (These particular questions involve the assumption of an ordering of the set of possible outcomes and, indeed, the assumption that the outcomes are numerical.)

4.1 Middle of a Sample

The *median* of a sample is a simply computed statistic that is intended to describe the location or center or middle of a set of observations. When the observations are lined up in numerical order, the median is defined to be the middle observation, if there is one. Thus, if the number of observations in a sample is *odd*, there is a middle number, and this is the median. If the sample size is *even*, there is no middle observation; there are *two* in the middle, and it is convenient and customary to take the number halfway between the two middle observations as the median.

The median can also be expressed in terms of *ranks*. The smallest observation in a sample is said to have rank 1, the second smallest to have rank 2, etc. The median is then the observation with the middle rank, or the average of the two observations with middle ranks, according as the number of observations is odd or even.

EXAMPLE 4.1

Suppose that a gym class has 75 boys in it. The instructor can determine their median height by having the boys line up along one side of of the gym in order of height, with the shortest boy at one end and the tallest boy at the other end. Counting from either end to the thirty-eighth boy in line, he will find the boy whose height is the median height. On each side of that boy are 37 boys, those on one side, who are all taller, and those on the other side, who are all shorter.

The *mean* of a sample is defined as the ordinary arithmetic average of the observations in the sample, their sum divided by their number. If the observations are $X_1, X_2, \ldots, X_n$, the sample mean is given by the formula

$$\overline{X} = \frac{X_1 + X_2 + \cdots + X_n}{n}.$$

The name $\overline{X}$ is used when the individual observations are referred to by the name X_i (for $i = 1, 2, \ldots, n$). The sigma notation is again very useful. A sum of X's of the form $X_1 + X_2 + \cdots + X_n$ is indicated by a capital sigma (which is the Greek S) preceding a typical summand, X_i; for instance,

$$\sum_{i=1}^{5} X_i \qquad \text{means } X_1 + X_2 + X_3 + X_4 + X_5,$$

or in words: Add up the quantities of the form X_i as the subscript index i goes through the integers from 1 to 5. In terms of this notation, the formula for the mean is

$$\overline{X} = \frac{1}{n}\sum_{i=1}^{n} X_i.$$

Other measures of "middle" are used. Some of these are constructed to eliminate the influence of extreme observations, with the thought that these might have resulted from errors of measurement or "freak" conditions. For example, if one omits the smallest and largest values in the sample, the mean of the remaining $n - 2$ observations is called a *trimmed* mean. Another location statistic is the *midrange*, which is the average of the smallest and largest observations, the middle of the interval over which the observations are spread. And a variant of this would be the average of the kth smallest and the kth largest, for some k.

EXAMPLE 4.2

The ages (in months) of the 30 students in Example 1.3, taken from Table 1.1 but arranged in numerical order, are as follows:

216	232	258
217	232	262
218	236	265
219	240	276
221	240	281
224	245	281
224	246	288
226	249	290
228	254	378
230	256	443

The fifteenth and sixteenth observations, which are the two in the middle, are 240 and 245, respectively; so the sample median would be defined to be 242.5. The sum of all the ages is 7675, and division by 30 (the sample size) yields a mean of $\bar{X} = 255.83$, a number considerably higher than the median. Examination of the data shows that $\bar{X}$ is high because of the two rather large numbers at the end of the list. These are not errors of measurement but simply the ages of two older students, ages 31 and 36, back in school after stints of family raising or military service or some such. Whatever the explanation, these large numbers have quite an influence on the sample mean. Subtracting the two lowest and the two highest observations from the sum of 7675 and dividing by 26 (the number of remaining observations), one obtains a trimmed mean:

$$\tfrac{1}{26}(7675 - 216 - 217 - 378 - 443) \doteq 247.$$

The midrange is $\frac{1}{2}(216 + 443) = 329.5$, but the average of the third smallest and third largest, say, is $\frac{1}{2}(218 + 290) = 254$.

All these numbers that have been computed are measures of the middle of the sample. No one can be said to be more "correct" than the others—at least, not without knowing more about the use to which the statistic will be put. Each measure has its flaws and its good points. The median is not influenced by the extreme observations in a sample; in some ways this is good and in others it is not so good. The mean *is* influenced by extreme observations; for some purposes this is good, and for others it is bad.

One motivation in determining a middle or center of a set of data is the urge to define a typical observation. The phrases "average man" and "typical family" are common in popular parlance. No doubt the reader has heard of the average American man, who has .82 wife, 2.3 children, 1.6 family pets, 1.3 automobiles, 1.7 television sets, and an income of $8796.32 per year. Of course, there never is a man with precisely these characteristics; but the numbers (*if* they were correct, which they are not since we just invented them) would carry some significance. For instance, the "2.3 children" would suggest something about family sizes, namely, that the distribution of the number of children per family is centered (in some sense) at a point between 2 and 3 children, even though the actual number varies from family to family from 0 to 25 or so. A statement that families in India average 3.1 children would mean that the distribution of the number of children in such families is "located" to the right of that for American families. This would say that Indian families generally include more children than American families (even though the comparison in some particular cases might be reversed). This in turn might suggest something about life styles or traditions in the two cultures.

EXAMPLE 4.3

Suppose that on a certain (perhaps unusual) block in a city, the values of the homes on the block are as follows:

$20,000	$26,000
$20,000	$34,000
$20,000	$40,000
$20,000	$155,000
$25,000	

(Perhaps the first four were put up by a builder following a common plan, and the last, owned by a wealthy man, is on a side of the block adjacent to an exclusive district.) The mean value of these houses is $\overline{X} = \$40,000$, whereas seven of them are worth less than that average. The median value is $25,000—four are worth less and four are worth more than this amount. But the most popular value is $20,000; this is referred to as the *mode*, or the *modal value*. The tax assessor would probably prefer to describe the "typical" house by the value $40,000, since multiplying this by the number of houses yields $360,000, the total value of these houses. For other purposes, the median value of $25,000 might be more appropriate.

Generally speaking (and such generalizations have their exceptions), the mean (i.e., $\overline{X}$) is useful in situations in which one is likely to multiply this typical value by a sample size to obtain a total amount for the whole sample. The median is simpler to compute and has a fairly clear interpretation in terms of the middle rank. Sometimes the choice is made, not with logic or reason, but with the devious purpose of proving some point. And then there is the story (which harks back to the days when money had greater value) of the Pullman porter who was asked by a passenger (wanting to get some idea of how much to tip) how much his average tip was. He replied: "My average tip is a dollar, but I don't very often get the average."

When numerical data are given in the form of a frequency distribution, the calculations of the mean can be effected in a way that exploits the tabulation in terms of frequencies. The idea is simple: Additions for *repeated* observations (such as occur in frequency tabulations) can be accomplished more simply by using multiplication. Thus to add together the numbers 7, 7, 7, 7, and 7, one can multiply the value 7 by its frequency, 5:

$$7 + 7 + 7 + 7 + 7 = 5 \cdot 7 = 35.$$

EXAMPLE 4.4

Three coins were tossed together 20 times. After each toss the number of heads that show among the three coins was counted and recorded. The sample is then $X_1, X_2, \ldots, X_{20}$, where X_i is the number of heads that show after the ith toss. The sample results can be, and were, recorded in a frequency tabulation; the possible values of the X's are 0, 1, 2, 3, and these occurred among the 20 observations with frequencies as follows:

Possible value (x_i)	0	1	2	3
Frequency(f_i)	2	8	6	4

The 20 results were, in some order:

0, 0, 1, 1, 1, 1, 1, 1, 1, 1, 2, 2, 2, 2, 2, 2, 3, 3, 3, 3,

and their sum is

$$\sum X_i = 0 + 0 + 1 + 1 + 1 + 1 + 1 + 1 + 1 + 1$$
$$+ 2 + 2 + 2 + 2 + 2 + 2 + 3 + 3 + 3 + 3 = 32.$$

The summation can be carried out by multiplying each possible value by the number of times it occurs in the sample and adding:

$$\sum X_i = 2 \cdot 0 + 8 \cdot 1 + 6 \cdot 2 + 4 \cdot 3 = 32.$$

The arithmetic mean is then $\frac{32}{20} = 1.6$.

The work is often put in tabular form to facilitate computation, as follows:

VALUE x_i	FREQUENCY f_i	PRODUCT $x_i f_i$
0	2	0
1	8	8
2	6	12
3	4	12
Sums	$20 = n$	$32 = \sum X_i$,

The idea illustrated in Example 4.4 leads to a general formula, to be given in terms of the following notational convention. Let $x_1, x_2, \ldots, x_k$ denote the *distinct, possible values* that can be observed. Let f_1 denote the frequency of x_1 among the observations X_i, f_2 the frequency of x_2 among the X_i's, etc. The sum of the observations can then be written in two forms:

$$X_1 + X_2 + \cdots + X_n = f_1 x_1 + f_2 x_2 + \cdots + f_k x_k.$$

(In Example 4.4, $n = 20$ and $k = 4$.) In terms of sigma notation,

$$\overline{X} = \frac{1}{n} \sum_{i=1}^{n} X_i = \frac{1}{n} \sum_{j=1}^{k} f_j x_j.$$

An arithmetic mean, or a weighted sum like the one above used to calculate the mean of grouped data, is often called an *average*. Although this word has a colloquial meaning somewhat broader than this—including any notion such a "typical", the word *average* as used here (and generally in mathematics and statistics) is synonomous with *mean*.

Problems

4.1. A weight-reducing drug given to nine overweight women resulted in the following weight losses in a 1-month period (in pounds):

14, 9, 17, 10, 10, 18, 11, 13, 6.

Determine

(a) The mean weight loss.

(b) The median weight loss.

(c) The midrange of the weight losses.

4.2. The number of letters in each of the 55 words of a paragraph picked at random from a newspaper column was counted, with these results:

Length of word	1	2	3	4	5	6	7	8	9	10	11	12
Frequency	2	8	13	11	6	7	4	0	2	0	0	2

Compute the average word length.

4.3. A typist, in 25 time trials, made errors at rates summarized in the following frequency table:

Errors per minute	0	1	2	3	4
Frequency	6	9	7	2	1

Compute the average number of errors per minute.

4.4. Figure 4.1 shows the numbers of hours devoted to physiology in 106 medical schools. Determine the mean number of hours required.

4.5. The attendance record at University Senate meetings, during a recent year, was as follows: 16 members attended no meetings, 13 attended one, 25 attended two, 38 attended three, 27 attended four, 25 attended five, and 14 attended all six meetings.

(a) Make a graphical representation of these data.

(b) Determine the average number of meetings attended per member.

(c) Determine the average attendance at the meetings.

4.6. A student with nothing better to do counted the number of cars between blue cars passing by his dorm window, with these results:

Number of nonblue in a row	0	1	2	3	4	5	6	7	8
Frequency	1	3	7	4	8	6	5	4	2

(a) Compute the average number of nonblue cars between blue ones.

(b) Compute the proportion of blue cars among those seen.

4.7. (a) (No calculator available.) Compute the mean birthweight of the 30 babies of Problem 2.6, first using the original data, and then using the frequency tabulation you made to construct a histogram. (Do they agree?)

(b) (Calculator available.) Compute the mean hemoglobin level for the data in Problem 2.9, first using the raw data, and then using the frequency tabulation you made in constructing a histogram.

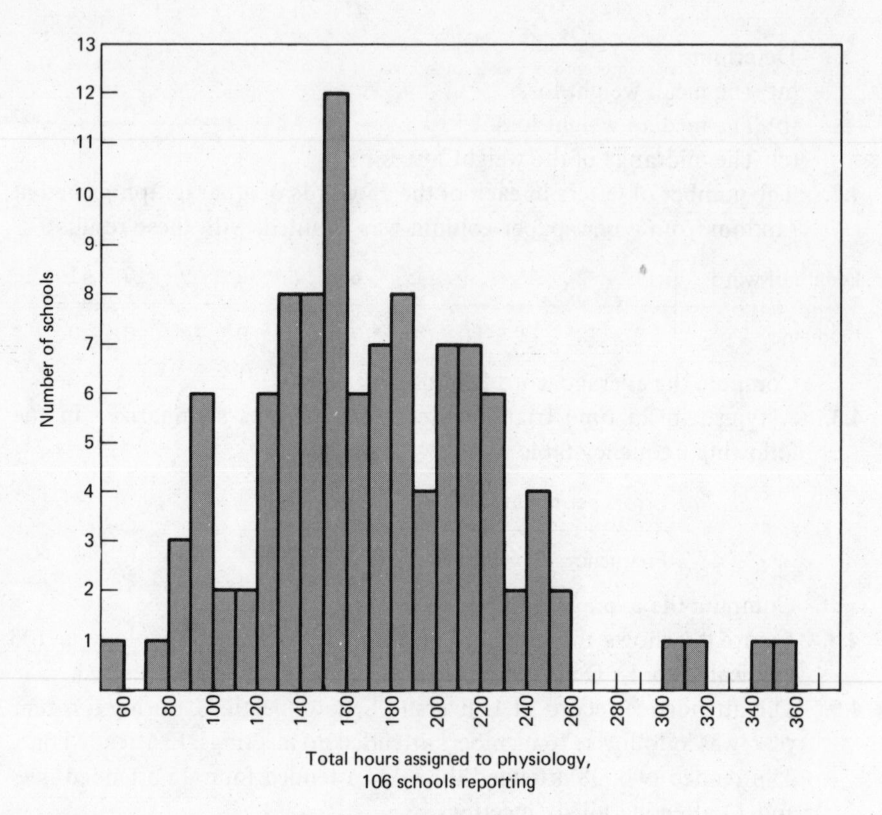

Figure 4.1
Histogram for Problem 4.4.

4.2 Dispersion in a Sample

Statistics intended to measure the characteristic of spread or dispersion or variability in a sample are almost as numerous as measures of location or middle. The simplest such statistic is the *range* of the sample, defined to be the difference between the extreme sample values, and denoted by R:

$$R = \text{range} \equiv \text{largest observation} - \text{smallest observation}.$$

It is just the width of the interval over which observations range in that particular sample. It is a nonnegative quantity.

Other measures of dispersion are constructed in terms of deviations from the middle. Thus, if $\overline{X}$ (the sample mean) is used as the middle, the deviation of an observation about that point is the difference $X_i - \overline{X}$, a quantity that is positive if X_i lies to the right of $\overline{X}$ and negative if it lies to the left of $\overline{X}$. Each observation in a sample has a deviation about the mean of that sample, and a measure of dispersion can be constructed by averaging these deviations in some fashion. The ordinary average or arithmetic mean of the deviations is not useful, since it always turns out to be zero. This is because the negative deviations of observations to the left of $\overline{X}$ exactly cancel the positive deviations of observations to the right of $\overline{X}$.

To avoid such cancelation, one can consider just the *magnitude* of the deviation, dropping the algebraic sign:

$$\text{absolute deviation of } X_i \text{ about } \overline{X} = |X_i - \overline{X}|.$$

This quantity is really more to the point of measuring dispersion, and the average of these absolute deviations is called the *mean deviation*:

$$\begin{aligned}\text{mean deviation} &= \frac{1}{n}(|X_1 - \overline{X}| + |X_2 - \overline{X}| + \cdots + |X_n - \overline{X}|)\\ &= \frac{1}{n}\sum_{i=1}^{n} |X_i - \overline{X}|.\end{aligned}$$

The same term, "mean deviation," is sometimes (and with some justification) applied to the average absolute deviation about the *median*; the two versions will not ordinarily agree.

Another way of avoiding the cancelation of positive and negative deviations is to square them before averaging. The average of the squared deviations is called the *variance* and will be denoted by S^2:

$$S^2 = \frac{1}{n}\sum_{i=1}^{n} (X_i - \overline{X})^2.$$

A convenient computation formula for the variance is based on the identity

$$\sum (X_i - \overline{X})^2 = \sum X_i^2 - n\overline{X}^2,$$

where the summation directions have been dropped to avoid cluttering, it being understood that sums are taken from $i = 1$ to $i = n$, i.e., over all the sample observations. The identity is easy to establish, by expanding each $(X_i - \overline{X})^2$ and adding like terms:

$$\begin{aligned}\sum (X_i^2 - 2X_i\overline{X} + \overline{X}^2) &= (X_1^2 - 2X_1\overline{X} + \overline{X}^2)\\ &\quad + (X_2^2 - 2X_2\overline{X} + \overline{X}^2)\\ &\quad + \cdots\\ &\quad + (X_n^2 - 2X_n\overline{X} + \overline{X}^2)\\ &= \sum X_i^2 - 2\overline{X}(\sum X_i) + n\overline{X}^2.\end{aligned}$$

Substitution of $n\overline{X}$ for $\sum X_i$ yields the asserted identity. Division by n then gives this result:

$$S^2 = \frac{1}{n}\sum_{i=1}^{n}(X_i - \overline{X})^2 = \frac{1}{n}\sum_{i=1}^{n} X_i^2 - \overline{X}^2$$

$$= \overline{X^2} - \overline{X}^2,$$

where $\overline{X^2}$ denotes the average of the squares of the observations, which will be referred to as the "average square." Thus the variance is the difference between the average square and the square of the average.

Because it does involve squares of observations, the variance is a bit hard to comprehend intuitively; its units would be the square of the units in which X itself is measured. If X is a number of inches, X^2 would be a number of square inches. It is common practice to work with the *square root* of the variance, called the *standard deviation*:

$$S \equiv \sqrt{\frac{1}{n}\sum(X_i - \overline{X})^2}.$$

This will have the same units as X; if X is a number of inches, then S will also be a number of inches and can be interpreted on the same scale as a kind of average deviation away from $\overline{X}$. It is not an uncommon type of average, having the name *root mean square*, encountered often in the physical sciences.

A word should be said about another version of the variance and standard deviation. Many texts and computer programs define the variance of a set of numbers to be

$$\tilde{S}^2 = \frac{1}{n-1}\sum_{i=1}^{n}(X_i - \overline{X})^2,$$

which is simply $n/(n-1)$ times what has been defined here to be the variance. Because dividing by $n - 1$ seems unnatural and has no real justification (despite the assertions of some textbooks), and because its use complicates the computation formulas, the version used here will be that in which the sum of squared deviations is divided by n.

When it is useful to indicate that the variance refers to a particular set of observations called X_i, the subscript x will be used: S_x and S_x^2.

EXAMPLE 4.5

To see the computation of standard deviation in a simple case, consider the four children in a certain family, of ages 2, 6, 7, and 9. The average age is 6, and the computations are perhaps best done as in Table 4.1. The average is $\overline{X} = \frac{24}{4} = 6$, the average deviation is

Table 4.1

	X_i	$X_i - \overline{X}$	$\lvert X_i - \overline{X}\rvert$	$(X_i - \overline{X})^2$	X_i^2
	2	−4	4	16	4
	6	0	0	0	36
	7	1	1	1	49
	9	3	3	9	81
Sums	24	0	8	26	170

zero (as predicted), but the mean deviation is $\frac{8}{4} = 2$. The average of the squared deviations is $\frac{26}{4} = 6.5$, and the average of the squares of the observations is $\frac{170}{4} = 42.5$. Thus

$$S^2 = 6.5 = 42.5 - 6^2 = \overline{X^2} - \overline{X}^2.$$

The standard deviation is $S = \sqrt{6.5} \doteq 2.55$.

Consider also another family with four children, ages 3, 5, 8, and 8. The table for computations is as shown in Table 4.2. Thus $\overline{X} = 6$, the mean deviation, is again $\frac{8}{4} = 2$, and the variance is

$$S^2 = \frac{18}{4} = 4.5 = \frac{162}{4} - 6^2.$$

Table 4.2

	X_i	$X_i - \overline{X}$	$\lvert X_i - \overline{X}\rvert$	$(X_i - \overline{X})^2$	X_i^2
	3	−3	3	9	9
	5	−1	1	1	25
	8	2	2	4	64
	8	2	2	4	64
Sums	24	0	8	18	162

Now compare these two families. The average ages of the children is the same, 6. The mean deviation is the same, because to get the second set of four numbers from the first, one can move the outside two numbers in by one and the inside two numbers each out by one (2 becomes 3, 9 becomes 8; and 6 becomes 5, 7 becomes 8). So the absolute deviations have the same sum (and therefore the same average). But because squaring is an operation that makes big numbers bigger, the average squared deviation about the mean is smaller in the second family than in the first. The standard deviation in the first instance is 2.55, and in the second it is $\sqrt{4.4} \doteq 2.12$. The *range* in the first sample is $9 - 2 = 7$, and in the second it is $8 - 3 = 5$. (All these are measured in *years*, whereas the variance would be a number of *square* years, which is not easy to interpret.)

One point should be made about the standard deviation, a point that will be helpful in providing a rough check on computations and may catch the gross blunder sometimes made by the novice. It is that the standard deviation is an average *deviation*; as such, it cannot exceed the largest single deviation about the mean (nor be smaller than the smallest). If the observations are represented graphically, the point one standard deviation from the mean (as computed) should be plotted to see if it falls in the midst of the data and not way out somewhere.

EXAMPLE 4.6

The nine weight losses given in Problem 4.1 were as follows (in numerical order):

$$6, 9, 10, 10, 11, 13, 14, 17, 18.$$

The mean loss is 12 lb, and corresponding deviations about this mean are

$$-6, -3, -2, -2, -1, 1, 2, 5, 6.$$

The average of the squares of these deviations is $\frac{120}{9}$, so the standard deviation is

$$S = \sqrt{\tfrac{120}{9}} \doteq 3.65.$$

The deviations range from 1 to 6 in magnitude, and the standard deviation lies between these.

Just as the computation of the mean in a frequency distribution was in terms of distinct values and their frequencies, so the mean deviation and variance can be calculated in terms of sample frequencies:

$$\text{mean deviation} = \frac{1}{n} \sum_{j=1}^{k} f_j |x_j - \bar{X}|$$

and

$$\text{variance} = \frac{1}{n} \sum_{j=1}^{k} f_j (x_j - \bar{X})^2.$$

(Here again, $x_1, x_2, \ldots, x_k$ are the distinct values in the sample and f_i is the frequency of occurrence of the value x_i in the sample.)

EXAMPLE 4.7

Consider again the data of Example 4.4, giving frequencies of 0, 1, 2, and 3 heads in 20 tosses of three coins. The computations are made systematic by using an extension of the frequency table, Table 4.3.

Table 4.3

POSSIBLE VALUE x_j	FREQUENCY f_j	DEVIATION ABOUT $\bar{X}$ $(x_j - \bar{X})$	ABSOLUTE DEVIATION $\lvert x_j - \bar{X}\rvert$	SQUARED DEVIATION $(x_j - \bar{X})^2$	PRODUCTS $f_j\lvert x_j - \bar{X}\rvert$	PRODUCTS $f_j(x_j - \bar{X})^2$
0	2	−1.6	1.6	2.56	3.2	5.12
1	8	−.6	.6	.36	4.8	2.88
2	6	.4	.4	.16	2.4	.96
3	4	1.4	1.4	1.96	5.6	7.84
Sums	20				16.0	16.80

Computations are as follows:

$$\text{mean deviation} = \frac{16.0}{20} = .80;$$

$$\text{variance} = \frac{16.8}{20} = .84;$$

$$\text{standard deviation} = \sqrt{.84} \doteq .92.$$

The variance can also be calculated as the difference between the average square and the square of the average, using Table 4.4. Thus

Table 4.4

x_j	f_j	x_j^2	$f_j x_j^2$
0	2	0	0
1	8	1	8
2	6	4	24
3	4	9	36
Sum			68

the average square is $\frac{68}{20} = 3.40$, and subtracting the square of the mean yields the variance:

$$S^2 = 3.40 - (1.6)^2 = 0.84,$$

which agrees with the earlier result.

The most common error made by students in calculating the variance in a frequency distribution is to square after, rather than before, multiplying by frequency. Remember that it is the values, or their deviations, that are squared, never the frequencies. (The role of the frequency multiplication is simply to collapse into one term what would otherwise be several terms, to simplify the addition when there are repeated values.)

4.3 Linear Transformations

One of the facts of life in connection with making measurements is that the choice of units, of a scale of measurement, is rather arbitrary. Different choices are convenient in different situations. Some data on lengths are conveniently recorded in centimeters, some in inches, some in miles. And it is often necessary to know the relationship between the measurements of a given quantity as read on different scales.

In most cases the relationship between scales of measurement is a linear one, and perhaps the most common instances are those of converting feet to inches, pounds to ounces, gallons to quarts. In each of these instances, the conversion is achieved by multiplying the number on one scale to obtain the corresponding number of units on the other scale. Another conversion of units commonly encountered (at least in arithmetic books) is that from degrees Fahrenheit to degrees Celsius. Here, not only the unit is different, but the reference origin is different; and the relationship is what is called *linear*:

$$C = \tfrac{5}{9}(F - 32),$$

or

$$F = \tfrac{9}{5}C + 32,$$

where F denotes the number of Fahrenheit degrees and C the number of Celsius degrees that identify a particular temperature. (Any relationship between variables x and y of the form $y = ax + b$ is called linear because of the fact that the graph of the function $ax + b$ is a straight line.)

The question will arise naturally, when measurements in a sample are converted from one scheme of units to another scheme related to the first by linear transformation, as to how to convert the statistics computed from the set of numbers on one scale to statistics that describe the corresponding set of numbers on the other scale.

The mean of a sample, i.e., the arithmetic average of the observations in the sample, is given by the following formula in the case of a frequency tabulation:

$$\overline{X} = \sum x_i \frac{f_i}{n}.$$

This can be interpreted as a weighted sum of the possible values, where the weights are the relative frequencies; for, each possible value is multiplied by that weight before summing. In physics it is shown that it is precisely this weighted sum that gives the center of gravity or balance point of a system of masses, in the amount f_1 units at x_1, f_2 units at $x_2, \ldots,$ and f_k units at x_k. This center of gravity is a characteristic of the particular masses and locations, and its location in that system does not depend on where distances are measured from or on the unit of measurement used. So, if the location of the zero point is moved and the scale of measurement is changed by a constant factor, the same change affects the coordinate of the center of gravity as affects each measurement.

The conclusion of the argument is then that if a new measurement system is introduced that is *linear*, of the type

$$y_i = ax_i + b,$$

then the mean value of the transformed data bears this same relationship to the mean value of the original data:

$$\overline{Y} = a\overline{X} + b.$$

It is also of interest to know what happens to the variance, and so what happens to the standard deviation, when a set of measurements is subject to a linear transformation. Because the variance is an average of squared *deviations* about the mean, a change in the reference origin does not affect it. More precisely, if $y_i = ax_i + b$, then

$$y_i - \overline{Y} = a(x_i - \overline{X}),$$

and it is seen that the translation amount b has disappeared in taking differences. And then, since

$$(y_i - \overline{Y})^2 = a^2(x_i - \overline{X})^2,$$

it follows that the average squared deviation of the y's about $\overline{Y}$ is a^2 times the average squared deviation of the x's about $\overline{X}$. Hence

$$S_y^2 = a^2 S_x^2 \qquad \text{or} \qquad S_y = aS_x,$$

provided that a is a positive quantity (as it is in most ordinary scale changes). Thus, if the scale is changed so that each observation is multiplied by a constant factor, then the standard deviation is multiplied by that factor.

EXAMPLE 4.8

Suppose that the data that give house prices in Example 4.3 are expressed in terms of the number of thousands of dollars above

\$20,000. That is, if X is the value in dollars as given there, then

$$Y = \frac{X - 20{,}000}{1000}$$

expresses the value in number of thousands above 20. Notice that this relation, and the reciprocal relation solved for X, are linear:

$$X = 1000Y + 20{,}000.$$

If the mean and standard deviation of the values given in terms of Y are computed, the mean and standard deviation of the values given in terms of X can be calculated therefrom:

$$\overline{X} = 1000\overline{Y} + 2000,$$

$$S_x = 1000S_y.$$

Table 4.5 gives old and new values, along with computations of mean and variance.

Table 4.5

X	$Y = \frac{X - 20{,}000}{1{,}000}$	$Y - \overline{Y}$	$(Y - \overline{Y})^2$
20,000	0	−20	400
20,000	0	−20	400
20,000	0	−20	400
20,000	0	−20	400
25,000	5	−15	225
26,000	6	−14	196
34,000	14	−6	36
40,000	20	0	0
155,000	135	115	13,225
	9)180	0	9)15,282
	$20 = \overline{Y}$		$1{,}698 = S_Y^2$

$$\overline{X} = 1{,}000\overline{Y} + 20{,}000 = 40{,}000; \qquad S_x = 1{,}000S_y = 1{,}000\sqrt{1{,}698} = 41{,}207.$$

As in this example, computations can often be simplified by a judicious choice of scale. However, the usefulness of such simplification has diminished with the advent of modern calculators and computers, which handle numerically complicated problems with as much relish and ease as numerically simple ones.

Problems

4.8. In Problem 4.1 the weight losses

14, 9, 17, 10, 10, 18, 11, 13, 6

were found to have mean 12 lb and median 11 lb.

(a) Determine the mean deviation about the mean and also the mean deviation about the median.

(b) Determine the standard deviation. (Use the divisor n and also $n - 1$ and note the difference it makes.)

(c) Determine the range.

4.9. (Calculator available.) Determine the standard deviation of the hourly earnings given in Problem 2.10 and repeated here:

$2.61, 2.81, 2.83, 2.37, 2.78, 3.25, 3.19, 3.30, 3.05

3.60, 2.79, 2.79, 2.64, 2.32, 3.42, 2.73, 2.47, 2.15

2.20, 2.59, 2.52, 2.36, 3.05, 2.90, 2.18.

4.10. Determine the range of the hourly earnings in Problem 4.9. Determine also the midrange (and compare with the mean).

4.11. Calculate the standard deviation of the typing errors per minute in the 25 time trials (Problem 4.3):

Errors per minute	0	1	2	3	4
Frequency	6	9	7	2	1

4.12. (Calculator available.) Compute the standard deviation of the hemoglobin level for the data in Problem 2.11, first using the raw data, and then using the frequency tabulation you made in constructing a histogram.

4.13. In Problem 2.1 the following frequency tabulation was obtained for the number of licorice "snaps" in a box:

Number	11	12	13	14	15
Frequency	3	3	10	8	1

Compute the mean and standard deviation.

4.14. Determine the mean weight loss and the standard deviation of weight losses in *kilograms* for the data in Problem 4.8.

4.15. Convert the credit loads in Problem 2.2, repeated here, to "number of credits above 18" (+ if above, − if below).

13, 18, 15, 16, 19, 18, 16, 14, 18, 18, 18, 20, 18, 16

16, 17, 20, 17, 19, 15, 17, 18, 22, 19, 14, 13, 18, 14

18, 18, 8, 19, 17, 18, 21, 13, 18, 9, 18, 13.

Make a frequency distribution for the new variable ($Y = X - 18$, where X = number of credits) and calculate its mean and standard deviation. From these compute the mean and standard deviation of the credit loads.

4.4 Parameters of Discrete Populations

A discrete random variable, or the population of its possible values and corresponding odds, can be defined by a table (as in Chapter 3) of this form:

Possible value	$x_1 \quad x_2 \cdots x_k$
Probability	$p_1 \quad p_2 \cdots p_k$

This, in turn, can be represented graphically by the same kind of picture as used for a sample from such a population, in which probabilities (rather than relative frequencies) are represented by vertical rods, as in Figure 4.2.

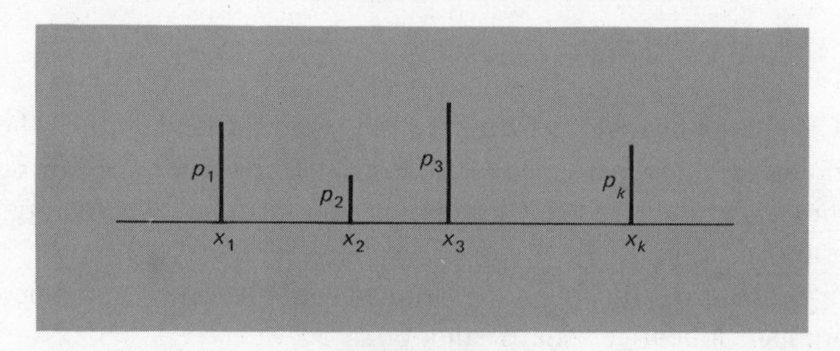

Figure 4.2
A discrete distribution.

Quantities devised to describe various aspects of such a distribution are called *parameters* of the population. (Admittedly, a "statistic," which describes some aspect of a sample, could have been called a "parameter" of the sample. But the convention used here is quite universal, *parameter* for population descriptive measures and *statistic* for sample descriptive measures.)

As in the case of a sample distribution, there are many ways to describe a probability distribution, many aspects that might be of particular interest. Indeed, because a probability table is not distinguishable in essence from a relative frequency table, one can use some of the same ideas in concocting descriptive measures. For instance, the *mode* of a sample distribution is a statistic that is defined simply as the most frequently observed (i.e., the most popular or "modish") value in a sample, among the values possible. Similarly, the mode of a population of discrete type is the *most probable value* among those that might be observed when the basic experiment is performed—the one (or ones) whose probability of occurrence is greatest.

Other population parameters are measures corresponding to various sample *averages.* The notion of averaging in a sample is that quantities are weighed according to the relative frequency of occurrence of the value from which they are computed, and then summed. In a discrete population, the relative frequencies are simply replaced by probabilities, their long-run limits, as weights. In making such replacement, the versions of sample averages with the n moved into the summation with the f are the starting point:

$$\overline{X} = \frac{1}{n}\sum x_i f_i = \sum x_i \frac{f_i}{n}$$

and

$$S^2 = \frac{1}{n}\sum (x_i - \overline{X})^2 f_i = \sum (x_i - \overline{X})^2 \frac{f_i}{n},$$

where each summation extends over $i = 1, \ldots, k$, i.e., over the collection of possible values $x_1, \ldots, x_k$.

To calculate or define corresponding population averages, then, one simply replaces the relative frequency f_i/n by the probability p_i. Thus the *population mean* is

$$\mu \equiv \sum x_i p_i,$$

and with this in place of $\overline{X}$ in the expression for variance, the *population variance* is

$$\sigma^2 \equiv \sum (x_i - \mu)^2 p_i.$$

These descriptive parameters (describing, respectively, location and dispersion, as in the case of the sample versions) are characteristics of the probability distribution of the model that describes what can happen in a single performance of the experiment.

In the case of a population, the averaging process is often indicated by the letter E in this notation, $\mu = EX$ and $\sigma^2 = E[(X - \mu)^2]$. The E stands for *expected value* (or sometimes *expectation*), and $E(X)$ is called the expected value of the random variable X; but it is just the mean or center (one type) of the population distribution of X.

Recall that the *probabilities* of outcomes, the long-run limits of relative frequencies, are thought of as characteristics of the basic experiment each time it is performed. That is, they are intended to describe certain aspects of what can happen and the corresponding "odds" when a single observation is to be obtained. The probability distribution, consisting of the possible outcomes and their probabilities, is thus a model for a single realization of the experiment—even though it is investigated empirically by performing the experiment many times. Therefore, the descriptive parameters computed using the ideal probabilities in place of sample relative frequencies are characteristics or descriptive parameters of the probability distribution—of the (ideal) model for the single performance of the experiment, or for the population.

EXAMPLE 4.9

The model for a fair die is defined by equal probabilities for the six faces, that is, for the six possible values of X, the number of dots on the face that turns up when the die is tossed. The center of this distribution, the expected value or population mean, is

$$\mu = E(X) = 1 \cdot \tfrac{1}{6} + 2 \cdot \tfrac{1}{6} + 3 \cdot \tfrac{1}{6} + 4 \cdot \tfrac{1}{6} + 5 \cdot \tfrac{1}{6} + 6 \cdot \tfrac{1}{6} = \tfrac{21}{6} = 3.5.$$

The value 3.5 is not expected, of course, in the sense that it could occur at a single toss; only integers can occur. It can be interpreted as an average payoff over a long series of tosses if a payoff of $\$x$ is offered for the outcome $X = x$. Thus the total income in N tosses is approximately

$$\frac{N}{6} \times (\$1) + \frac{N}{6} \times (\$2) + \cdots + \frac{N}{6} \times (\$6) = \frac{N}{6} \times (\$21).$$

Spread over the N tosses, this would be \$21/6, or \$3.50 per toss. Incidentally, \$3.50 would be the entry fee for playing such a game if it were perfectly fair (no advantage to the "house").

The calculation of the mean μ, as in the case of the mean $\overline{X}$, is often facilitated by an extension of the table of probabilities; see Table 4.6. Observe that here there is *no division by n*, simply because there is no n! There is no sample size because the computation is not being performed for a sample but for a population model; the probabilities refer to *one* observation and are already relative quantities. Thus

$$\mu = \sum x_j p_j = \tfrac{21}{6} = 3.5.$$

Table 4.6

POSSIBLE VALUE x_j	PROBABILITY p_j	PRODUCT $x_j p_j$
1	$\frac{1}{6}$	$\frac{1}{6}$
2	$\frac{1}{6}$	$\frac{2}{6}$
3	$\frac{1}{6}$	$\frac{3}{6}$
4	$\frac{1}{6}$	$\frac{4}{6}$
5	$\frac{1}{6}$	$\frac{5}{6}$
6	$\frac{1}{6}$	$\frac{6}{6}$
Sums	1	$\frac{21}{6} = \mu$

This parameter is characteristic of the die or, more precisely, of the result of the toss of a fair die. It relates to the experiment, to its mathematical representation by a discrete probability distribution.

The mean of a distribution, whether sample or population, is perhaps most easily comprehended as its *center of gravity*. This will be illustrated in the following example, going back to an earlier example of a discrete sample distribution.

EXAMPLE 4.10

The sample of Example 4.4 reported 20 tosses of three coins. The resulting numbers of heads included two 0's, eight 1's, six 2's, and four 3's. Imagine each observation represented as a block piled at the appropriate point on a weightless teeter-totter, as in Figure 4.3.

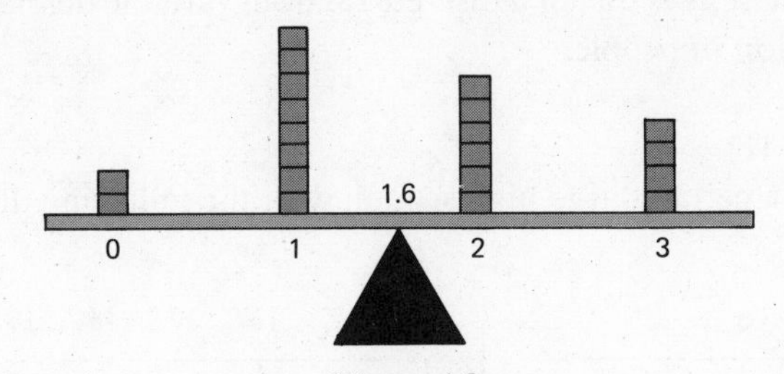

Figure 4.3
Balance at a sample mean.

This collection of blocks would exactly balance with a fulcrum at the point 1.6 on the scale, i.e., at the center of gravity:

$$1.6 = \tfrac{2}{20} \cdot 0 + \tfrac{8}{20} \cdot 1 + \tfrac{6}{20} \cdot 2 + \tfrac{4}{20} \cdot 3 = \overline{X}.$$

Similarly, the mean of a discrete probability model for this same experiment can be interpreted as a center of gravity. The ideal or fair toss of three coins (which was the basic experiment of Example 4.4) has this model:

Number of heads	0	1	2	3
Probability	$\frac{1}{8}$	$\frac{3}{8}$	$\frac{3}{8}$	$\frac{1}{8}$

This distribution "balances" at its mean:

$$\mu = 0 \cdot \tfrac{1}{8} + 1 \cdot \tfrac{3}{8} + 2 \cdot \tfrac{3}{8} + 3 \cdot \tfrac{1}{8} = \tfrac{3}{2}$$

as illustrated in Figure 4.4. (These figures are approximate, for the blocks spread out too much and might actually alter the balance. A more precise representation would employ thin rods.)

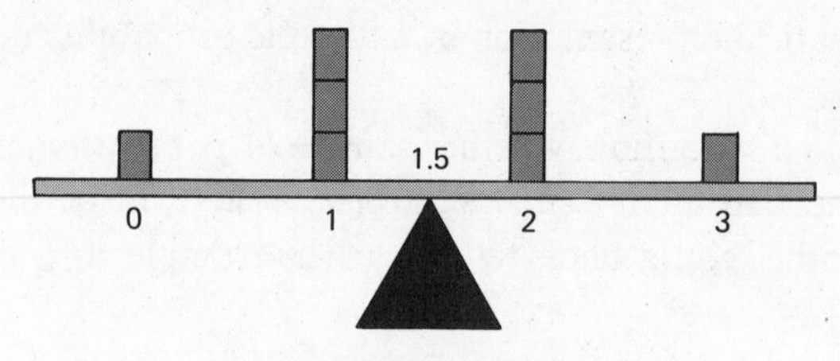

Figure 4.4
Balance at a population mean.

One feature of the ideal distribution in Example 4.10 is its symmetry about 1.5. There is precisely the same pattern of weight to the right as to the left of 1.5. So it is quite natural to find that 1.5 is the balance point.

The next example is one of a discrete random variable defined on an actual finite population of people.

EXAMPLE 4.11

Consider a certain large high school, with the following distribution of ages:

Age	14	15	16	17	18	19
Number of students	12	384	608	596	398	2

If a student is selected "at random" from this population of students, the selection is done so that all 2000 students have the same chance of being drawn; and the a priori model in which each student is assigned probability $\frac{1}{2000} = .0005$ is thus assumed to describe the experiment. In terms of *ages*, the population can be thought of as consisting of the possible values 14, 15, 16, 17, 18, 19, and corresponding probabilities:

$$P(14) = \frac{\text{number of students of age 14}}{2000} = .006,$$

$$P(15) = \frac{\text{number of students of age 15}}{2000} = .192,$$

etc. The probability distribution for the discrete random variable, the age of the student drawn, is given by Table 4.7. The population

Table 4.7

AGE (x_i)	PROBABILITY (p_i)	PRODUCT ($x_i p_i$)
14	.006	.084
15	.192	2.880
16	.304	4.864
17	.298	5.066
18	.199	3.582
19	.001	.019
Sums	1	16.495

mean μ, which here is just the average age of students in the population of 2000, is then

$$\mu = \sum x_i p_i = 16.495.$$

Suppose now that a sample of 20 students is selected (by a process that is irrelevant at this point) from the population of 2000 students, with results as given in Table 4.8. The average age in this sample is then

$$\begin{aligned} \overline{X} = \tfrac{1}{20}(X_1 + X_2 + \cdots + X_{20}) &= \tfrac{1}{20} \sum_{i=1}^{6} x_i f_i \\ &= \tfrac{324}{20} = 16.2. \end{aligned}$$

But another sample of 20 from the student body would have a different frequency distribution, and a correspondingly different sample mean. Thus, although there is just *one* population mean, descriptive of the typical age in the population (and relating to what can happen in a single drawing of a student from the 2000), the various possible

Table 4.8

AGE x_i	FREQUENCY f_i	PRODUCTS $x_i f_i$
14	0	0
15	5	75
16	8	128
17	5	85
18	2	36
19	0	0
Sum	$20 = n$	324

samples of 20 students will have various average ages. That is, the statistic $\overline{X}$ varies from sample to sample. The population mean, on the other hand, is a constant, characteristic of the basic population (or the experiment of drawing a student from the population).

The last example illustrates vividly a reason that confusion sometimes exists as to what is the population and what is the sample, and hence as to the difference between a statistic and a corresponding population parameter. For, if one were studying young people of high school age in a certain large city, he might take this group of 2000 students in a particular high school as a *sample* from the larger population of students in the city. In that case, the value 16.495 that was the population mean when the school's enrollees were assumed to constitute the reference population would be interpreted as the mean of the sample of those 2000 students. For, the probabilities involved in picking one student from the 2000 become the relative frequencies of ages in the same group of students interpreted as a sample from a larger population. The designation of what is the population is arbitrary, to the extent that the object of an investigation is arbitrary.

(To understand what the reference population is or should be, in a given study, is important but is not always easy. It will often require understanding of the field under investigation. Moreover, it is usually a nontrivial matter to obtain a sample from the right population in a structured fashion so that one can intelligently analyze the results.)

As in the case of the sample variance, the computation of a population variance is sometimes easier with the use of this relation between the variance and the average square:

$$\sigma^2 = \sum (x_i - \mu)^2 p_i = \sum x_i^2 p_i - \mu^2$$
$$= E(X^2) - [E(X)]^2.$$

To show this, that the variance is the difference between the average square and the square of the average, is to go through exactly the same steps used in Section 4.3 in showing the corresponding relation for sample quantities.

EXAMPLE 4.12

The probability table for the number of points showing on a fair die (Example 4.9) can be extended further to yield the variance; see Table 4.9. The variance is

$$\sigma^2 = E(X^2) - [E(X)]^2 = \tfrac{91}{6} - (\tfrac{21}{6})^2 = \tfrac{35}{12}$$

Table 4.9

POSSIBLE VALUE x_i	PROBABILITY p_i	SQUARE x_i^2	PRODUCT $p_i x_i^2$
1	$\frac{1}{6}$	1	$\frac{1}{6}$
2	$\frac{1}{6}$	4	$\frac{4}{6}$
3	$\frac{1}{6}$	9	$\frac{9}{6}$
4	$\frac{1}{6}$	16	$\frac{16}{6}$
5	$\frac{1}{6}$	25	$\frac{25}{6}$
6	$\frac{1}{6}$	36	$\frac{36}{6}$
Sum	1		$\frac{91}{6}$

and the standard deviation is

$$\sigma = \sqrt{\tfrac{35}{12}} = 1.68.$$

(Again, notice that probabilities are not squared—as relative frequencies were not, in the case of a sample. *Values* are squared, and then weighted with probabilities, to determine the average squared value.)

Problems

4.16. Compute the mean and standard deviation of the random digits distribution (0, 1, . . . , 9, each with probability $\frac{1}{10}$).

4.17. Compute the mean and standard deviation of X, the number of points assigned in the Goren system of bidding to a card drawn at random from a standard deck of playing cards. (The system assigns 4 points to an ace, 3 to a king, 2 to a queen, 1 to a jack, and 0 to anything else.)

4.18. (a) What would be a *fair* amount to bet on red if red pays \$10 in a roulette game? (Compute the expected value of the amount won at a given trial. See Problem 3.8.)

(b) Suppose that the "house" agrees to pay $10 for a double zero and $5 for a single zero in a roulette game. What would be a fair amount to bet?

4.19. The a priori model for the sum Y of the points on two pair of dice is defined by

$$P(Y = k) = \frac{\{6 - |7 - k|\}}{36}.$$

(You may want to check these probabilities with calculations like those in Problem 3.9.) Make a probability table and calculate the mean and variance of Y. (It is interesting to look at the mean and variance for each die in conjunction with the answers here. Any comment or conjecture?)

4.20. Referring to Problem 3.27, compute the "population" mean. (Notice that this would be a "sample mean" *if* the 24 students were thought of as selected from a larger population—the population of all kindergarten students in a city, or perhaps the population of all conceivable boys and girls who have entered kindergarten.) Compute also the mean of the sample of 10 that you drew (in Problem 3.27) from the population of 24. What is an essential difference between the population mean and the sample mean? (Answer this after reflecting on how your respective means would compare with those of your fellow students.)

4.5 Parameters of Continuous Populations

In collecting and recording the observations comprising a sample from a continuous distribution, values must be rounded off, because of limitations in measuring and recoding schemes. The resulting frequency distribution really summarizes a sample from a distribution that is only an approximation, a discrete approximation, to the continuous model for the experiment. Hence the replacement of relative frequencies by probabilities in formulas for averages (or equivalently, taking larger and larger samples) would result in descriptive parameters for that approximating discrete distribution. To define such parameters for a continuous model, a further limiting process is needed, one in which the degree of round-off becomes ever less, so that measurements are made more and more precise. (Of course, this has to be a conceptual or theoretical process, since physical instruments do have limitations.)

The definition and calculation of a population average in the case of a continuous population, given its density function, requires either numerical methods or the methods of infinitesimal calculus, both of which are beyond the scope of the present discussion. What will be done here, then, is to assume (on intuitive grounds) that such parameters can be defined and calculated and that they have properties corresponding and similar to those of parameters of discrete populations. Moreover, certain appealing results about these parameters can be given that are not so hard to comprehend as to prove mathematically.

The mean of a continuous population, or continuous random variable, is perhaps most easily understood as its *center of gravity*, considered earlier in the case of a discrete population and a sample therefrom.

For a continuous population, imagine a template cut out in the shape of the density function of the population and placed on a weightless teeter-totter, complete with scale. The coordinate of the physical balance point is the mean of the distribution, the *population mean* (see Figure 4.5). This balance point is a kind of *middle* of the continuous distribution analogous to the middle defined as the average of a discrete distribution. The Greek letter mu (μ) will ordinarily denote this mean. It is a population parameter; it describes an aspect of the model that represents what might happen at any given performance of the experiment.

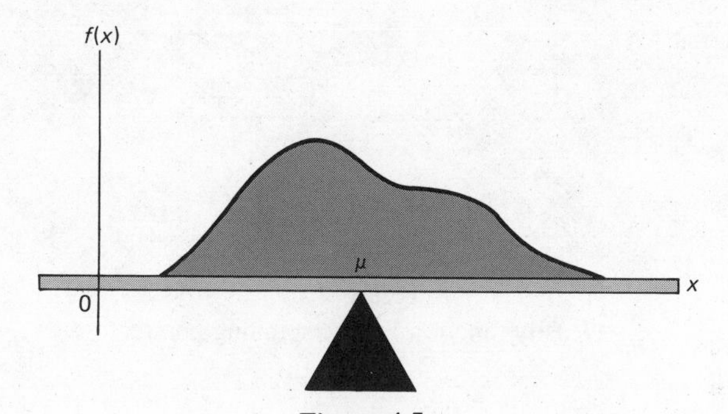

Figure 4.5
Mean of a continuous distribution.

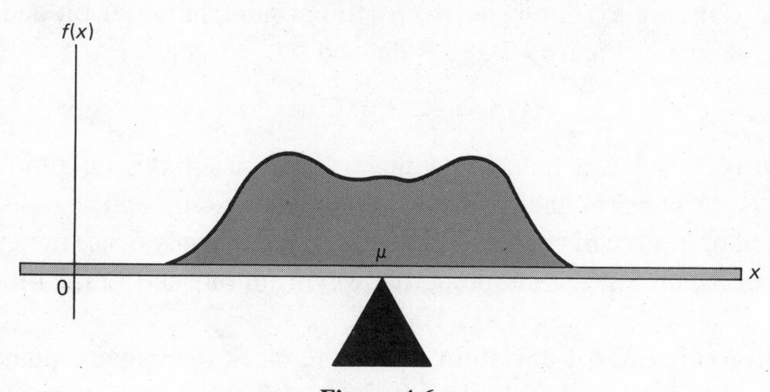

Figure 4.6
Mean of a symmetrical distribution.

Even if a simple mathematical function can be found to represent the curve $f(x)$, it usually requires mathematical tools beyond our scope to calculate μ from $f(x)$. (Of course, approximate methods, such as actually making a template and balancing it, or making some kinds of discrete approximation, can be used to come close to μ.) There is at least one important case, however, in which the balance point or mean is obvious, the case of a *symmetrical* distribution. If the density $f(x)$ is symmetrical about some point, then that point is the balance point, as it was in the case of a discrete symmetrical distribution, Example 4.2 (see Figure 4.6).

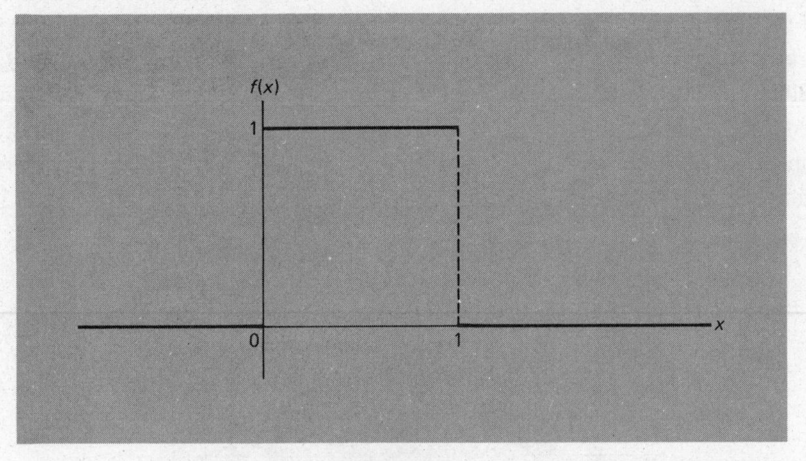

Figure 4.7
Density function for the spinning pointer.

EXAMPLE 4.13
In Chapter 3 the ideal spinning continuous wheel (or spinning pointer) was defined by a model with *constant* density over the range of the scale used. If the scale is numbered from 0 to 1 around the wheel, the density is as shown in Figure 4.7 and is defined by

$$f(x) = 1, \qquad \text{for } 0 < x < 1.$$

Clearly, $x = \frac{1}{2}$ is a point of symmetry. Hence for this distribution, $\mu = \frac{1}{2}$. (A physical analog would be something long and thin, like a pencil or a piece of chalk; if made out of a uniformly dense material, it will balance at the midpoint, halfway from one end to the other.)

The average squared deviation about the mean (or balance point), as in the case of a discrete distribution, is again defined to be the variance, the *population variance*. Its calculation, for a given continuous model, again would involve either numerical approximations (using some sort of discretizing scheme) or the methods of calculus. And it is not easy to get an intuitive feel for the variance, even with the analog of a continuous mass distribution. The corresponding quantity for mass distributions is called an *average moment of inertia*. Although numerical results are not readily obtained, one *can* often make correct qualitative observations by inspection. Thus, because the density in Figure 4.8(a) is high near the mean and the density in Figure 4.8(b) is high at points far from the mean, it should be evident that the variance of the former distribution is small and that of the latter is large.

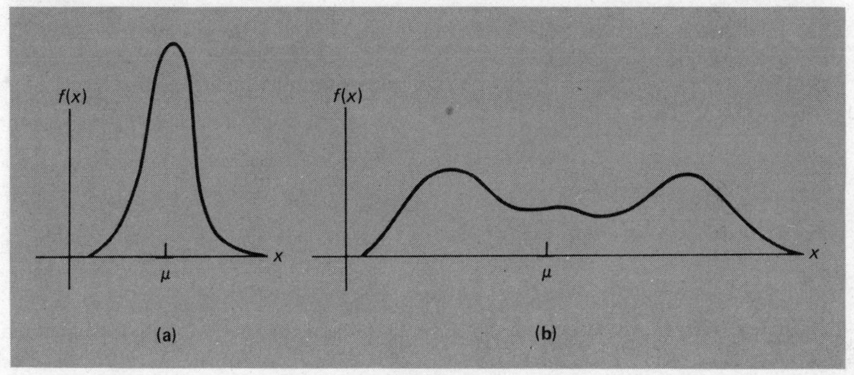

Figure 4.8
(a) Small variance; (b) large variance.

As before, the square root of the variance, whose units would be the same as those of the population variable, is called the (population) *standard deviation* and is often denoted by σ. Figure 4.9 gives a number of simple continuous distributions, as defined by density functions, together with their means and standard deviations. Because σ is a ("root-mean-square") type of average deviation about the mean, it is marked on the graphs as a deviation from μ.

As in the case of samples, descriptive measures for populations other than averages are also useful. The *population median* is a direct analog of the notion of sample median. The sample median is a value that divides the observations so that one-half are to the left and one-half are to the right; so the population median is a value of a random variable that divides the distribution so that one-half of the probability is to the left and one-half is to the right. Since probability for a continuous distribution is represented by the area under the density function graph, the median divides the total area (of 1) into equal parts (of $\frac{1}{2}$ each). And then

$$P(X > \text{median}) = P(X < \text{median}) = \tfrac{1}{2}.$$

In particular, if a distribution is symmetrical about, say, $x = a$, then that value (a) is the population median as well as the mean.

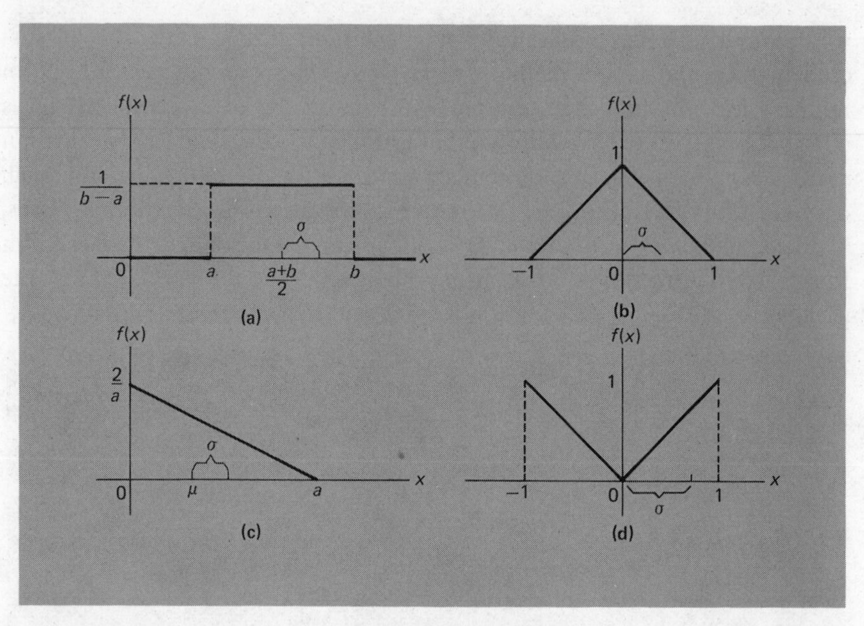

Figure 4.9

(a) Uniform distribution on $a < x < b$: $\mu = (a + b)/2$; $\sigma = (b - a)/\sqrt{12}$. (b) Triangular distribution on $-1 < x < 1$: $\mu = 0$; $\sigma = 1/\sqrt{6} = .408$. (c) $\mu = a/3$; $\sigma = a/\sqrt{18} = .236a$. (d) $\mu = 0$; $\sigma = 1/\sqrt{2} = .707$.

EXAMPLE 4.14

The density in Figure 4.9(c) is shown again in Figure 4.10 but with the median shown also. The median M has to be located so that the area

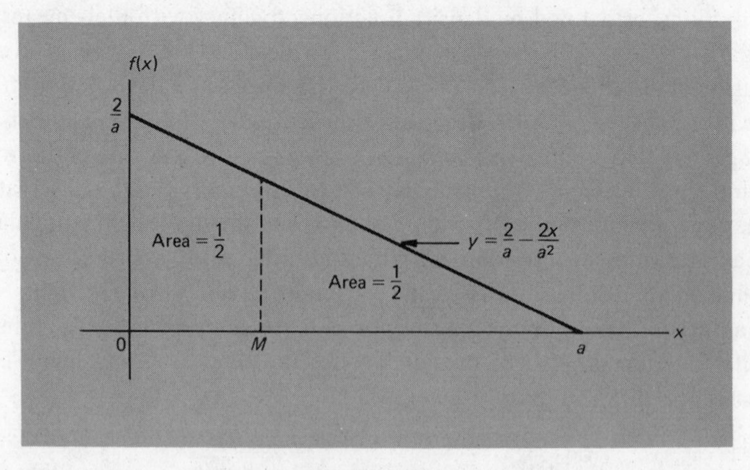

Figure 4.10

Median of a triangular distribution.

of the triangular piece from M to a is $\frac{1}{2}$:

$$\frac{1}{2}(a - M) \cdot \left(\frac{2}{a} - \frac{2M}{a^2}\right) = \frac{1}{2}.$$

Solving for M, one obtains

$$M = a\left(\frac{1 - 1}{\sqrt{2}}\right) = (.293)a,$$

which is the median; the probability that an observation from this population exceeds this value is the same as the probability that it is less than this value. [Note that the mean is $a/3$ and is not the same as the median.]

The population *mode* can be defined (by analogy with the discrete case) as the value at which the probability is most dense, that is, at which $f(x)$ has a maximum. The concept is not often useful, except as a basis for the adjectives "unimodal" and "bimodal," to describe distributions that have, respectively, one hump and two humps or peaks [see Figure 4.8(a) and (b), respectively].

The idea of a median that divides a probability distribution into two equal parts can be generalized. The *quartiles* of a distribution denoted by Q_1, Q_2, and Q_3 divide the probability into four equal parts: one fourth below the "first" quartile (Q_1), another fourth between the first and second quartiles (Q_1 and Q_2), another between the second and third quartiles (Q_2 and Q_3), and another to the right of the third quartile (Q_3) (see Figure 4.11). It should be clear that Q_2 is the same as the median, since $\frac{2}{4} = \frac{1}{2}$.

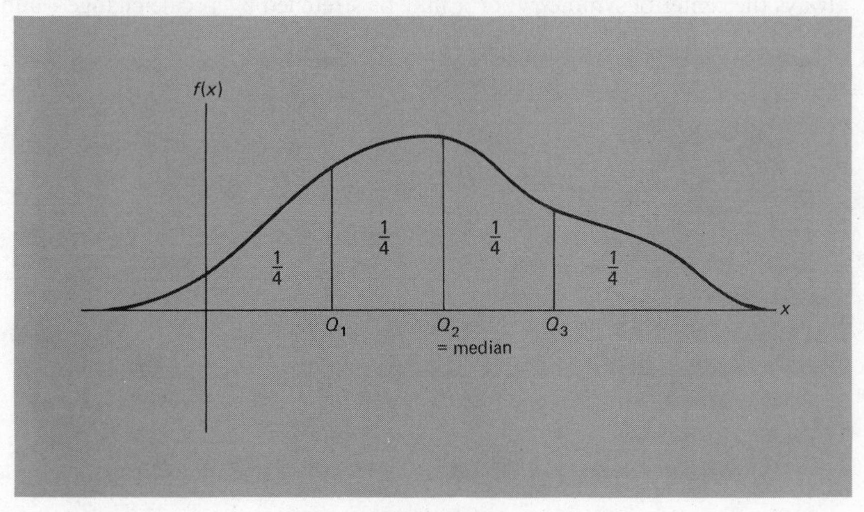

Figure 4.11
Quartiles of a distribution.

Similarly, percentiles divide a distribution into 100 equal parts. The probability to the left of the forty-third percentile, for instance, is .43:

$$P(X < k\text{th percentile}) = \frac{k}{100}.$$

The percentiles of a distribution are commonly used to table a distribution function. Thus tables of any function $y = F(x)$ can be constructed *either* by giving y-values for conveniently chosen (usually equally spaced) values of x *or* by giving x-values for conveniently chosen y-values.

4.6 Parameters of the Normal Distribution

An important continuous model, defined in Section 3.6, is the normal distribution. The standard normal distribution, with density

$$f(x) = Ke^{-x^2/2},$$

is symmetric about $x = 0$. Its mean is therefore this value, $\mu = 0$. Its variance can be shown to be $\sigma^2 = 1$, and this combination of parameter values ($\mu = 0$, $\sigma^2 = 1$) is what the adjective "standard" refers to. It can also be shown mathematically that the two points of inflection (where a curve changes from concave down to concave up or vice versa) occur at $x = \pm 1$, or one standard deviation on either side of the mean (see Figure 4.12).

The more general normal distribution may be relocated—but the mean is always the center of symmetry; or it may be stretched by a scale change—and

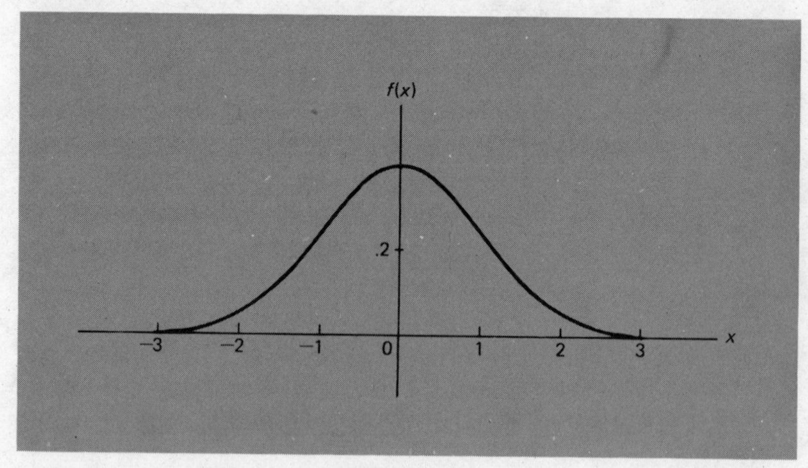

Figure 4.12
Standard normal density.

the distance from the center to the point of inflection is always one standard deviation. Figure 4.13 shows a normal density that is not standard. It can be standardized by a linear transformation. Thus, if X is a normal variable with mean μ and standard deviation σ, then

$$Z \equiv \frac{X - \mu}{\sigma}$$

has mean 0 and standard deviation 1 (that is, Z is standard normal.) This linear relationship defining Z can be thought of as simply a *coding*, which assigns a code value to X which tells how many times the standard deviation σ fits into the distance from the center μ to X. Thus σ becomes one unit and μ becomes the reference origin of the coded values. In this way everything is referred to the *standard* normal distribution, which is the distribution of the code Z.

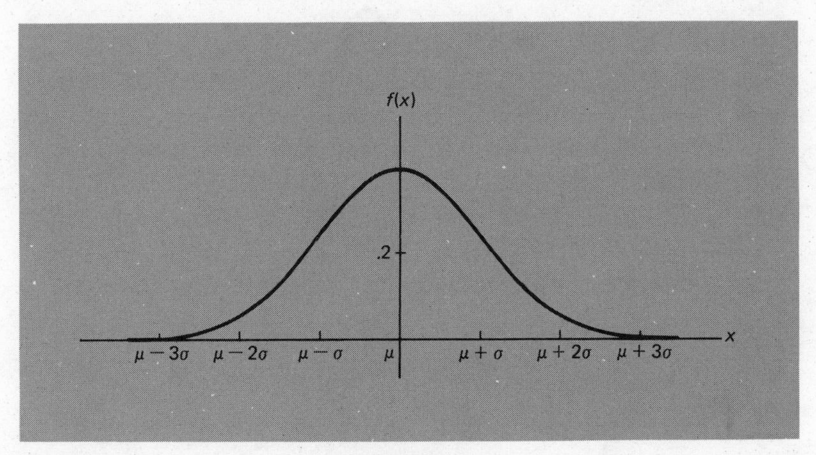

Figure 4.13
General normal density.

In particular, probabilities for a general normal X can be found using the table of probabilities of the standard normal Z, for the area to the left of an X-value is equal to the area under the Z-density to the left of the *code* for that X-value.

EXAMPLE 4.15
A certain viscosity measurement is found to have a pattern of variation that is representable by the ideal model of a normal distribution. If the mean of that distribution is $\mu = 32$ and the standard deviation is $\sigma = 1.8$, the probability that a measurement will fall in the range,

say, from 31 to 34 is computed as follows. The code values for 31 and 34 are, respectively,

$$\frac{31 - 32}{1.8} = -.555\ldots$$

and

$$\frac{34 - 32}{1.8} = 1.111\ldots.$$

That is, 31 is a little more than a half a standard deviation to the left, and 34 is a little over 1 standard deviation to the right of the center, $\mu = 32$. The area (or probability) between 31 and 34 under the density

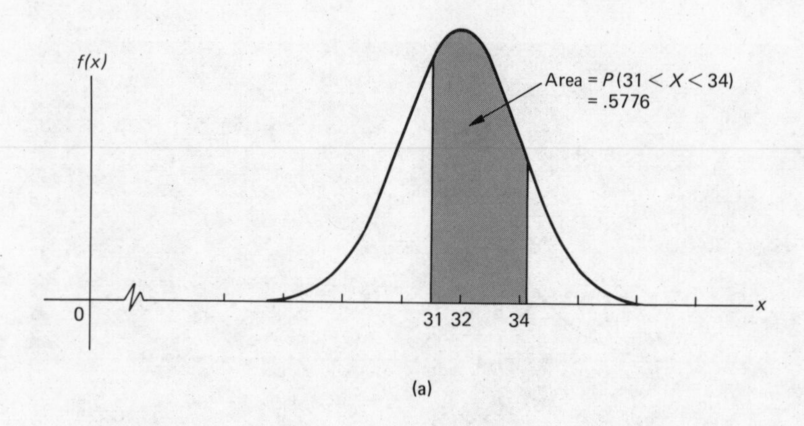

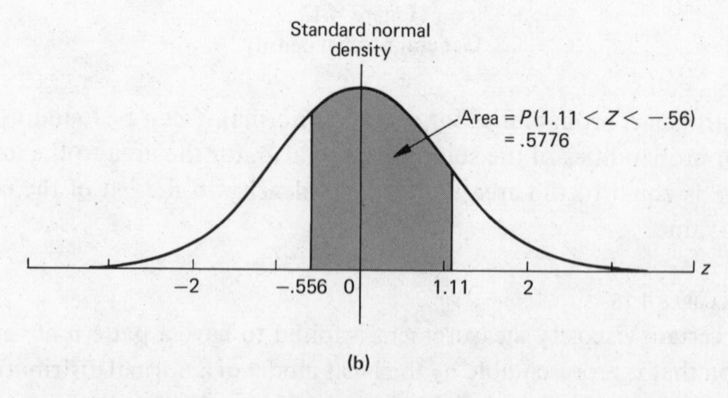

Figure 4.14
Use of the standard normal curve for normal probabilities.

curve representing viscosity variation is just the area under the standard normal curve between $-.555\ldots$ and $1.111\ldots$. If X denotes a viscosity measurement, and $\Phi(z)$ is the standard normal c.d.f.,

$$\begin{aligned} P(31 < X < 34) &= \Phi(1.111) - \Phi(-.556) \\ &= .8667 - .2891 = .5776 \end{aligned}$$

(see Figure 4.14).

Problems

4.21. A spinning pointer has a uniform scale from 0 to 10 marked on the path of its tip. If Y denotes the value on the scale at which the pointer stops, what are the mean and standard deviation of Y?

4.22. The life of a certain type of home light bulb is found to have a distribution that is approximately normal with mean 700 hr and standard deviation 50 hr.
(a) Draw a rough sketch of the density function of this distribution using facts about symmetry and points of inflection.
(b) Determine the probability that a bulb will last more than 800 hr.
(c) Determine the probability that a bulb will last less than 650 hr.
(d) Determine a life length such that 40 per cent of the bulbs last at least that long.

4.23. The weight of a 15-cent candy bar is found to have an approximately normal distribution with mean 1.6 oz and standard deviation .04 oz. What is the probability that a bar is at least as heavy as the weight, 1.5 oz, printed on the wrapper?

4.24. A certain aptitude test has been given to thousands of students; the results can be thought of as pretty well defining the population or model for the aptitude score and say that it is (approximately, at least) normal with mean 450, standard deviation 50.
(a) What is the *percentile* score of a student who achieves a mark of 535?
(b) If a student is told that his score is at the 37th percentile, what is the score?

4.25. Figure 4.15 shows the distribution function of family incomes in a large community.
(a) What is the median income?
(b) What proportion of incomes exceed $10,000?
(c) What is the mode?

4.26. In Figure 4.16(a)–(c) are shown pairs of distributions, one for X and one for Y. In each case decide whether $\mu_x = \mu_y$ or $\mu_x > \mu_y$ and whether $\sigma_x = \sigma_y$ or $\sigma_x > \sigma_y$.

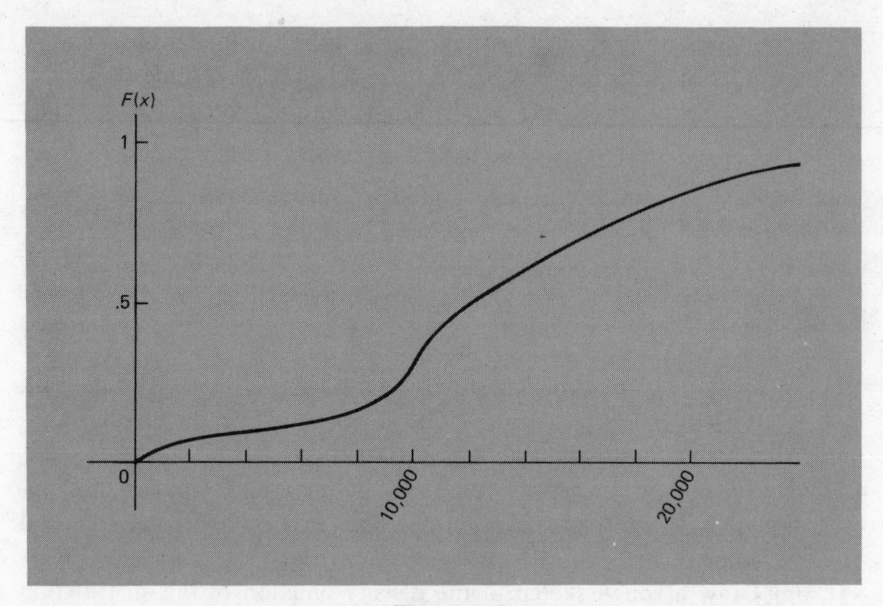

Figure 4.15
Distribution function of incomes for Problem 4.24.

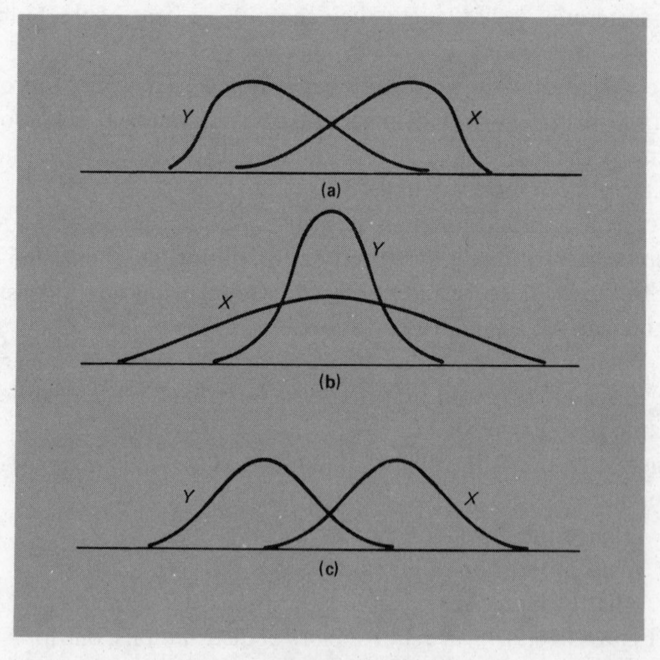

Figure 4.16
Density functions for Problem 4.25.

4.7 Parameters as Limits of Statistics

Certain population parameters were defined by analogy with certain sample statistics, by replacing relative frequencies in the latter by probabilities. In such instances, the approach of relative frequencies to probabilities suggests that the sample statistic would approach the corresponding population parameter as more and more observations are made and included. For example, since

$$\overline{X} = \sum x_i \frac{f_i}{n}$$

and

$$E(X) = \sum x_i p_i,$$

the fact that f_i/n tends to p_i suggests that, in some sense,

$$\overline{X} \to E(X) \qquad (\text{as } n \to \infty).$$

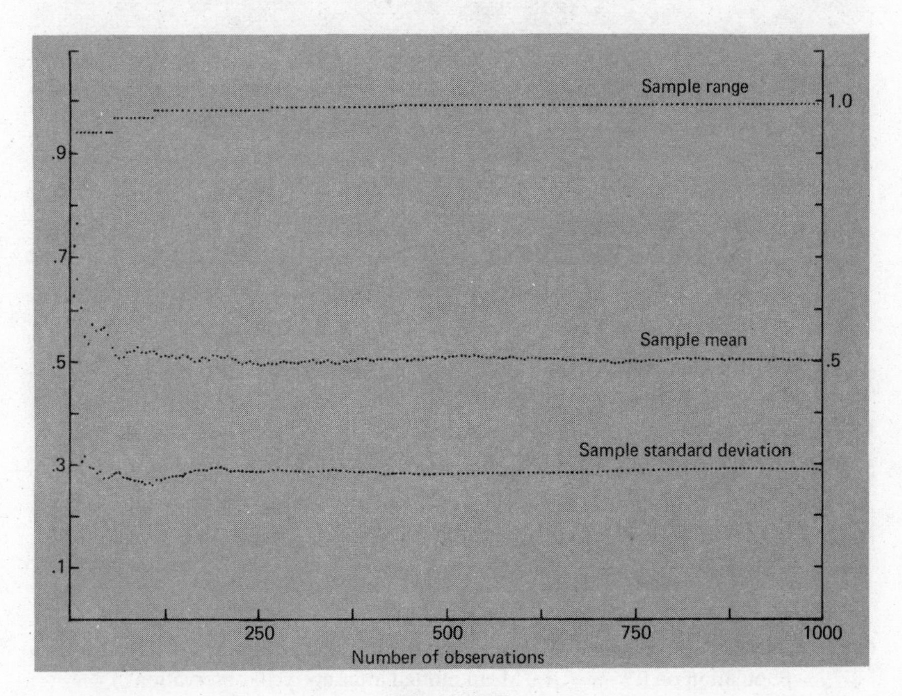

Figure 4.17
Convergence of sample statistics to population parameters: 1000 observations on a uniform population on $0 < x < 1$. (Values plotted after every five observations.)

There are mathematical theorems* justifying such a notion, but these are beyond our scope. On the other hand, the tendency can be illustrated by the technique of artificial sampling. The following example shows the stabilizing of $\overline{X}$ and S, in a certain case, and also the convergence of the sample range, which is not so obvious from a consideration of relative frequencies.

EXAMPLE 4.16

A sequence of random observations of a spinning pointer, uniform on $0 < X < 1$ (see Example 4.13), was generated on a calculator. After every five observations the mean, standard deviation, and range of all observations generated up to that point were calculated and plotted in Figure 4.17. Notice that all three statistics appear to be converging. The mean $\overline{X}$ is approaching the population mean,

$$\overline{X} \to E(X) = \tfrac{1}{2};$$

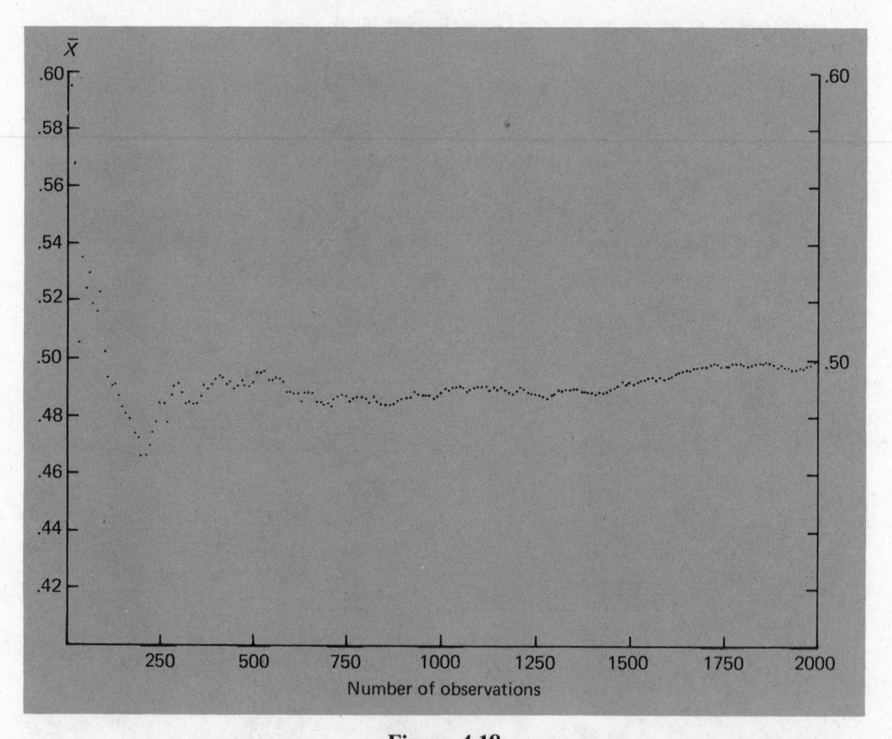

Figure 4.18

Convergence of sample mean to population mean: 2000 observations on a uniform population on $0 < x < 1$. (Mean plotted after every 10 observations.)

* One theorem says that $\overline{X}$ will have $E(X)$ as a limit, with probability 1. A sequence in which $\overline{X}$ does *not* approach $E(X)$ can occur theoretically but never will in practice.

the standard deviation is approaching the population standard deviation,

$$S_x \to \sigma_X = \frac{1}{\sqrt{12}} \doteq .289;$$

and the sample range is approaching the population range,

$$R \to 1.$$

But even after 1000 observations, $\bar{X}$ is not exactly *equal* to $E(X)$. The limit statement means that $\bar{X}$ will be as close as one might specify [within ε of $E(X)$, where ε is any positive number, no matter how small] if one takes enough observations. Figure 4.18 shows the approach of $\bar{X}$ in a sequence of 2000 observations, with an expanded scale for $\bar{X}$ to show better the fluctuations in $\bar{X}$ as it settles down.

5 Sampling Distributions

In order to know what to do with the various statistics that have been introduced so far, and the countless others that might be proposed, one must gain some understanding of their nature and the way this relates to the population being investigated. The first point that must be understood is that the value of a statistic is *variable*.

For a given sample, a statistic has a particular value and is not varying. But it must be realized that the particular sample that has been obtained is one of many different samples that might have been obtained. Hence the particular value of the statistic is but one of many values that might have been obtained—and would be obtained by other investigators, or from other samples however generated.

Consider the familiar experiment of tossing a die: When you toss a die, you get just one outcome. But you know that it is only one of six outcomes you could have got and would get if you were permitted other tosses. To understand the significance of a particular outcome, or to know how to bet before the die is tossed, you would need to know the probability model for the "experiment of chance" that is the toss of a die. This probability model incorporates a knowledge of all the things that can happen when a die is tossed, as well as a lot of experience (at least on the part of mankind) with this and with similar experiments. What makes representation of the toss of a die by a mathematical model feasible is that the experimental conditions are controlled in such a way that certain long-run tendencies emerge. (It would certainly be possible to undermine the relevance of the model of equal probabilities for the six faces by "loading" the die, by not shaking and rolling it thoroughly enough, or perhaps by other sneaky devices. But for a die to be really useful, it is necessary that it be constructed and tossed in such a way as to permit representation by a mathematical model.)

Similarly, when you obtain the observations in a sample and use them to calculate a statistic, you get just one value for that statistic. But that value is only one of many values that might have been obtained and would have been obtained if other samples were permitted. Understanding the significance of a particular value of a statistic, or knowing how to bet before it is obtained, would require knowledge of an appropriate probability model for the experiment which consists of getting the sample and the corresponding value of the statistic. This model would have to incorporate a knowledge of the various samples that can occur and a lot of experience with experiments of the same or similar type. For the model to be useful, experimental conditions would have to be controlled so that the model represents what happens.

EXAMPLE 5.1

To learn about the quality of a lot of manufactured articles, it is common practice to inspect a sample from the lot and so to determine the sample "fraction defective." This fraction is a statistic, being a quantity calculated from the sample. In order to study the nature of the variation in this statistic and how this sampling variation relates to the *lot* fraction defective, it is necessary to have a sampling method that is at least controlled or well defined. However, when the articles in the lot are in some order (in a package, or according to order of manufacture), it would not be smart just to take the first 10, or the last 10, or the top 10. (Have you ever compared the top layer with other layers in a crate of fruit?) A model for the sampling can be constructed if there is a kind of *randomization*, if the articles are drawn in a manner thought of as at random from the lot, with all articles having the same chance of being drawn in a given selection. Such a method of selection does yield samples that are informative as to the quality of the lot.

The methods of sampling that make it possible to represent a sampling process by a mathematical model, the kind of model to use for a given method of sampling, and the various techniques for reducing sampling errors and sampling variability constitute the subject of "sampling," which has an extensive literature. Here only the simplest (yet quite common) kind of sampling will be used, *random sampling*, in which successive observations are from the same basic experiment (or population) and in which the probabilities for any set of observations are not altered by a knowledge of other observations. (A variant of this for finite populations will be considered also.)

5.1 Empirical Sampling Distributions

Some experience with random samples has been gained in examples in Chapters 2 and 3, where artificial sampling with the aid of a computer was used to illustrate sampling fluctuations and large-sample tendencies in frequency distributions. The same device can be used to study the variation in the value of a statistic from sample to sample of a given size, and the way this variational pattern changes with a changing sample size.

It should be emphasized that in practice one must deal with only a single value of a statistic. What will now be done with artificial sampling *is not done in practice*; it is being done here to provide experience with sampling that will help illuminate and point out facts about statistical variation that will be needed for intelligent inference.

EXAMPLE 5.2
A random number generator was used to generate 100 samples of five observations each from a continuous population with the uniform density shown in Figure 5.1. (This is the model for the continuous spinning pointer discussed in Section 3.7.) The *mean* of each of the 100 samples was calculated, and the results are given in Table 5.1. It is evident from this tabulation that the sample mean is a *variable*; its value varies from sample to sample. Careful inspection of the list of 100 values would also suggest that, although variable, the sample mean is not as variable as the observations themselves. This fact and other features of this "sample" of sample means are made more visible if one constructs a frequency distribution and histogram of the 100

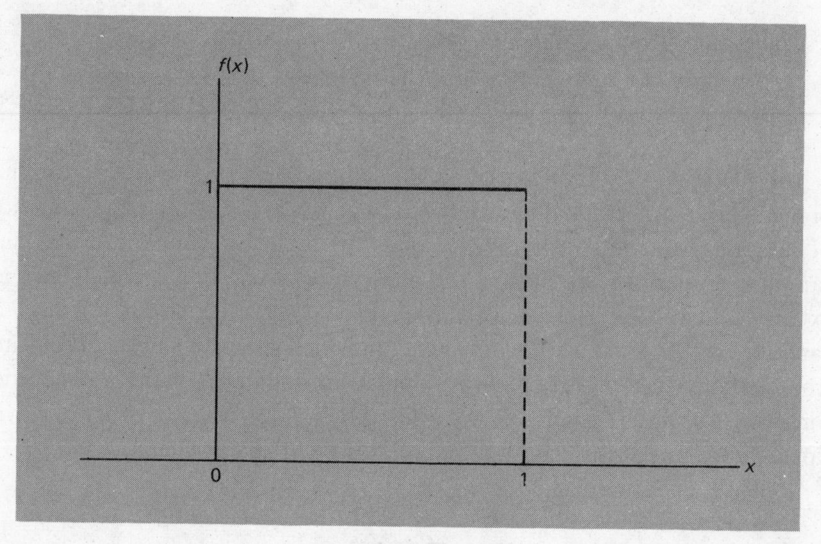

Figure 5.1
Uniform population.

means, in the same way that a frequency distribution and histogram are constructed for any list of 100 numbers. The histogram is shown in Figure 5.2.

Table 5.1

.214	.584	.491	.650	.383
.518	.194	.625	.389	.504
.544	.378	.516	.675	.348
.697	.731	.373	.516	.445
.473	.368	.332	.494	.499
.475	.389	.689	.687	.438
.290	.402	.306	.314	.643
.595	.553	.238	.327	.308
.418	.465	.541	.279	.524
.666	.472	.346	.525	.344
.617	.275	.478	.468	.388
.561	.256	.608	.832	.621
.583	.488	.491	.498	.582
.615	.484	.279	.465	.423
.547	.697	.293	.519	.362
.629	.395	.479	.505	.716
.476	.628	.532	.620	.257
.267	.498	.594	.695	.404
.276	.546	.390	.424	.531
.458	.547	.470	.445	.456

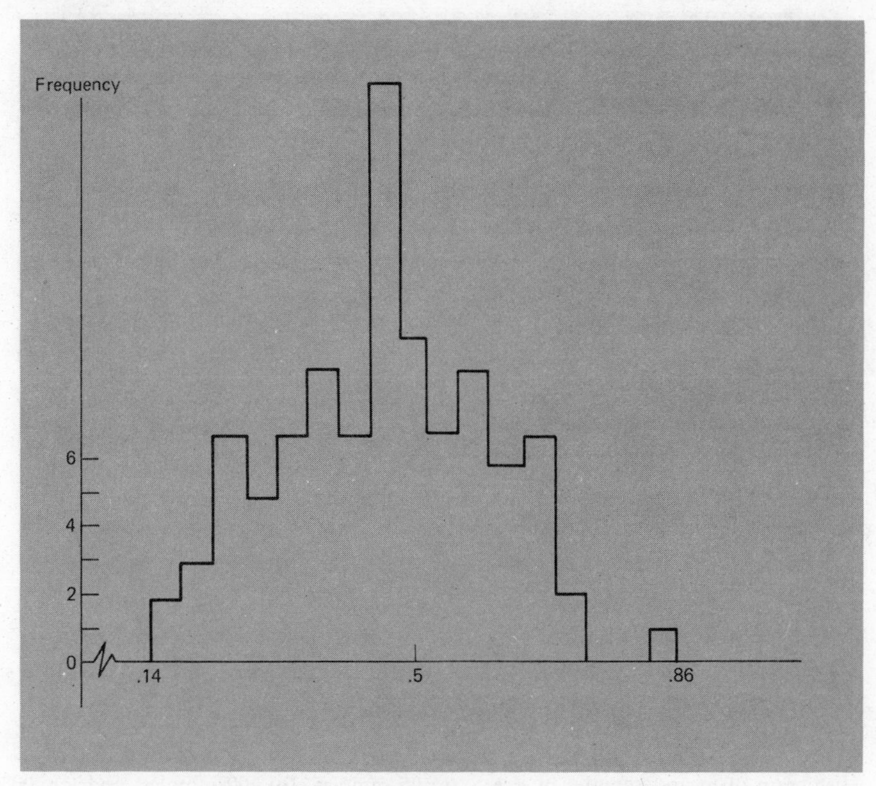

Figure 5.2
Histogram of means of 100 samples of size 5 from uniform populations.

In the study of histograms in Chapter 3, it was shown how the histogram evolved into a probability model as the sample size increases without limit. Similarly, a histogram of sample means will approach a density function of a probability model for the sample mean as a random variable. This will be illustrated in the following example.

EXAMPLE 5.3
Histograms for "samples" of sample means are shown in Figure 5.3—in 5.3(a) for the means of 25 samples of size 5, in 5.3(b) for the means of 100 samples of size 5, and in 5.3(c) for the means of 400 samples of size 5. The trend is apparent; a limiting distribution is emerging which is an ideal distribution for the variable $\overline{X}$, the mean of a sample of five observations on the spinning pointer. In Figure 5.4 the histogram of Figure 5.3(c) is repeated, with a continuous distribution superposed, a free-hand sketch of what the limiting density appears to be.

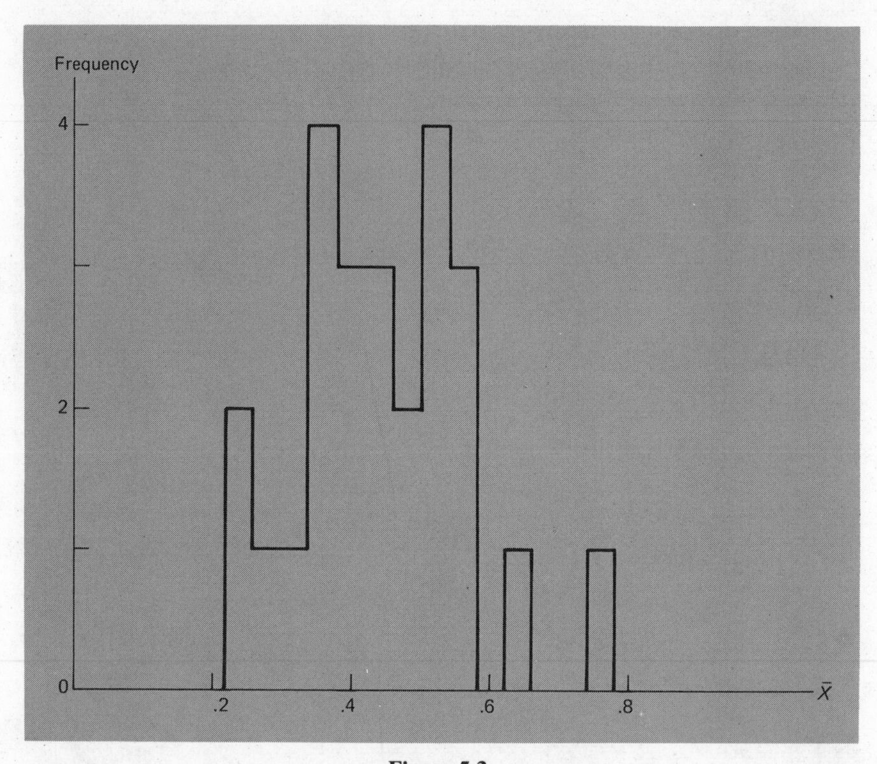

Figure 5.3

Histogram of means, samples of size 5. (a) 25 samples; (b) 100 samples; and (c) 400 samples.

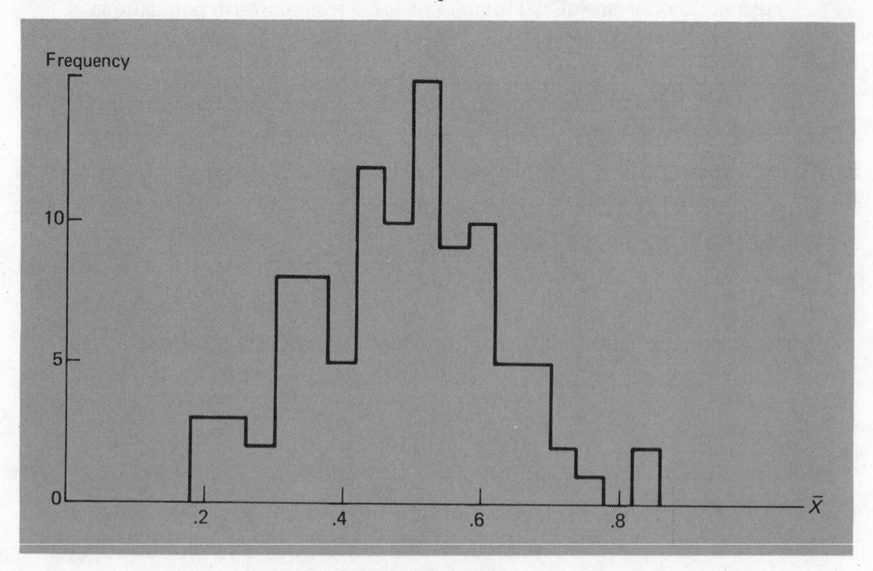

Figure 5.3(b)

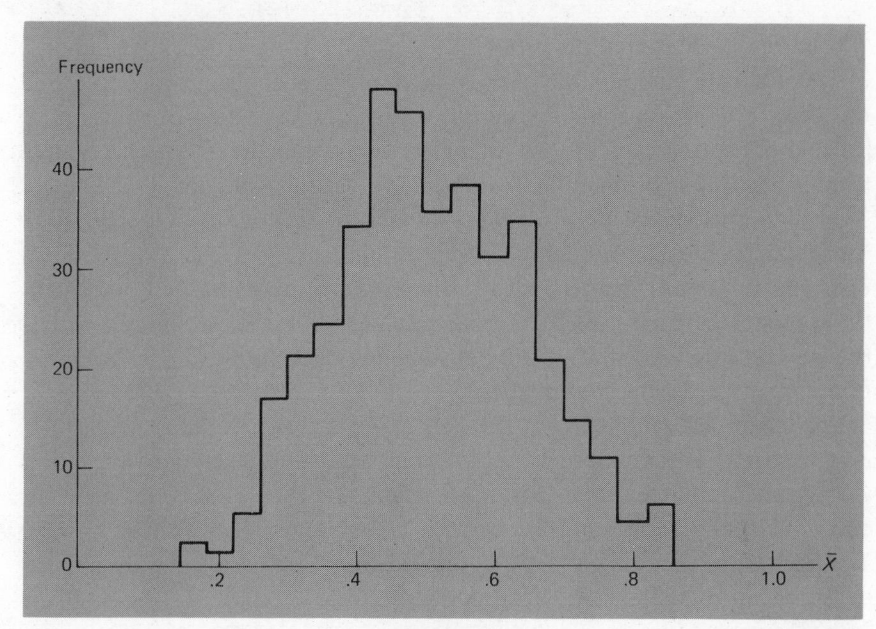

Figure 5.3(c)

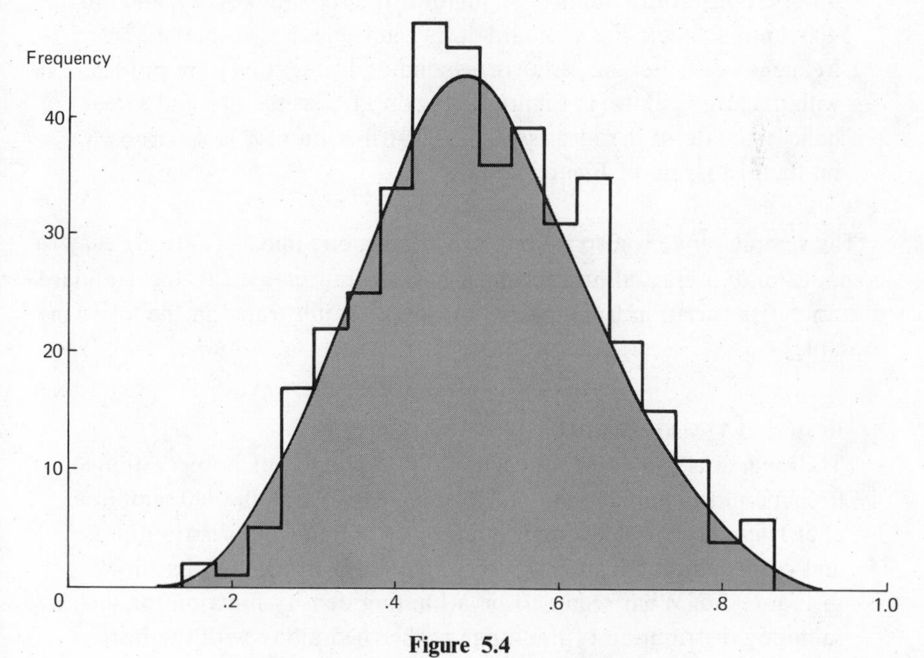

Figure 5.4
Histogram of sample means and its idealization.

It bears repeating that the process of obtaining many samples (a "sample" of samples from a population of possible samples) is an artificial one, carried out here to illustrate some theoretical truths. The collection of values of $\overline{X}$ (calculated from each of the many samples) can be thought of as itself a sample from a *population* of possible $\overline{X}$-values. The density obtained as a limit of $\overline{X}$-histograms defines the probability model for $\overline{X}$ (for the $\overline{X}$-population). This model is often called the *sampling distribution* of $\overline{X}$. Generally, this ideal distribution can only be derived by a theoretical analysis beyond our scope; the approach here has been to learn something about ideal sampling distributions by an empirical study of (finite) sampling distributions in certain cases of artificially generated observations.

The mean of a sample is not the only statistic that is a random variable; every statistic is a random variable. Thus the sample standard deviation varies from sample to sample, although, again, in an actual sampling problem one would not *see* this variation because he would use only a single sample. To learn about what *might* have happened in obtaining a sample standard deviation, an empirical study is again instructive.

EXAMPLE 5.4 (continuing Example 5.3)
The *standard deviation* was computed for each of the 25 samples of 5 observations from a uniform population, the 100 samples of 5, and the 400 samples of 5. These standard deviation values were summarized in frequency distributions, and corresponding histograms were plotted, with the results shown in Figure 5.5. Again a trend emerges, and a freehand estimate of the ideal sampling distribution of S is superposed on the histogram of Figure 5.5(c).

The sample range is also a random variable, one that is relatively easy to compute and useful, although often not as informative as the standard deviation. Its variation from sample to sample is illustrated in the following example.

EXAMPLE 5.5 (continuing Example 5.4)
The *range* was computed for each of the 25 samples of 5 observations from a uniform population, the 100 samples of 5, and the 400 samples of 5. These range values were summarized in frequency distributions, and corresponding histograms were plotted, with the results shown in Figure 5.6. What seems to be a limiting density function for the sampling distribution of the range is sketched along with the histogram of Figure 5.6(c).

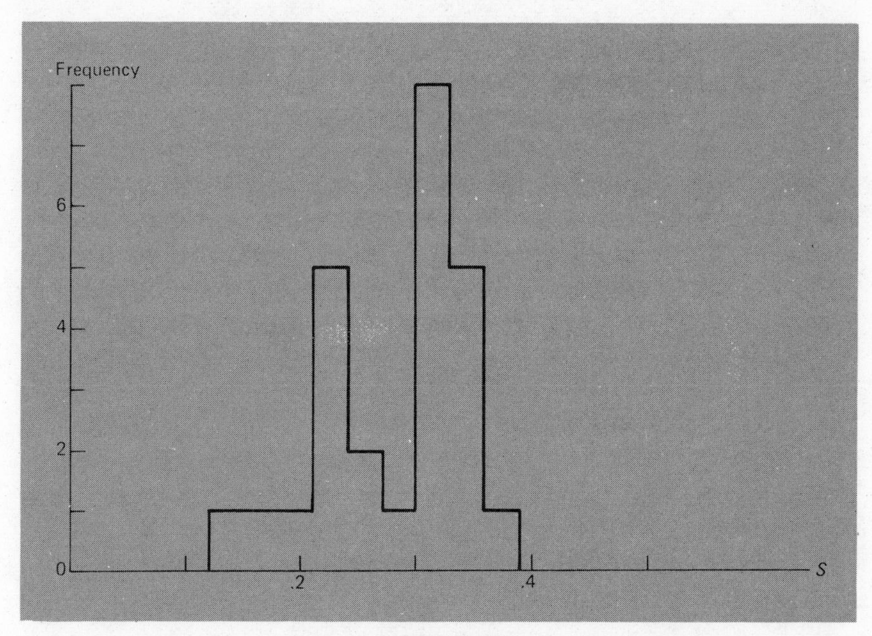

Figure 5.5
Histogram of standard deviations of samples of size 5. (a) 25 samples: (b) 100 samples; and (c) 400 samples (and idealization).

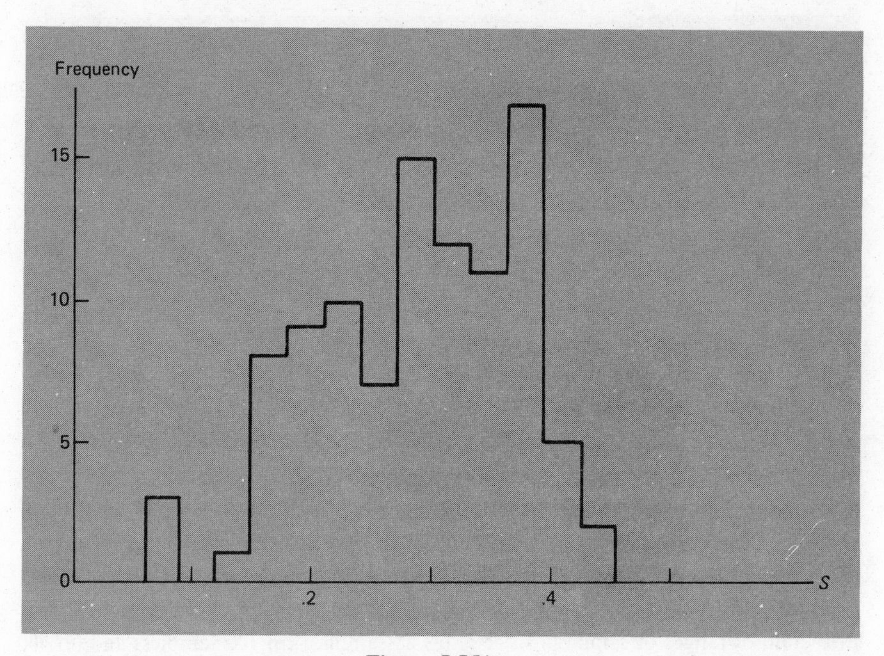

Figure 5.5(b)

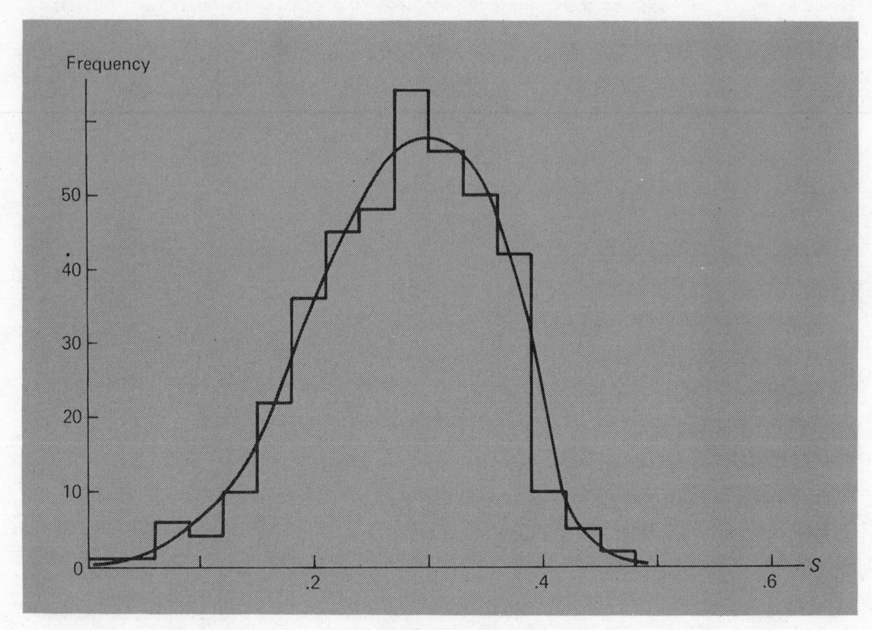

Figure 5.5(c)

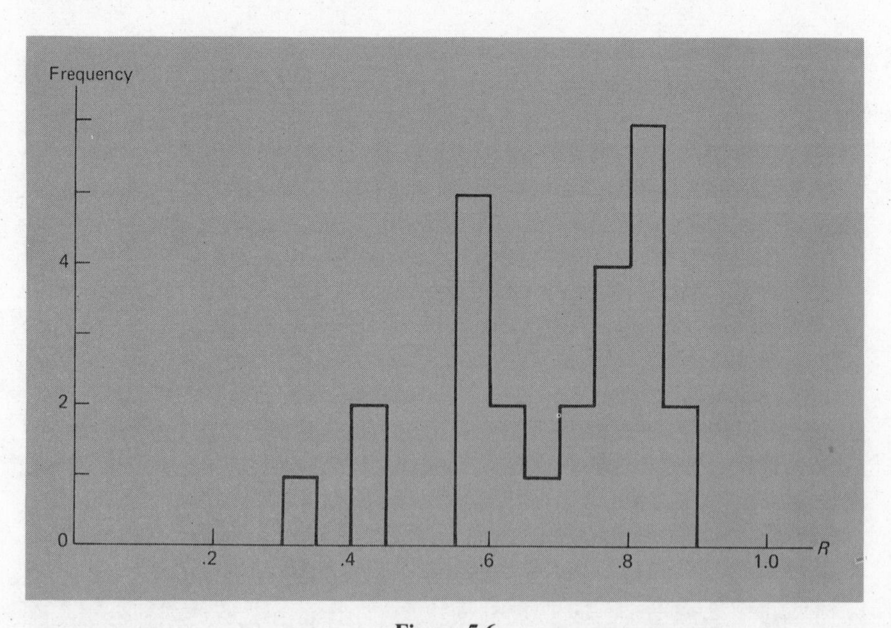

Figure 5.6
Histogram of ranges of samples of size 5. (a) 25 samples; (b) 100 samples; and (c) 400 samples (and idealization).

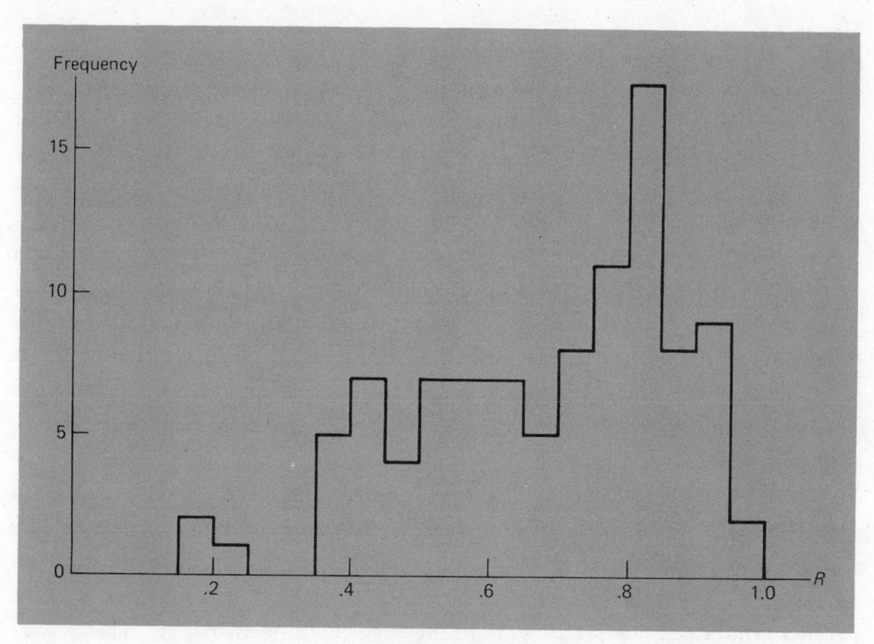

Figure 5.6(b)

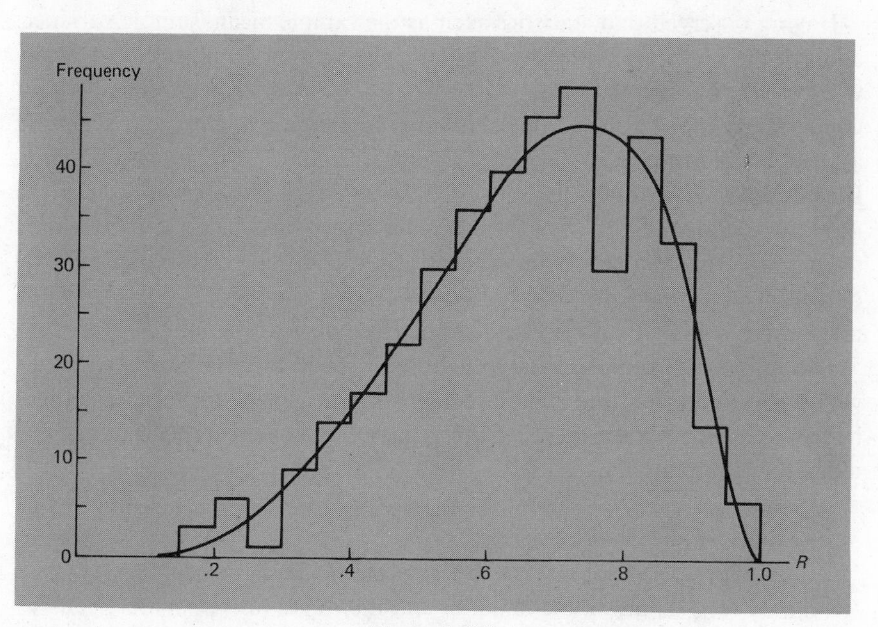

Figure 5.6(c)

5.2 Theoretical Sampling Distributions

A statistic is a *variable*, in the sense that it was produced by a sampling process whose result or outcome is unpredictable; and replicating the sample (repeating the whole sampling process, were this possible) would yield different values of the statistic each time it is done. Moreover, if the sampling process, the gathering of data, is carried out in a way that can be represented by a probability model, the value of a statistic computed from the sample is a *random variable*; i.e., it can itself be represented by a probability model consisting of a set of possible values and an idealized pattern of variation. It is the goal of the present discussion and the sampling experiments reported here to provide some understanding of the nature of this model for the variation in a statistic from sample to sample.

A notable difference between the population model and the model for a statistic lies in their impact on the statistician in a practical problem. The sample observations give *many* visible realizations of a population random variable, but a sample statistic will have only a *single* value in a given statistical problem that does not even seem, on the surface, to be a variable. This is why the device of artificially obtaining many samples is useful, for one can obtain many values of the statistic by such a process, values that *could* have been obtained (rather than the one that *was* obtained) and that might have been obtained by others who obtained similar samples.

Having observed that statistics such as the sample mean, sample variance, and sample range are variables and have distributions, one would naturally wonder whether anything can be learned about these distributions without going through the tedious, expensive, and often impossible process of getting a great many samples, as has been done artificially in the above examples. Indeed, it would be a bit silly to draw 100 samples of 50 observations each to learn about the randomness of the mean of a sample of 50 and then to use only the mean of one sample of 50 in a statistical problem; for there would be 5000 observations available for use, and the mean of a sample of 5000 ought to be much more reliable than the mean of only 50 observations.

The following example, though a little artificial and unnaturally simple, will illustrate the fact that sampling distributions of statistics can sometimes be *derived* from a knowledge of the population being sampled and of the method of sampling that is used.

EXAMPLE 5.6

Let us consider a population that consists of the six boys in a certain family. The names of the children and their ages (in years) are given in the following list.

Arthur	4	Dave	8
Bill	6	Ed	9
Charles	6	Frank	9

A stranger is asked to estimate the average age, after he is allowed to see three of the children chosen at random and ask them their ages. He places the six names in a hat and draws three: Bill, Ed, and Frank. The average age in his sample of three is $\frac{1}{3}(6 + 9 + 9) = 8$. If he takes this as his guess, how reliable is it?

"Population" here is the set of six equally probable boys' names; or it could be taken to be the set of possible ages and probabilities:

Age	4	6	8	9
Probability	$\frac{1}{6}$	$\frac{2}{6}$	$\frac{1}{6}$	$\frac{2}{6}$

The population average,

$$4 \times \tfrac{1}{6} + 6 \times \tfrac{2}{6} + 8 \times \tfrac{1}{6} + 9 \times \tfrac{2}{6} = 7,$$

is the average age the stranger is to guess. Figure 5.7 exhibits this distribution and its average (or balance point).

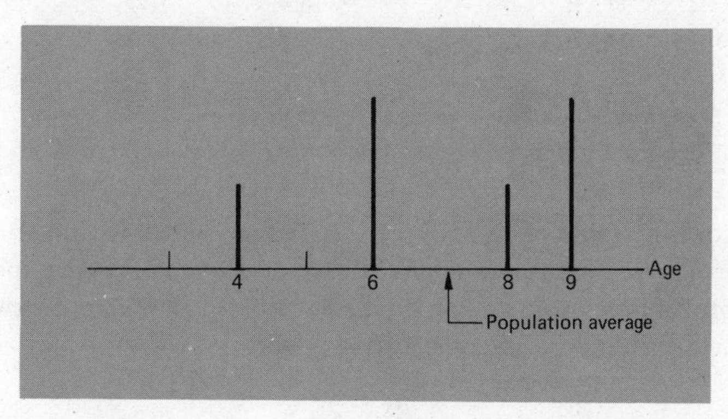

Figure 5.7
Population distribution of ages for Example 5.6.

The samples that might turn up can be enumerated completely, and it is implied in the phrase "at random" that every one of the 20 possible samples had the same chance of being drawn. Table 5.2 presents these 20 equally probable samples, the ages in each sample, and the sum of the observations in each sample. (The sample actually obtained is indicated by an asterisk.)

Table 5.2

SAMPLE (BOYS)	PROBABILITY	SAMPLES (AGES)	SAMPLE SUM
A B C	$\frac{1}{20}$	4, 6, 6	16
A B D	$\frac{1}{20}$	4, 6, 8	18
A B E	$\frac{1}{20}$	4, 6, 9	19
A B F	$\frac{1}{20}$	4, 6, 9	19
A C D	$\frac{1}{20}$	4, 6, 8	18
A C E	$\frac{1}{20}$	4, 6, 9	19
A C F	$\frac{1}{20}$	4, 6, 9	19
A D E	$\frac{1}{20}$	4, 8, 9	21
A D F	$\frac{1}{20}$	4, 8, 9	21
A E F	$\frac{1}{20}$	4, 9, 9	22
B C D	$\frac{1}{20}$	6, 6, 8	20
B C E	$\frac{1}{20}$	6, 6, 9	21
B C F	$\frac{1}{20}$	6, 6, 9	21
B D E	$\frac{1}{20}$	6, 8, 9	23
B D F	$\frac{1}{20}$	6, 8, 9	23
*B E F	$\frac{1}{20}$	6, 9, 9	24
C D E	$\frac{1}{20}$	6, 8, 9	23
C D F	$\frac{1}{20}$	6, 8, 9	23
C E F	$\frac{1}{20}$	6, 9, 9	24
D E F	$\frac{1}{20}$	8, 9, 9	26

From the table of samples, it is a simple matter to construct a table of average ages that might be encountered and the corresponding probabilities of encountering them; see Table 5.3. Thus the center of the distribution of average ages is 7 years:

$$\mu_{\overline{X}} = 7.$$

The distribution of $\overline{X}$ is shown graphically in Figure 5.8. Observe that it is rather narrower than the population distribution in Figure 5.7. Recall, too, that the *population* average age (i.e., the average age of the six children in the family) is also 7 years:

$$\mu = 7.$$

So, sample mean ages vary from $5\frac{1}{3}$ to $8\frac{2}{3}$ years, with an average of 7, whereas *individual* ages vary from 4 to 9 years, but also with average 7.

Table 5.3

AVERAGE AGE IN SAMPLE, $\overline{X}_i$	PROBABILITY p_i	PRODUCT $\overline{X}_i p_i$
$\frac{16}{3}$	$\frac{1}{20}$	$\frac{16}{60}$
$\frac{18}{3}$	$\frac{2}{20}$	$\frac{36}{60}$
$\frac{19}{3}$	$\frac{4}{20}$	$\frac{76}{60}$
$\frac{20}{3}$	$\frac{1}{20}$	$\frac{20}{60}$
$\frac{21}{3}$	$\frac{4}{20}$	$\frac{86}{60}$
$\frac{22}{3}$	$\frac{1}{20}$	$\frac{22}{60}$
$\frac{23}{3}$	$\frac{4}{20}$	$\frac{92}{60}$
$\frac{24}{3}$	$\frac{2}{20}$	$\frac{48}{60}$
$\frac{26}{3}$	$\frac{1}{20}$	$\frac{26}{60}$
Sum		$\frac{420}{60} = 7$

The lesser variation apparent among the $\overline{X}$-values is seen numerically in a comparison of *variances*. The population variance is

$$\tfrac{1}{6}(4-7)^2 + \tfrac{2}{6}(6-7)^2 + \tfrac{1}{6}(8-7)^2 + \tfrac{2}{6}(9-7)^2 = \tfrac{30}{9}.$$

The variance of $\overline{X}$ comes, similarly, from the probability table for $\overline{X}$:

$$\tfrac{1}{20}[(\tfrac{16}{3}-7)^2 + 2(\tfrac{18}{3}-7)^2 + 4(\tfrac{19}{3}-7)^2 + (\tfrac{20}{3}-7)^2 + 4(\tfrac{21}{3}-7)^2$$
$$+ (\tfrac{22}{3}-7)^2 + 4(\tfrac{23}{3}-7)^2 + 2(\tfrac{24}{3}-7)^2 + (\tfrac{26}{3}-7)^2] = \tfrac{6}{9}.$$

Although it was possible to determine completely the distribution of $\overline{X}$, in this example, because of its simplicity, it would have been

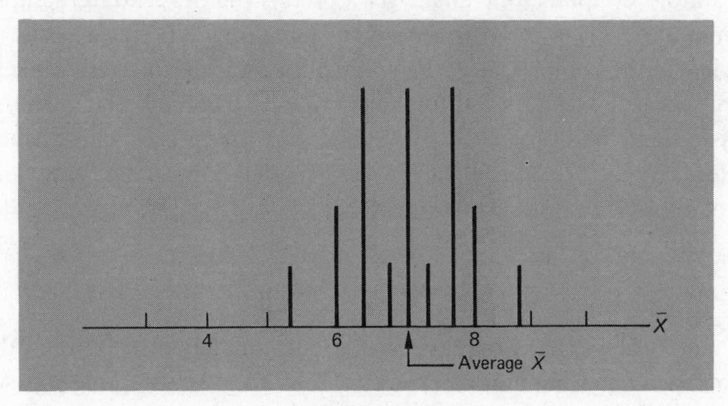

Figure 5.8
Distribution of $\overline{X}$ for Example 5.6.

possible to learn about $\overline{X}$ by empirical means as well. A computer-generated series of 100 samples of the same kind, each drawn at random from the six boys, resulted in the empirical distribution of $\overline{X}$ shown in Figure 5.9.

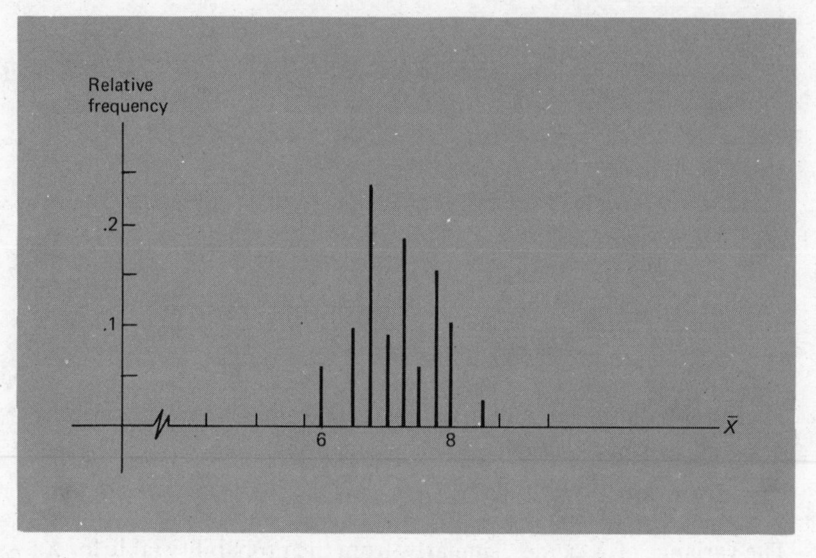

Figure 5.9
Empirical distribution of average ages.

The above example is quite simplistic; it had to be, to make the enumeration of all possible samples feasible. But it should help make the point that even with random sampling, one might get a pretty wild (misleading) sample, and he would never know it. Drawing conclusions from data has its risks.

In view of the complexity of the enumerative approach, even in the simple example considered, the need for mathematical theorems that avoid such a process should be apparent. The case is even more compelling when one considers sampling that is done from a conceptual population, where enumeration of possible samples is impossible.

Projects

No matter how much you have *read* about sampling distributions, the best way to appreciate what and why they are is to do some sampling yourself and to observe first hand the kind of variation that is encountered in obtaining a value of some statistic. On this account we suggest here some projects or laboratory exercises designed to give you such essential experience. They are not "problems" with "answers," and they will only serve their purpose if carried out thoughtfully at each step. Naturally, there will be a degree of artificiality involved, and part of the thought process is to see that what you do tells you something about more practical problems. It will be instructive to compare notes with classmates, and it will be even more instructive if you can pool your results with those of others to achieve a still larger collection of values of the statistic being studied. (Access to a desk calculator will ease the burden of computation considerably, but the tasks can and should be done even if you do not have one.)

Project 1

Use the random number table (Table VIII of the Appendix) to give you observations from the uniform spinning pointer, using the blocks of five random digits to define a five-digit decimal on the range $0 < x < 1$.

(a) Calculate $\overline{X}$ for each of 20 (or more, if you use a calculator!) samples of size 5. (It may be convenient to take a 5 by 5 block of digits as a column of 5 observations comprising a sample.)

(b) Make a frequency tabulation of your $\overline{X}$'s, using as class intervals .14–.22, .22–.30, etc., and from this construct a histogram. Sketch in what you think might be the ideal model for $\overline{X}$, and compare with the population model. Is the model for $\overline{X}$ shaped differently? Shifted in location? Narrower or wider?

(c) From the frequency distribution of $\overline{X}$ values, compute the mean and standard deviation. [If you do not have a calculator, this job is simpler if you relabel thus: $Y = (\overline{X} - .5)/.8$, compute $\overline{Y}$ and S_y, and then transform back to the $\overline{X}$ scale.] Observe their relationship to the population mean and standard deviation, and save for reference in the next problem set.

(d) Construct a sample distribution function for the original 20 $\overline{X}$-values. Superimpose a smooth curve that is suggested by your data as the underlying model, and save for reconsideration in the next problem set.

Project 2

Obtain the ranges of 40 samples of 5 as in the preceding project. (The range is easier to compute than $\overline{X}$, so you can afford to take more samples.) Make a frequency table and histogram for the 40 R-values. Repeat with 40 samples of size 10 (or even 20 samples of 10 if your enthusiasm is waning). Compare the results. Is the apparent shift of the distribution to be expected? (That is, when you take 10 observations instead of 5, do you anticipate a larger R, generally?)

Project 3

Use the table of normal random numbers (Table IX of the Appendix) as a source of data to obtain 40 samples of size 5. Calculate the *median* and *midrange* (average of largest and smallest) for each sample. Construct a histogram for each of these empirical sampling distributions and sketch in a smooth density that (you think) is close to the ideal density. How does it compare, in each case, with the density for a single observation, i.e., the population density, which is standard normal? (Compare with respect to location and degree of variability.)

Project 4

Using at least 25 random normal numbers (Table IX) and class intervals -3 to -2.5, -2.5 to -2.0, etc., make a histogram for an empirical version of the population from which the random normal numbers are taken. Then take 20 samples of 5 and compute $\overline{X}$ for each sample. Construct a histogram for these $\overline{X}$-values and compare with the population distribution.

Project 5

Consider the 40 fraternity students whose credit loads were reported in Problem 2.2 as a (finite) *population*. The credit loads are

13, 18, 15, 16, 19, 18, 16, 14, 18, 18, 18, 20, 18, 16

16, 17, 20, 17, 19, 15, 17, 18, 22, 19, 14, 13, 18, 14

18, 18, 8, 19, 17, 18, 21, 13, 18, 9, 18, 13.

The (population) average credit load is 16.625. Use the random number table to obtain 20 samples of 10 fraternity members and calculate the mean of each sample. Compare the distribution of these means with the distribution of individual credit loads (the population) obtained in Problem 2.2.

5.3 The Sampling Distribution of $\overline{X}$: Some Theorems

Example 5.6 illustrated certain facts that are true in greater generality, facts that will aid in studying the performance of certain inferential procedures. These facts—mathematical theorems in probability theory—relate in particular to the use of the sample mean as a tool in inference. Other statistics (such as the sample variance, sample range, etc.) also have distributions, with moments that are of interest, but each case requires techniques and theorems of its own. Since the main emphasis in this introductory treatment is on problems relating to population means, and the principal technique will involve $\overline{X}$, we do not burden the discussion at this point with theorems about other statistics.

THEOREM 5.1 *The distribution of the mean of a random sample is centered at the same value as is the population:* $\mu_{\overline{X}} = \mu$.

The term *random sample* used here must be given a technical mathematical meaning in order to be able to prove such a mathematical theorem. The proof will not be presented, so the practical definition of random sample given earlier will suffice. Successive observations must all be from the same basic experiment or population, and probabilities for any set of observations must be the same with as without any specific knowledge of the values of the remaining observations.

EXAMPLE 5.7

The mean μ of a variable that is uniformly distributed for $0 < x < 1$ (the continuous spinning pointer) is $\frac{1}{2}$, which is the center of symmetry and hence the balance point of the distribution. Hence, according to Theorem 5.1, the center of the distribution of the mean of a random sample from this uniform population is also $\frac{1}{2}$. Observe that the empirical distribution of $\overline{X}$ for samples of size 5 in Figure 5.3 would balance at about (but not exactly at) $\frac{1}{2}$. [In the case of 400 samples, Figure 5.3(c), the center value is .5015.]

Variability in $\overline{X}$ is covered by the following result, again stated without proof:

THEOREM 5.2 *The variance of the distribution of the mean $\overline{X}$ of a random sample of size n is $1/n$ times the population variance:*

$$\sigma_{\overline{X}}^2 = \frac{\sigma^2}{n}.$$

This theorem says that in a large sample there is little variation among the means (from sample to sample), as compared with the variation of the individual observations (from observation to observation).

EXAMPLE 5.8

Again, for the case of sample means from a uniform population on $0 < x < 1$, the population variance (see Figure 4.9) is $\sigma^2 = \frac{1}{12}$.

Hence, according to Theorem 5.2, the variance of $\overline{X}$ is

$$\text{var } \overline{X} = \frac{1}{12n}.$$

For the sample size $n = 5$, this is $\frac{1}{60} = .01666\ldots$; this might be compared with the variance of the empirical distribution of 400 sample means in Figure 5.3(c), which is .0163. (The comparison of standard deviations would be $\sigma_{\overline{X}} = \sqrt{\frac{1}{60}} = .129$ against the empirical result $\sqrt{.0163} = .1275$.)

An important kind of sampling does not fall under the category of random sampling as described above. It is used in the case of finite populations of actual objects, and the requirement is that the n observations of the sample are obtained in such a way that all possible selections of n objects from the population are equally likely. This is referred to as *simple random sampling* and can be achieved by selecting the objects for the sample one at a time so that as each one is selected, all remaining objects have the same chance of being chosen. In such cases, the following results can be established:

THEOREM 5.3 *The distribution of the mean of simple random sample of size n from a population of size N has these parameters:*

$$\mu_{\overline{X}} = \mu, \qquad \sigma_{\overline{X}}^2 = \frac{\sigma^2}{n} \cdot \frac{N-n}{N-1},$$

where μ and σ^2 are, respectively, the population mean and variance.

EXAMPLE 5.9

In Example 5.6 the 20 possible selections of three boys from the six were assumed to be equally likely, the result of simple random sampling. It was shown there that

$$\mu_{\overline{X}} = \mu = 7$$

and that

$$\sigma_{\overline{X}}^2 = \frac{2}{3}, \qquad \sigma^2 = \frac{10}{3}.$$

With $N = 6$, $n = 3$, the relationship of Theorem 3 can be verified:

$$\frac{\sigma^2}{n} \cdot \frac{N-n}{N-1} = \frac{\frac{10}{3}}{3} \cdot \frac{6-3}{6-1} = \frac{2}{3} = \sigma_{\overline{X}}^2$$

Thus it is possible to compute the mean and variance of the random variable $\overline{X}$ if the population mean and variance are known, which is generally not the case. The mean and variance of $\overline{X}$ can be approximated from the empirical distribution of $\overline{X}$ shown in Figure 5.9. The frequency distribution for those 100 sample means is as follows:

$\overline{X}$	$\frac{16}{3}$	$\frac{18}{3}$	$\frac{19}{3}$	$\frac{20}{3}$	$\frac{21}{3}$	$\frac{22}{3}$	$\frac{23}{3}$	$\frac{24}{3}$	$\frac{26}{3}$
Frequency	6	10	24	7	19	6	16	10	2

The mean and standard deviation of this empirical sampling distribution should be approximately equal to $E(\overline{X})$ and $\sigma_{\overline{X}}^2$, respectively:

$$E(\overline{X}) \doteq 6.88,$$

$$\sigma_{\overline{X}}^2 \doteq 0.79.$$

(The theoretical values are 7 and $\frac{2}{3}$, respectively.)

5.4 Large-sample Distribution of $\overline{X}$

Statistical procedures based on large samples are generally easier to analyze than those that employ small samples. This is largely because of facts like that contained in Theorem 5.4. This particular result is of such importance, playing a central role in inference, that it is known as the *central limit theorem*:

THEOREM 5.4 *Let $\overline{X}_n$ denote the mean of the first n of a sequence of independent observations from a population with mean μ and variance σ^2. Then the random variable*

$$Z = \frac{\overline{X}_n - \mu}{\sigma/\sqrt{n}}$$

has a distribution that approaches the standard normal distribution as n becomes infinite.*

At the end of Chapter 4 it was demonstrated how a sample mean $\overline{X}$ converges to the population mean as the sample size increases. But for any given sample size, no matter how big, the mean $\overline{X}$ is still a *random variable*; i.e., it has a distribution. This distribution gets bunched closer and closer around μ as n increases, but Theorem 5.4 tells about the *shape* of the distribution. It is a

* "Approaches" means the probability that $Z \leq z$ can be made as close as one pleases to $\Phi(z)$, in Table I, by taking n big enough.

bell-shaped distribution similar to the standard normal, differing from it only by a change of center or location and a change of scale. Values of $\bar{X}$ are clustered about its mean μ (so the deviations $\bar{X} - \mu$ would be clustered about zero). And the values of $\bar{X}$ are *tightly* clustered about μ when n is large. However, when $\bar{X} - \mu$ is measured according to the scale factor $\sigma/\sqrt{n}$, this measure,

$$Z = \frac{\bar{X} - \mu}{\sigma/\sqrt{n}},$$

has standard deviation 1. As discussed in Section 4.8, the term "normal" is applied not only to the distribution in Table I but also to a distribution that differs from it only in location and/or scale. Thus we say that in large samples $\bar{X}$ is approximately normally distributed.

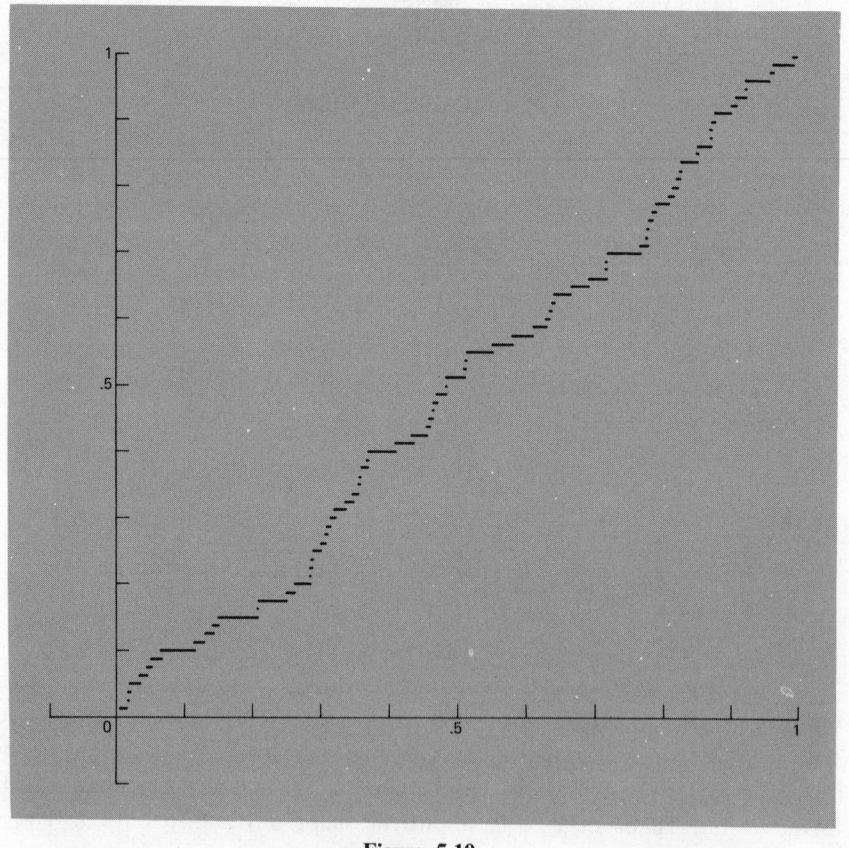

Figure 5.10
Empirical distribution function of a sample mean: 80 samples of size 1, from uniform population.

EXAMPLE 5.10

Samples of size 1, 4, 20, and 100—eighty samples for each size—from the spinning pointer population (see Examples 4.13 and 4.16), were generated on a computer. The mean of each sample was computed, and for each sample size a distribution of the 80 means was plotted in a distribution function graph. The graphs are shown in Figures 5.10 through 5.13. The first, for $n = 1$, is essentially an empirical distribution for the population itself (see Figure 5.1). Recall that to interpret this kind of plot you simply note that an observation occurred at each value above which there is a jump. Thus in Figure 5.10 the observations are spread more or less uniformly throughout the interval $0 < x < 1$. In the graphs for larger n, the observations (values of $\overline{X}$) become more and more clustered about the value $x = \frac{1}{2}$,

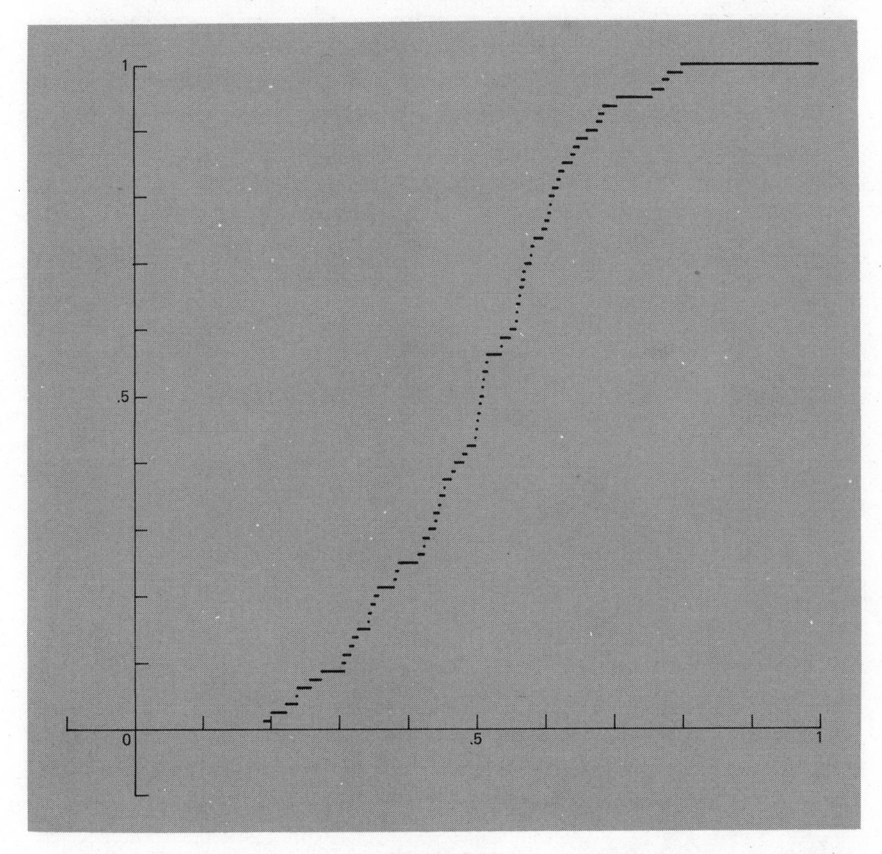

Figure 5.11
Empirical distribution function of a sample mean: 80 samples of size 4, from uniform population.

the population mean. But the *shape* of the distribution also has a tendency, observable more in Figure 5.14, in which the x-scale has been expanded. In that figure the limiting normal distribution (promised in Theorem 5.4) has been superimposed.

EXAMPLE 5.11

Two hundred samples of size 50 were taken from an artificial "coin-like" population with a probability .6 for head; see Table 5.4. The mean of each of the 200 samples was computed and the results plotted as a frequency plot in Figure 5.15. (Notice that the outline of the relative frequency rods suggests a normal curve.) The cumulative plot corresponding to these frequencies is shown in Figure 5.16, with a normal curve superimposed—the c.d.f. of a normal distribution with

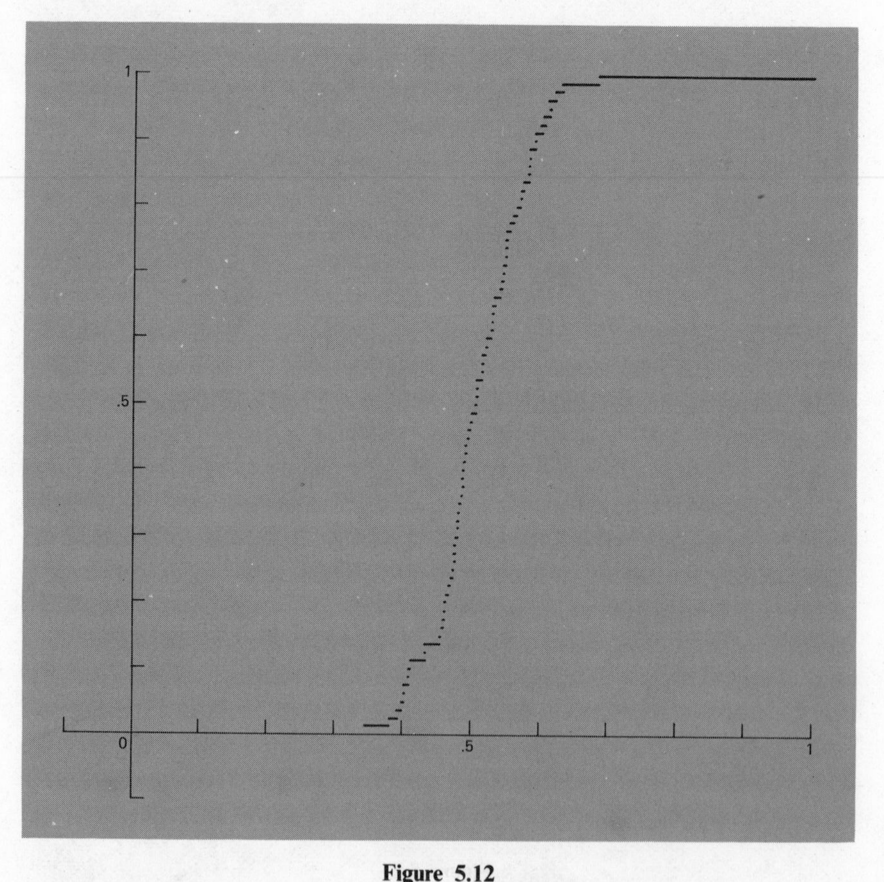

Figure 5.12
Empirical distribution function of a sample mean: 80 samples of size 20, from uniform population.

mean $\mu = .6$ and standard deviation $\sigma/\sqrt{50}$. Thus the step function—an empirical version of $P(\overline{X} < k)$ based on 200 samples—is, as it should be, not far from the limiting normal distribution promised by the central limit theorem.

Table 5.4

OUTCOME	NUMERICAL CODE, X	PROBABILITY
Heads	1	.6
Tails	0	.4

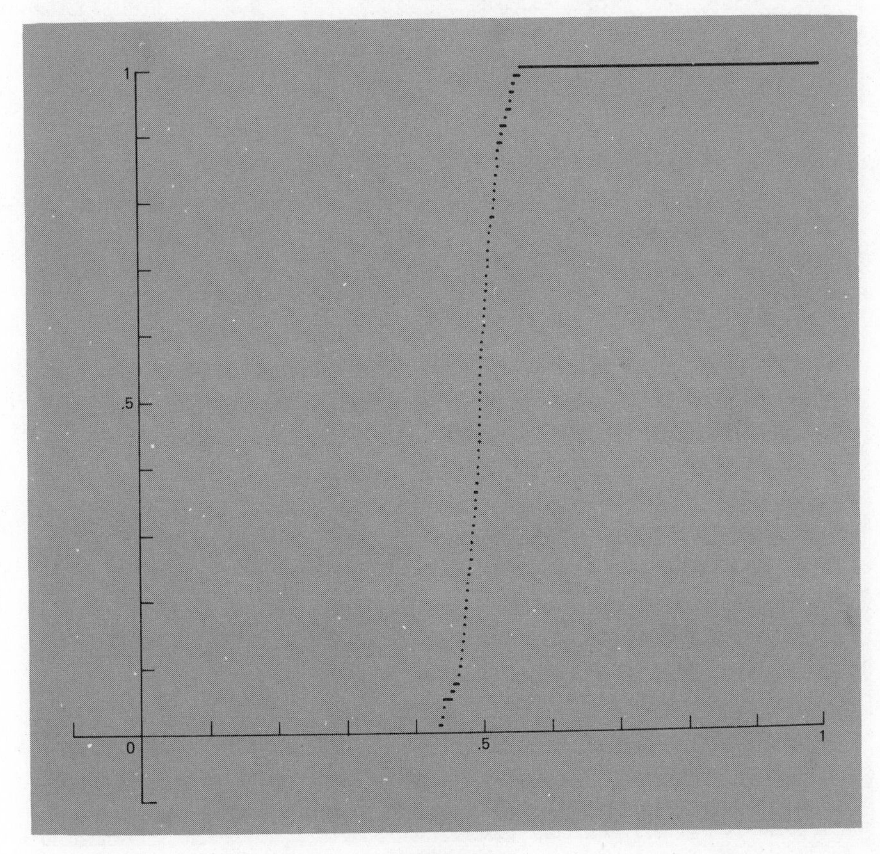

Figure 5.13
Empirical distribution function of a sample mean: 80 samples of size 100, from uniform population.

Knowing the approximate shape of the distribution is more than just aesthetic, of course, it is practical. With this knowledge one can compute probabilities for $\overline{X}$:

THEOREM 5.4—Practical Version: *Let $\overline{X}$ denote the mean of a large random sample from any population with mean μ and standard deviation σ. Then probabilities for $\overline{X}$ can be approximated as follows:*

$$P(\overline{X} < k) \doteq \Phi\left(\frac{k - \mu}{\sigma/\sqrt{n}}\right)$$

where $\Phi(z)$ is the standard normal cumulative distribution function (tabulated in Table I).

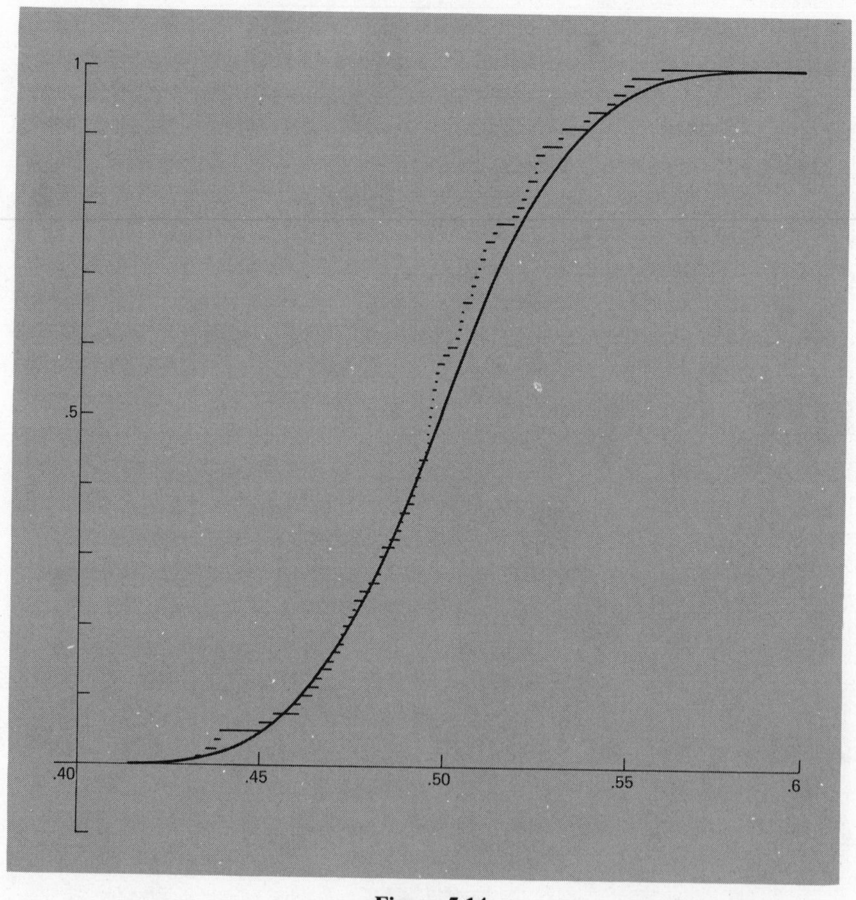

Figure 5.14
Empirical distribution function of Figure 5.13 with expanded scale and superimposed limiting c.d.f.

EXAMPLE 5.12

An elevator has a sign "Maximum load: 2800 lb, Capacity: 17 persons." If the population of people using the elevator has a mean of 150 lb and standard deviation 35 lb, what is the probability that when filled to "capacity" it will be overloaded in weight?

If $\overline{X}$ denotes the average weight of a load of 17 persons, the total weight is $17\overline{X}$. The problem, then, is to calculate

$$P(17\overline{X} > 2800) = P(\overline{X} > \tfrac{2800}{17}).$$

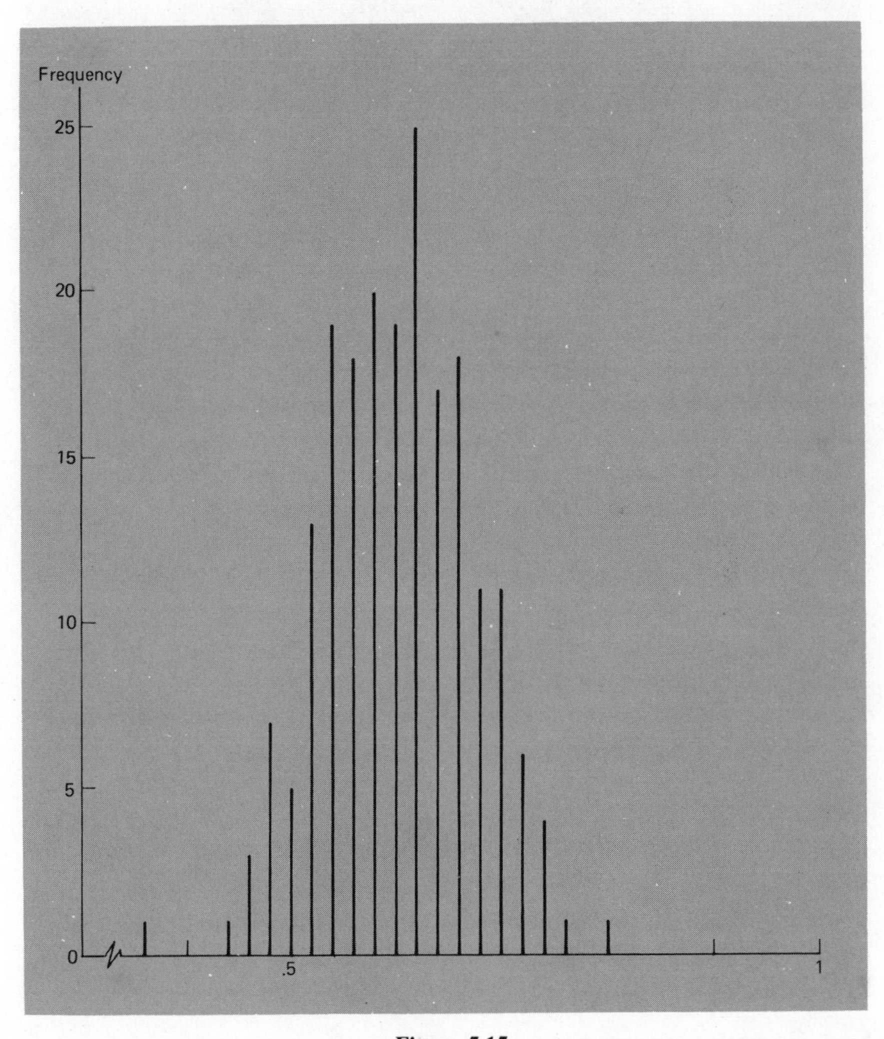

Figure 5.15
Frequency diagram for 200 sample means—samples of size 50 from 0–1 population.

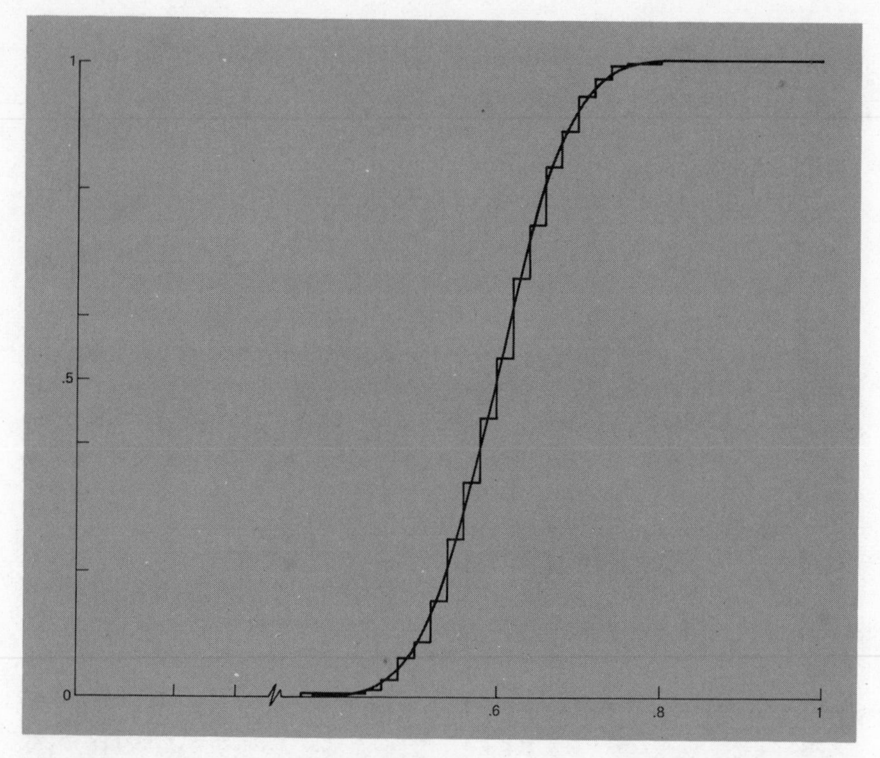

Figure 5.16
Cumulative frequency version of Figure 5.15 with limiting c.d.f. superimposed.

Now, $\bar{X}$ is approximately normal with mean 150 and standard deviation 8.5:

$$\mu_{\bar{X}} = \mu_X = 150, \qquad \sigma_{\bar{X}} = \frac{\sigma_X}{\sqrt{17}} = 8.5.$$

Therefore,

$$P\left(\bar{X} > \frac{2800}{17} = 164\right) = -\Phi\left(\frac{164 - 150}{8.5}\right) = -\Phi(1.65).$$

That is, 164 is 1.65 standard deviations of $\bar{X}$ to the right of the mean of $\bar{X}$. Table I shows that the probability of exceeding such a value is 5 per cent.

Problems

5.1. Go back to your work on Project 1(c) and note the values of the mean and standard deviation of your empirical $\bar{X}$-distribution. These should be not too far from the theoretical values, μ and $\sigma/\sqrt{n}$, respectively. Make the comparison. Check also your sketch of a distribution function for $\overline{X}$. Although the sample size was only 5, is it beginning to look normal?

5.2. Going back to Project 4, compare the empirical version of $E(\overline{X})$, namely, the mean of your distribution of 20 sample means, with the true value of $E(\overline{X})$.

5.3. Given a population of students with average IQ of 100 and standard deviation of 20, what is the probability that a random sample of 25 students will have an average IQ exceeding 110? That the sample mean will be less than 95?

5.4. The weight of a child entering kindergarten is a random variable whose standard deviation has been found to be 5 lb. What is the probability that the mean weight in an incoming class of 64 pupils differs from the population mean by more than 1 lb (either way)?

5.5. The scores on an algebra placement exam have been found, over the past several years, to be distributed with mean 62 and standard deviation 10. This year a group of 36 from a school system that has dropped the "new math" and gone back to the old averages 65 on the placement exam. What is the probability, if the new group could be considered as a sample from the old population, that a sample of 36 would average 65 or more?

6 Statistical Inference

To "infer" is to draw a conclusion, and statistical inference denotes the drawing of conclusions from statistical evidence, i.e., from data. These conclusions are concerned with the way things are, with the state of affairs, the state of what we call "nature," which governs the generation of data and results in the particular kind of data that constitute the evidence.

The problem of inference is a hard one because it amounts to reasoning from the particular to the general, and this is always a hazardous process. But it is a process that we are constantly engaged in, and there is really no alternative. We are continually presented with data, with experiences, with experimental results; and to plan our lives and conduct our affairs intelligently, we have no choice but to form our concepts of governing laws on the basis of those data, those experiences and experimental results. Consider the following questions, typical of those that demand answers in daily life.

How many television viewers are watching Channel 4 at this moment? Certainly, advertisers want to know, but no one can know with complete accuracy. What people who conduct surveys do is to ask a limited number of viewers if they are watching, and from this information they generalize—draw conclusions or make inferences about the whole population.

Is Crest Toothpaste effective? Consumers would like to know, and even the makers of Crest would like to know, although if the answer were negative they might not publicize the fact. It is difficult to determine whether a toothpaste is effective, because other factors enter into the development of tooth decay. Even if controllable factors such as brushing, fluoridation, and geography were held constant, individual differences make results variable. But if one is content with some average measure of effectiveness, he can conduct experiments on a limited number of subjects and draw general conclusions therefrom, at some risk of being wrong.

How well will a certain incoming freshman do in college? Here, what is wanted by the counselor, the parents, and the student himself is a *prediction*. The prediction might be based on such available information as high school rank, college board exams, etc., but so many other factors affect the student's progress that students with the same high school rank and exam scores will not usually perform equally well in college. From a sample of past students, their precollege status and final college grade-point averages, it is necessary to infer a general model characterizing something about all students and then to apply this to predicting success for the incoming freshman.

How reliable is this piece of equipment? Users want to know this, and past experience with supposedly comparable equipment gives them data from which to infer its reliability; but the data exhibit variability, and again the inference cannot be guaranteed. Some years ago, a large communications company wanted to know if a certain new type of transistor would stand up for 50 years in an underwater cable. The statistician was rather helpless, because of course he had no data on which to base an inference. The transistor was new and had not been exposed to *any* environment for a 50-year period.

Similar questions are legion: How dangerous is this anesthetic? Who is going to win the election? What is the best seed corn for a given location? How many toll gates are needed at this bridge? Is business getting better? What kind of housing is needed? Is this production line operating within acceptable standards? And so on.

6.1 Data and the State of Nature

Because experimental results derive from certain governing laws, and because those laws therefore determine the nature of the data, it is assumed that it is possible to learn something about the laws from the data—that the data contain some information about the state of nature. This will prove to be the case provided that the data are gathered intelligently, in such a way that one knows the mechanism by which the laws or the state of nature gives rise to the data. Data whose origin and method of collection are completely unknown are of little value. There is a story about some blind men each of whom encountered different parts of an elephant (one his trunk, one his leg, etc.); each one, of course, reported an entirely different picture of what an elephant is like. Whenever one encounters data, he must beware lest the relationship of his data to the actual state of affairs is like the relationship of the nature of the trunk to that of the whole elephant.

EXAMPLE 6.1
Everyone I spoke to about a certain city election said he was going to vote against the incumbent mayor. Yet it turned out that he won by an overwhelming majority. My data were not gathered in a way that permitted a proper conclusion about the outcome of the election.

EXAMPLE 6.2
In a group of five economists this morning I found that none was smoking. In a similar group of political scientists this afternoon I found that three of five were smoking. What can I infer about the state of nature from these particular data? Can I infer that people smoke more in the afternoon than in the morning? Or that economists smoke less than political scientists? These may sound like silly conclusions, but sad to say, one can find conclusions in scientific literature that are just about as ill-founded.

The term *state of nature* refers to certain aspects that relate to a particular situation, not to every conceivable aspect of nature. The actual proportion of voters who will vote for a certain candidate is a state of nature. The relative merits of two brands of seed corn would constitute a state of nature. The laws of planetary motion that Kepler labored to uncover constitute a state of nature. The actual characteristics of a population of people (or of animals, or birds) constitute a state of nature. The existing conditions of poverty in a certain city constitute another state of nature. The shape and balance of a die that is to be tossed, and the mechanism used to toss it, define a state of nature.

In all these situations, the data that are gathered to learn about the true or actual state of affairs involve variability, either because the nature of the experiment is (like the toss of a die) what is thought of an inherently random, or because extraneous factors enter into the process of experimentation to corrupt the readings. In order to know how to make inferences from data, a general model for the state of nature is needed, one that includes the element of randomness. Moreover, it is necessary to know what the model is, to know how the state of nature determines the generation of data. Thus in specifying a state of nature, one specifies a certain probability model for the gathering of observations. For practical purposes, then, that probability model is a state of nature.

EXAMPLE 6.3

The Library Board in a certain city asked the chief librarian to determine the proportion of transactions that involved people living in the suburbs so that negotiations between the Board and the county library board, which operates the suburban libraries, could be carried on intelligently. To make this determination, the librarian proposed to take a sample of 1500 transactions from the 2,500,000 or so transactions over the preceding year. The state of nature consists of the actual (but unknown) proportion of transactions involving suburban patrons, or the probability that a transaction drawn at random would involve a suburban patron. The actual population from which this transaction would be drawn is not totally clear; the Library Board undoubtedly did not make it clear. But including the transactions that take place over a year's time would eliminate the seasonal fluctuation of the proportion, whereas taking those that occur over several year's time might erroneously bias the inference about the present state of affairs, if there had been a gradual change in the proportion of interest. In summary, the state of nature would be the probability distribution given in Table 6.1. The quantity p is not

Table 6.1

TYPE OF TRANSACTION	PROBABILITY OR PROPORTION
Suburban patron	p
City patron	$1 - p$

known but is assumed to exist and to define the model. It is either the probability that a single transaction drawn at random from a year's record of transactions represents a suburban patron, or the proportion of such transactions among the year's record of transactions.

In this example, the probability models that can be considered are of a very special type. They belong to a restricted class of models that are each defined by the value of a *parameter* (the probability p is an example). Each state of nature can be identified with a value of the parameter p, and the job of inference about nature is that of drawing conclusions about the value of that parameter. More generally, it is often possible to define a class of admissible states of nature for a given problem that can be defined by one or more parameters, and inference about the state of nature is accomplished in terms of inference about the values of those parameters. The degree of simplicity of this class of admissible states determines in part how easy the problem of inference is.

EXAMPLE 6.4

A laboratory makes measurements of viscosity and has found through experience that the model describing the variation in measurements, although having a different mean for each specimen tested, has the same pattern of variation about the respective mean for each specimen. That is, the density function defining the model has the same shape in each case but is simply translated to different mean points for different specimens. Thus the density is of the form

$$f(x;\theta) = g(x - \theta),$$

where $g(x)$ is a function describing the variation in the measuring process and θ is a parameter that fixes the location of the pattern on the viscosity scale. (Indeed, the parameter θ may be the mean.) Thus a state of nature corresponds to, or *is*, a value of θ, and the problem is to make inferences regarding θ. (The parameter θ, in this case, may be thought of as the actual viscosity of the specimen, with variation about θ explained as measurement and sampling errors.)

EXAMPLE 6.5

In certain physical and other problems, it is known (rather, it is assumed) that variables are linearly related: $y = ax + b$. It is desired to determine from experimentation the values of a and b. Ordinarily, since two points determine a line, it would suffice to fix the variable x at each of two values and measure the corresponding y, but it is often the case that random errors creep in and spoil the measurements. Thus it may be desirable to use several x-values. For each of them the measured response is a random variable Y, composed of the actual linear response and the measurement error:

$$Y = ax + b + Z,$$

where Y is the measured value and Z is a random variable. The model for the experiment is defined by the fixed and known x-values that are used, the unknown values of a and b, and the unknown distribution of the random error component Z. If it is assumed (as it often is) that Z is normal with mean zero and variance σ^2, the model is completely defined by the three parameters a, b, and σ^2, provided an assumption of independence of the measurements is also made. So each admissible state of nature corresponds to a triple of parameter values: (a, b, σ^2).

6.2 How Much Data?

The amount of data used in a process of inference varies all the way from none to thousands and occasionally millions of observations. It would seem clear intuitively that the more data one has, the better the inferences he can make. Yet it may be that after obtaining a certain amount of data, there is little left to be learned by gathering more. It is also generally true that observations are not obtained free of charge, and sometimes the degree of improvement in inference made possible with more data is not worth the cost of the additional data.

EXAMPLE 6.6

A television commercial in the early 1960s asserted that a certain product had been "tested in five nationally known laboratories." The viewer was apparently expected to infer, on the basis of *no data whatever*, that the product was a superior one.

EXAMPLE 6.7

In some scientific laboratories measurements are made in duplicate or triplicate. That is, samples of size 2 or 3 are used. Although possibly adequate in some cases, the degree of accuracy possible with samples of this size is not great.

[It is sometimes suggested that if three measurements are obtained, the two closest together should be used, the other being discarded. This procedure has no justification and would likely be less accurate than just taking two observations in the first place. Several years ago a well-known instrumentation laboratory, in testing a long-range inertial guidance system, flew the system across country three times to a specified target, to test the system's accuracy. The one flight that missed by a considerable amount was thrown out, and the system was evaluated on the basis of the two that came closest to the target. On this basis the system was found to have met the specifications!]

EXAMPLE 6.8

A letter to the editor of a metropolitan newspaper is here quoted in part to show people's concern with sample size: "I'd like to know how [Dr. H.] ever got the idea that 'a national survey of 2000 college seniors' could ever show anything. His survey might have some significance if he had interviewed 200,000 college seniors."

It is true that many national surveys are conducted with sample sizes in the vicinity of 1000 to 2000. The National Nielsen Television Index, for example, involves residents of 1200 households (of 60,100,000 homes that have television sets). The American Research Bureau conducts surveys with varying sample sizes. In one local survey they used 1125 households in 57 counties, and in a metropolitan survey they used 590 households (compared with Nielsen's 531 in the same area). "Overnights" are conducted by ARB on special order and involve between 300 and 500 phone calls during the course of a television program.

It is true that such surveys are based on samples that are selected by techniques considerably more sophisticated than simple random sampling, techniques that are devised to reduce sampling errors and biases. However, even with simple random sampling, it turns out that in most cases almost as much can be learned from 2000 as from 200,000.

Although a large sample is inherently more reliable than a small one, it is often not practical, or even possible, to obtain a large number of observations. The statistician must then figure out how to make the best possible use of the data, to make the best possible inference from the data he has or can get.

The *Constitution of the United States* calls for a complete count of our population every 10 years, and this is a very costly operation. The Census Bureau uses the occasion to gather data about us, and these data constitute a 100 per cent sample, i.e., a sample that consists of the whole population. Such sampling is usually subject to errors involved in gathering data from such a large group, errors that may well be as great as the errors arising because of the sampling fluctuations in a less than 100 per cent sample. In other words, a complete tabulation of a population is not usually successful in eliminating uncertainty about its characteristics.

In summary, it is usually adequate and often necessary to settle on a sample smaller than the whole population. And the more observations the better, up to a point where costs outweigh gain of information. Thus the question "how much data?" has not been answered. It is a question that investigators always ask their statistical consultants, but the answer usually is, "it depends." It depends on how much the data costs to obtain and on how precise or definite the conclusions must be. Later chapters will study the effect of sample size on precision and so provide a basis for determining a sample size for a particular problem.

EXAMPLE 6.9

A shipment of goods is often spot-checked for quality by both the shipper and the receiver or consumer. The state of nature is the actual lot quality, as defined by some kind of average over the lot (which, then, is the population), and this quality might be determined by inspecting the entire lot. But a complete inspection is always costly, and in the case of destructive testing (flashbulbs, bullets, etc.), it is obviously impractical. What is done instead, usually, is to take a sample from the lot and test the articles in the sample; the lot quality is reflected to some extent in a judiciously chosen sample, and the choice between accepting the lot or rejecting it can be made with some degree of confidence on the basis of the sample information.

It must be recognized that not all problems that challenge us come to us as statistical problems, and sometimes they defy attempts to cast them in a statistical mold. It is not always clear what data to collect or how to interpret what is collected. In particular, it is not always clear how to construct a model, probability or otherwise, for the data or for the population (if a population can be identified).

EXAMPLE 6.10

One current problem is how to resolve the issue of capital punishment. Data are sometimes put forward, such as "those states that have (or have had) capital punishment have the highest crime rates." But does this really prove anything? One might like to propose a model that establishes, for states with and for states without certain penalties, probabilities that a person faced with a provocation to commit a crime will, in fact, do so. Perhaps someday someone will propose reasonable models and collect pertinent data that will furnish a statistical guide in this area—and then again, perhaps they will not.

6.3 Statistical Problems

Problems of inference are statistical when an inference about the state of nature is to be made from data. Sometimes the subject of statistics is described as the making of inferences or decisions under uncertainty; the uncertainty is one's incomplete knowledge of the state of nature.

In some situations the statistician's work is to provide an educated basis for making a decision or taking action. The next two examples illustrate such problems. In other situations the statistical inference is to be made for the purpose of shedding light on or refining old theories or suggesting new ones; scientific work involves such problems. It may be that when findings are published, persons unknown may take action or make a decision based on those findings, and Example 6.11 illustrates this.

EXAMPLE 6.11

A major study was conducted* to determine whether or not the anesthetic halothane was resulting in more fatalities than earlier types of anesthetic, as feared by some medical authorities. Conclusions were based on hospital data collected nationwide. The relevant state of nature here might be defined by the true fatality rates and how they compare with each other. The action contemplated was the discarding of halothane, despite its good characteristics; and the alternative action was to continue its use.

EXAMPLE 6.12

The weight guesser at the state fair has a problem, to estimate a person's weight within a given number of pounds. On the surface it may seem that he must do this, quickly, on the basis of no data in the traditional sense. But of course he is rapidly making at least one observation, at best crude, on each of several variables that he has found in the past to be correlated with weight: height, girth, ratio of flab to muscle, and who can say what else? His procedure is of necessity rough, intuitive, and difficult to analyze.

There are some classes of problems whose common statistical aspects permit generalizations and unity of approach. One such problem is the problem of estimation, touched on in the preceding example. *Statistical estimation* deals with the estimating of population parameters on the basis of

* Reported in more detail in J. M. Tanur et al., *Statistics: A Guide to the Unknown* (San Francisco: Holden, 1972), p. 14.

samples. When data can be obtained that relate to the parameter being estimated, these data can be presumed to contain at least some information about that parameter and so can be used to educate one's guess. Statistical estimation, then, is a process of combining or reducing data in some way to produce an estimate.

Another recurring statistical problem, one that will be considered in many contexts in subsequent chapters, is that called *hypothesis testing*. The situation often occurs in which one has some hypothesis about a state of nature that needs to be checked out, or tested. The hypothesis may come from some theory, from hunches, from past experience that may or may not be relevant to the present, or from a manufacturer's claim (or, equally well, from a consumer advocate's claim)—to mention a few sources.

Some statistical hypotheses relate to scientific models—theories in sociology, medicine, education, etc., that want verification or refutation. Others, as in Example 6.9, are hypotheses about the state of affairs in an operational setting, where decisions must be made according to the best available judgment as to what is true.

6.4 Statistical Procedures

An orderly procedure for statistical inference is a rule that prepares one with a pronouncement to make, or an action to take, for every foreseeable result of the sampling—a rule of the form: If outcome x is observed, I shall do this; if y is observed, I shall do that; and so on, for all possible outcomes.

Having prepared himself with such a rule, the statistician collects and summarizes the data and consults his rule to see what to do—corresponding to the outcome that is actually observed.

It must be admitted that statistical practice, the analysis and interpretation of data, is not always so orderly as required by the kind of statistical procedure just outlined, and it need not be. It is often necessary to study preliminary data to get a better feel of what the problem is and what kinds of models for nature to propose. In many instances, moreover, an investigator will come to a statistician, having already collected masses of data, and say "what does all this mean?"

The reason for establishing the kind of orderly procedure described is so that conclusions can be drawn that have some foundation other than pure intuition. This is not to say that intuition is not extremely useful, but it is hard to defend against skeptics or those whose intuition leads them to still different conclusions. When a rule is set forth that prepares the statistician for whatever data may arise (as opposed to the kind of rule that is set up after looking at the data), and when the data are collected so that a prescribed probability model can be employed to represent the collection process, *then it is possible to make probability statements about errors and precision that are achieved with the given rule.*

In the case of *estimation*, a statistical procedure is a rule whose input consists of the observations in the sample and whose output is a number that is to be announced as the estimate of the parameter. This language contains the idea of what mathematicians call a *function*: An *estimator* of a parameter is a function of the observations in a sample. But it will be recalled that this is precisely what a *statistic* is—a function of the observations in a sample. Thus an estimator is a statistic. The statistic used in estimation should be chosen to do a good job of estimating. Any statistic could be considered to be a candidate to be an estimator of a given population parameter, but some statistics will be better than others at the job of estimating.

It would be fortuitous indeed if there were some guiding principle that would enable one to choose a best estimator from the host of statistics that might be used. Various principles for selecting estimators have been proposed, but none seems to afford the assurance of producing a good estimator in every kind of problem. Very often one's intuition is as good a thing to follow as any more sophisticated principle. (Indeed, the only principles that have been proposed have themselves been proposed on intuitive grounds!)

One obvious, but rather vague, principle to follow in estimating some characteristic of a population is to use a sample statistic that measures the same kind of characteristic. Thus, to estimate the population mean, which is a measure of the middle of a population, it seems intuitively reasonable to use one of the various measures of the middle of a sample—the sample mean, the sample median, the mid-range, or a trimmed mean, etc. And if it is desired to estimate a population variance, which measures variability or dispersion, it would seem reasonable to base an estimate on the sample variance or on the sample range, quantities that measure dispersion in a sample.

Another idea, which almost suggests itself in the case of large samples, is applicable when the population parameter to be estimated is the long-run limit of a sample statistic. The idea, of course, is to use that statistic as an estimator.

However an estimator is proposed, there remain questions. How well does it perform? Does it yield an estimate that is *close* to the correct value? The second question does not have a definite answer, because an estimator (which is a statistic) is a random variable; its value varies from sample to sample. If one decides to use the sample mean as an estimator for the population mean, in some cases (i.e., some samples) it comes close and in others it does not. About all one can do is to measure success of estimation by some kind of *average* "miss," which is what will be done in Chapter 7.

In the *testing of hypotheses*, a statistical procedure is a rule that calls for rejecting a proposed hypothesis or not rejecting it, on the basis of data collected for the test. The test is usually couched in terms of a *test statistic*, a quantity summarizing that aspect of the data that is informative about the validity of the hypothesis being tested. The rule will then be of the form: If the test statistic has a value that lies anywhere in such and such region of values, reject the hypothesis. That region, calling for rejection, is called the *critical region* of the test. The proper choice of a critical region will be seen to depend on what might be true if the hypothesis being tested were not true; such alternatives, taken together, constitute the *alternative hypothesis*.

It is tempting to formulate the alternative hypothesis and critical region, and perhaps even the null hypothesis, after looking at the experimental results, but this should be avoided. The reason is that every finite amount of data can be found to be peculiar in some way, and in a way that is not generally true of the next sample one might obtain. If inspection of data reveals some peculiarity, which peculiarity is used to formulate the alternative hypothesis, and if the same data are then used to carry out an appropriate test, one is almost sure to reject the null hypothesis. The test is "rigged" for that data.

What should be done, if (as is often quite in order) one looks at data to form preliminary judgments, is to gather *more* data for use in testing between hypotheses formulated in view of the early data.

EXAMPLE 6.13

After the draft lottery of 1970, people examined the resulting sequence of numbers and claimed, on the basis of what they found, that the lottery was not fair. (It was noticed, for example, that an unexpectedly large number of early draft numbers were assigned to December dates.) One could only mean by "fair" that the 366! arrangements of 366 numbers were equally likely to have resulted. Any one such sequence would have some kind of peculiarity. It is not logical to look for a peculiarity and, having found one, to claim unfairness. The only way to test for fairness (other than looking for flaws in the way that the selection of numbers was carried out) would be to use the same method of generating draft numbers to repeat the process and to observe whether an observed peculiarity repeated itself.

It is natural to ask in connection with whatever test be proposed: How successful is the test procedure? "Success" would mean, of course, that the test will call for rejection whenever it is applied to a situation in which the hypothesis being tested is not correct and will not call for rejection, whenever applied to a situation in which the hypothesis is true. But a test based on data, incomplete information about the population, cannot be expected to function perfectly every time. It can err, indeed, in *two* ways, and the reader will be asked in Question 6.10, to identify the two types of error.

Since a test statistic is a random variable, one can at best lay odds on its making one type of error or the other, i.e., define probabilities that can be used to assess the overall performance of a test.

Questions

Being intended to provide motivation and background for what follows, Chapter 6 has not given specific techniques that permit testing knowledge by the usual sort of problem. However, the following questions should serve to check on comprehension as well as to stimulate more thinking than may have accompanied a preliminary reading of the chapter. Not all the questions have "pat" answers, and they should perhaps be discussed in class.

6.1. What is *inference*?

6.2. What is *statistical* inference?

6.3. Why does one take samples rather than make a complete survey of an entire population?

6.4. If it is true that the larger the sample, the more reliable the inference, why are statisticians content to take samples that do not seem very large?

6.5. Why are samples taken in *random* fashion?

6.6. A television commercial claims that a certain brand of power lawnmower will start most of the time on the first try. To back up this claim, they show 100 mowers lined up with operators, who, at the wave of a flag, attempt to start them; 92 of 100 succeed. Is the right population being sampled for you who are going to buy, possibly, just one lawnmower? Could you devise a more convincing or more relevant experiment?

6.7. If you never know the *true* value of a parameter you want to estimate, how is it possible to tell whether an estimator is any good?

6.8. Give an example of a situation in which an estimate of a population parameter is important.

6.9. If you did Project 5.3, compare your distribution of sample medians with the distribution of sample mean from Project 5.1. Which statistic, mean or median, would you prefer to use as an estimate of the population center (the population mean and median coincide in this case) on the basis of your empirical results, and why?

6.10. For the decision problem faced by a jury in a criminal case, what is the hypothesis being tested? What is the alternative? What are the two types of *error*—and is one more important than the other? (Is this decision problem a *statistical* problem?)

6.11. If you are in a poker game and get a hand consisting of 4 aces and a king, do you declare a misdeal because the outcome is so very unlikely to have occurred? Suppose that you get a hand consisting of the 2 and 4 of hearts and the 5, 6, and 7 of spades; is there anything unusual about this hand? If so, is it grounds for rejecting the hypothesis of a fair deal? Think about how you might devise a test for fairness and what the alternative might be. Should you wait until you gather the data to decide on your alternative?

6.12. Refer to the accompanying reprint of a popular comic strip.

"Hi and Lois" by Walker and Browne. (c) King Features Syndicate, Inc., 1972.

(a) What is the basic chance phenomenon being discussed, and what is the probability model for it (i.e., the population)?

(b) What is the hypothesis the twins set out to test, and what is the alternative as proposed by panel 2?

(c) What test statistic are they apparently planning to use? (What sample size?)

(d) What critical region do you suppose they have in mind? (That is, what value or values of the test statistic will they use to choose between the null and alternative hypotheses?)

(e) Would you say that the test is for the purpose of taking action or to advance scientific knowledge?

7 Inference for Numerical Populations

Populations and samples of values—of numbers, as opposed to attributes—may be described, by virtue of the numerical identification of outcomes with its inherent ordering, by devices that exploit the properties and relationships of numbers. Chapter 2 dealt mostly, and Chapter 4 wholly, with such populations. In particular, one can define the notions of "center" and "variability" and measure these in various ways.

This chapter will treat problems of inference in numerical populations, the estimation of parameters, tests of hypotheses about parameters, and a test for the validity of a specific model. The populations may, in general, be either continuous or discrete. But some techniques and conclusions will be given that apply only to a continuous model, and some, more specifically, only to a *normal* model.

7.1 Estimating a Population Mean

Many problems call for estimating the population average, or the center of gravity of the population distribution, denoted by μ, or by $E(X)$ when the population variable is X. Thus one wants to know average income, or average IQ, or average age, or average yield, or average quality, in each case for a specified population. But if, for any of various reasons we have discussed, the whole population cannot be surveyed to obtain the average, the best one can do is to take a sample and base an *estimate* of the population average on the data in the sample.

Because the population mean is a long-run limit of sample means (as more and more observations are used), it is natural to take the value of the mean of a sample, of whatever size can be economically obtained, as an estimate of μ.

The mean $\overline{X}$ of a sample has been seen to be a random variable, varying from sample to sample. So if $\overline{X}$ is used to estimate μ, different samples would yield different estimates. Having but one sample to work with in any practical situation, one might well wonder how reliable the estimate from that sample is. If the actual value of the population mean is μ and the estimated value from the sample is $\overline{X}$, the *error* in estimation can be defined to be the difference:

$$\text{error in estimation} \equiv \overline{X} - \mu.$$

This will never be known, because μ is not known. (If μ were known, why would one be estimating it?) However, one *can* say something about how small this error is *likely* to be, in terms of the distribution of the estimator $\overline{X}$.

It is known that the distribution of $\overline{X}$ is centered at μ:

$$E(\overline{X}) = \mu,$$

and that its standard deviation is inversely proportional to the square root of the sample size:

$$\sigma_{\overline{X}} = \frac{\sigma}{\sqrt{n}}.$$

When n is large, the distribution tends to concentrate around μ, and it will be unlikely that $\overline{X}$ strays far from μ. But this is interpreted only in the sense of repeated trials or in the sense of betting odds for a single trial. A *particular* sample *can* be "way out" and its mean far from μ.

The standard deviation of $\overline{X}$ is often taken as a measure of its success in estimating μ, since it characterizes variability in the distribution of $\overline{X}$:

$$\sigma_{\overline{X}} = \sqrt{E[(\overline{X} - \mu)^2]}.$$

This average deviation is a root-mean-square (rms) average of the deviation of $\bar{X}$ about μ, previously called the *error* of estimation:

$$\text{rms error of estimation} \equiv \sigma_{\bar{X}} = \frac{\sigma}{\sqrt{n}}.$$

Of course, when σ is not known, the rms error cannot be computed. However, because σ is a long-run limit of sample standard deviations (as $n \to \infty$), the replacement of σ by S from the sample gives an approximation to the rms error, called the *standard error of estimate*:

$$\text{standard error} \equiv \frac{S}{\sqrt{n}}.$$

Again, this measures the variability of $\bar{X}$ about μ, and so the reliability of $\bar{X}$ as an estimator of μ.

To get a "feel" for reliability of an estimator from its standard error is a matter of experience. But in the case of $\bar{X}$, the distribution is approximately normal (a *good* approximation, in most cases, for samples of 20 or more), and the standard deviation of a normal distribution can be readily interpreted: One standard deviation away from the mean takes one out to the point of inflection on the density; and two standard deviations on either side of the mean include about 95 per cent of the area. Three standard deviations on either side (i.e., from $\mu - 3\sigma$ to $\mu + 3\sigma$) would include 99.74 per cent of the area. It is *extremely unlikely*, therefore, that $\bar{X}$ will deviate more than three standard errors on either side of the true mean.

EXAMPLE 7.1

A sample of 200 observations was generated (on a computer) from a population whose distribution is normal with mean $\mu = 31.7$ and standard deviation $\sigma = 1.8$. (We consider first an artificial example in order to have one in which μ and σ are known. This would not be the case in a real-life example.) The sample moments turned out to be as follows:

$$\bar{X} = 31.688, \qquad S = 1.677.$$

The rms error in using $\bar{X}$ to estimate μ is

$$\text{rms error} = \frac{\sigma}{\sqrt{n}} = \frac{1.8}{\sqrt{200}} = .127.$$

The approximation to this called the *standard error*, obtained by using the sample standard deviation in place of σ, is

$$\text{standard error} = \frac{S}{\sqrt{n}} = \frac{1.677}{\sqrt{200}} = .119.$$

The actual error in the observed $\overline{X}$ as an estimate of μ was

$$\overline{X} - \mu = 31.688 - 31.7 = .012,$$

which happens to be quite small compared to the standard error. Other samples would involve errors, some larger, some smaller, the "standard" error being a kind of average (an rms average) error.

EXAMPLE 7.2

The average number of hamburger boxes in a carton supposedly containing 800 is of interest both to the supplier and to the hamburger shop. The data in Problem 2.9 consisted of the counts in 30 cartons, with mean $\overline{X} = 807.23$ and standard deviation 20.02. The estimate of the population average number would be 807.23, and the standard error is

$$\text{standard error} = \frac{S}{\sqrt{n}} = \frac{20.02}{\sqrt{30}} = 3.66.$$

So $\overline{X}$ is almost sure to fall within about 11 boxes of the true average number per carton.

7.2 Estimating a Population Standard Deviation

The population variance is the long-run limit of sample variances, as more and more observations are taken and included, and so again it is natural to take the sample statistic S^2 as an estimate of the population parameter σ^2. Before going to the obvious estimation of σ by S, a comment is in order about another possible definition of the sample variance.

Many statisticians and textbooks, as well as "canned" computer programs, use the following definition of sample variance:

$$\tilde{S}^2 = \frac{1}{n-1}\sum_1^n (X_i - \overline{X})^2 = \frac{n}{n-1}S^2.$$

The reason they give for this is that (as can be shown)

$$E(\tilde{S}^2) = \sigma^2,$$

whereas

$$E(S^2) = \sigma^2 - \frac{\sigma^2}{n}.$$

The language used is that S^2 is *biased* and $\tilde{S}^2$ is *unbiased* in estimating σ^2. (An estimator is called unbiased if its distribution is centered at the parameter being estimated.) However, this is really no justification for the modification, because the actual center of the distribution of an estimator is usually irrelevant as long as the distribution concentrates most of its probability near the parameter being estimated.

Moreover, in estimating σ by the sample standard deviation, it turns out that both S and $\tilde{S}$ are biased, so bias or the lack of bias does not serve to choose between them. For this reason, and because division by n is more natural than division by $n - 1$, we continue to use S^2 as the sample variance and S as the sample standard deviation.

The mean-squared error in S as an estimate of σ is difficult to compute. For large samples it is inversely proportional to n, and for the case of a *normal population*, it is approximately $\sigma^2/(2n)$. Hence we define

$$\text{standard error of } S = \frac{S}{\sqrt{2n}} \qquad (n \text{ large, normal population}).$$

EXAMPLE 7.3

The sample of Example 7.1 taken from a normal population with $\mu = 31.7, \sigma = 1.8$, turned out to have a standard deviation $S = 1.677$. The error in using this as an estimate of σ is

$$S - \sigma = 1.677 - 1.8 = -.123.$$

The standard error of estimate is

$$\text{standard error} = \frac{S}{\sqrt{2n}} = \frac{1.677}{\sqrt{400}} = .084$$

and the rms error (using σ rather than S) is .090. Thus the observed error of $-.123$ is not out of line; indeed, it is quite close to the average (rms) error, although some samples will yield better, and others worse, estimates of σ.

It might be worth noting that the estimate of average error (rms) in $\bar{X}$ provided by $S/\sqrt{n}$ is in error because of the error in S as an estimate of σ. An idea of how far off it is apt to get (for large samples) can be obtained from the standard error of S itself. For S is approximately normal and falls within 2 standard deviations of its mean with 95 per cent probability. Thus S will fall in the range $1.8 \pm 2 \times .09$, or 1.62 to 1.98 with probability .95, and the standard error of the mean can easily be off by as much as $.18/1.8 = 10$ per cent. But for the purposes to which the standard error of the mean is put, such an error is not worth worrying about.

Another measure of variability in a sample introduced in Chapter 4 is the sample range:

$$R = \text{largest observation} - \text{smallest observation}.$$

This can sometimes be used to provide an estimate of σ. In the case of *normal populations*, in particular, the necessary tables are available. The estimate used is a fraction of the range that depends on the sample size:

$$\text{estimate of } \sigma = \frac{R}{a_n},$$

where a_n is given in Table 7.1 for normal populations. The standard deviation of this estimator is

$$\sigma_{R/a_n} = \frac{b_n \sigma}{a_n},$$

where b_n is given also in Table 7.1 and σ, of course, is the parameter being estimated, which is not known. If one replaces σ by R/a_n, which is an estimate thereof, he obtains what again can be called a *standard error of estimate*:

$$\text{standard error} = \frac{b_n R}{a_n^2}.$$

Table 7.1

n	5	6	7	8	9	10	12	15
a_n	2.326	2.534	2.704	2.847	2.970	3.078	3.258	3.472
b_n	.864	.848	.833	.820	.808	.797	.778	.755
b_n/a_n^2	.160	.132	.114	.101	.092	.084	.073	.063

Table entries are given for only a limited number of sample sizes. The reason for this is that the usefulness of R in estimating σ lies primarily in this range of sample sizes. For large samples the estimator R/a_n becomes increasingly inefficient in estimating σ when compared to the sample standard deviation.

EXAMPLE 7.4
The nine weight losses given in Problem 4.1 were (in pounds) 14, 9, 17, 10, 10, 18, 11, 13, and 6. The range of these losses is

$$R = 18 - 6 = 12.$$

Assuming that the weight loss under study can be represented by a normal distribution, the standard deviation σ of that distribution can be estimated by

$$\frac{R}{a_9} = \frac{12}{2.97} = 4.04 \text{ (lb)}$$

with standard error

$$\frac{b_9 R}{a_9^2} = .092R = 1.1 \text{ (lb)}.$$

The estimate provided by S, by way of comparison, is

$$S = 3.65 \text{ (lb)},$$

with standard error

$$\frac{S}{\sqrt{n}} = \frac{3.65}{3} = 1.22 \text{ (lb)}.$$

Problems

7.1. The number of siblings for each of 24 students was recorded, with results as follows:

Number of siblings	0	1	2	3	4	5	6	7
Frequency	0	3	7	4	7	2	0	1

Give an estimate of the expected number of siblings for such students and the standard error of estimate.

7.2. Twelve female students in a class gave their weights (in pounds) as follows: 110, 128, 135, 115, 150, 100, 95, 145, 113, 115, 134, and 131. Given that the mean weight is $\bar{X} = 122.58$ and the standard deviation is $S = 16.52$, what is your estimate of the mean of the population from which these are drawn? What is the standard error?

7.3. Given the data in Problem 7.2, give estimates of the population standard deviation σ based on (a) the sample standard deviation and (b) the sample range. Give a standard error for each estimate. (What assumption about the population is needed to do this?)

7.4. Suppose you know that the standard deviation of IQ measurements is $\sigma = 20$. How large a sample from a large high school's junior class would be needed so that the standard error of estimate of the average IQ in the class given by the sample average IQ does not exceed 2 (i.e., 2 points on the scale on which IQ is measured)?

7.5. The birth weights of 30 babies given in Problem 2.8 are repeated here; assume that they are from a normal population.

8.2	7.5	7.0	5.5	7.1
8.0	8.3	6.5	6.75	5.9
8.3	7.75	5.25	8.25	6.7
9.0	5.8	7.1	7.25	7.5
8.0	8.25	6.75	6.7	5.4
7.7	7.25	5.75	6.6	6.75

(a) Given $\overline{W} = 7.093$ and $S_W = .970$, give estimates of μ_W and σ_W together with standard errors.

(b) Although n is too large to use R successfully, one can do the following. Divide the sample into six subsamples of five and compute the range of each. Their mean $\bar{R}$ divided by a_5 gives a good estimate of σ, with standard error $\bar{R}\, b_5/a_5^2\sqrt{6}$ (more generally, for k samples of m, it is $\bar{R}\, b_m/a_m^2\sqrt{k}$). Follow this procedure to obtain an estimate of σ and a standard error.

7.3 Interval Estimate for μ

The kind of estimate provided by a statistic such as $\bar{X}$ (for the population mean μ) is called a *point estimate*. It gives a single value as an estimate but involves an element of error or fuzziness. That is, even though a single value is announced, it is recognized that the actual value of the parameter being estimated can only be claimed to be somewhere in the vicinity of the announced value.

It would be desirable, perhaps, to be able to give an interval of values in which the parameter could be guaranteed to lie. Actually, the most that could be hoped for is an interval that would be highly likely to contain the parameter. There are two approaches to this: One is to act as though the parameter itself were a random variable, with a probability distribution, and to give the odds (in view of the sample results) that the parameter lies in any given interval. The other approach is to treat the parameter as a fixed (albeit unknown) constant and to construct an interval for each sample which has a specified chance of trapping the parameter value. The latter approach is the more popular, but oddly enough, the resulting interval tends to be interpreted as though it had been obtained using the first approach.

Because $\bar{X}$ is nearly normal, for large samples, probabilities are computed using the normal table (Table I of the Appendix). From that table it is seen that any normally distributed variable falls within (say) two standard deviations of its mean with roughly 95 per cent probability. Since $E\overline{X} = \mu$ and $\sigma_{\overline{X}} = \sigma/\sqrt{n}$ (where σ is the population standard deviation), it follows that

$$P\left(\mu - \frac{2\sigma}{\sqrt{n}} < \overline{X} < \mu + \frac{2\sigma}{\sqrt{n}}\right) \doteq .95.$$

That is, $\overline{X}$ falls within $2\sigma/\sqrt{n}$ units of the mean μ, 95 per cent of the time. But to say that $\overline{X}$ is within $2\sigma/\sqrt{n}$ of μ is equivalent to saying that μ is within $2\sigma/\sqrt{n}$ units of $\overline{X}$, and so

$$P\left(\overline{X} - \frac{2\sigma}{\sqrt{n}} < \mu < \overline{X} + \frac{2\sigma}{\sqrt{n}}\right) \doteq .95.$$

The quantities $\overline{X} \pm 2\sigma/\sqrt{n}$ (when σ is known) depend just on $\overline{X}$, and hence on the sample. The interval from $\overline{X} - 2\sigma/\sqrt{n}$ to $\overline{X} + 2\sigma/\sqrt{n}$ is therefore a random interval; and the probability that this random interval includes the actual μ is about 95 per cent. It is called a *95 per cent confidence interval for* μ. A number of points should be noted:

1. A 95 per cent confidence interval will *vary from sample to sample*. If you were to obtain 1000 samples (which you only do in speculation or in simulation), about 95 per cent of them would cover or include the value μ, the rest would not. But you will never know, *in practice*, whether or not your single confidence interval covers μ. All you know is that the procedure you use to obtain it has a 95 per cent chance of providing an interval that includes the actual mean.
2. To obtain an interval with a confidence coefficient other than 95 per cent, it is simply necessary to replace the 2 with a corresponding value. Since about 90 per cent of the area under a normal curve falls within 1.64 standard deviations of the mean, the 90 per cent confidence limits for μ are $\overline{X} \pm 1.64\sigma/\sqrt{n}$. Notice that the more confident you want to be, the wider the confidence interval must be.
3. If σ is *not* known, and this is the more common situation, it can be replaced by the sample standard deviation S if n is large. Thus the 95 per cent limits in a large sample would be $\overline{X} \pm 2S/\sqrt{n}$. (What is "large"? This is hard to specify, since it depends on the accuracy desired. Perhaps $n = 100$ is a good lower limit, but $n = 50$ may be enough for two-place accuracy.)

EXAMPLE 7.5

The sample of 200 observations from a normal population with mean $\mu = 31.7$ and standard deviation $\sigma = 1.8$, exploited in two preceding examples, yields the following 95 per cent confidence intervals for μ computed from the sample mean $\bar{X} = 31.688$ and the sample standard deviation, $S = 1.677$:

$$\sigma \text{ known } (= 1.8): \quad 31.688 \pm 1.96 \times \frac{1.8}{\sqrt{200}}, \qquad \text{or } 31.44 < \mu < 31.94.$$

$$\sigma \text{ unknown}: \quad 31.688 \pm 1.96 \times \frac{1.677}{\sqrt{200}}, \qquad \text{or } 31.46 < \mu < 31.92.$$

The actual population mean (31.7) is contained in these intervals; but in any practical problem one would not know whether this would be the case. All that can be said is that the method of construction used has a 95 per cent chance of producing a confidence interval that traps the mean. Equivalently, there is a 95 per cent chance of getting sample whose $\bar{X}$ is close enough to the actual μ so that the confidence interval centered at $\bar{X}$ manages to enclose μ.

EXAMPLE 7.6

The hemoglobin levels of 50 cancer patients given in Example 1.2 have a mean of 13.362 and a standard deviation of 1.747. The 95 per cent confidence limits for the mean hemoglobin level of such patients would be

$$13.362 \pm 1.96 \frac{1.747}{\sqrt{50}},$$

so that the confidence interval is

$$12.88 < \mu < 13.85.$$

An 80 per cent confidence interval would be obtained by using 1.28 (the 90th percentile of the standard normal distribution) in place of 1.96, and is thus

$$13.04 < \mu < 13.68.$$

Observe that with less confidence, a narrower interval can be given.

EXAMPLE 7.7

The concept that a 95 per cent confidence interval for the population mean is the result of using a method that produces, "95 times out of 100," an interval covering the mean can be demonstrated by artificial sampling. Two hundred samples of size 25 from a uniform population were generated on a computer. (*Note:* Although $n = 25$ is not

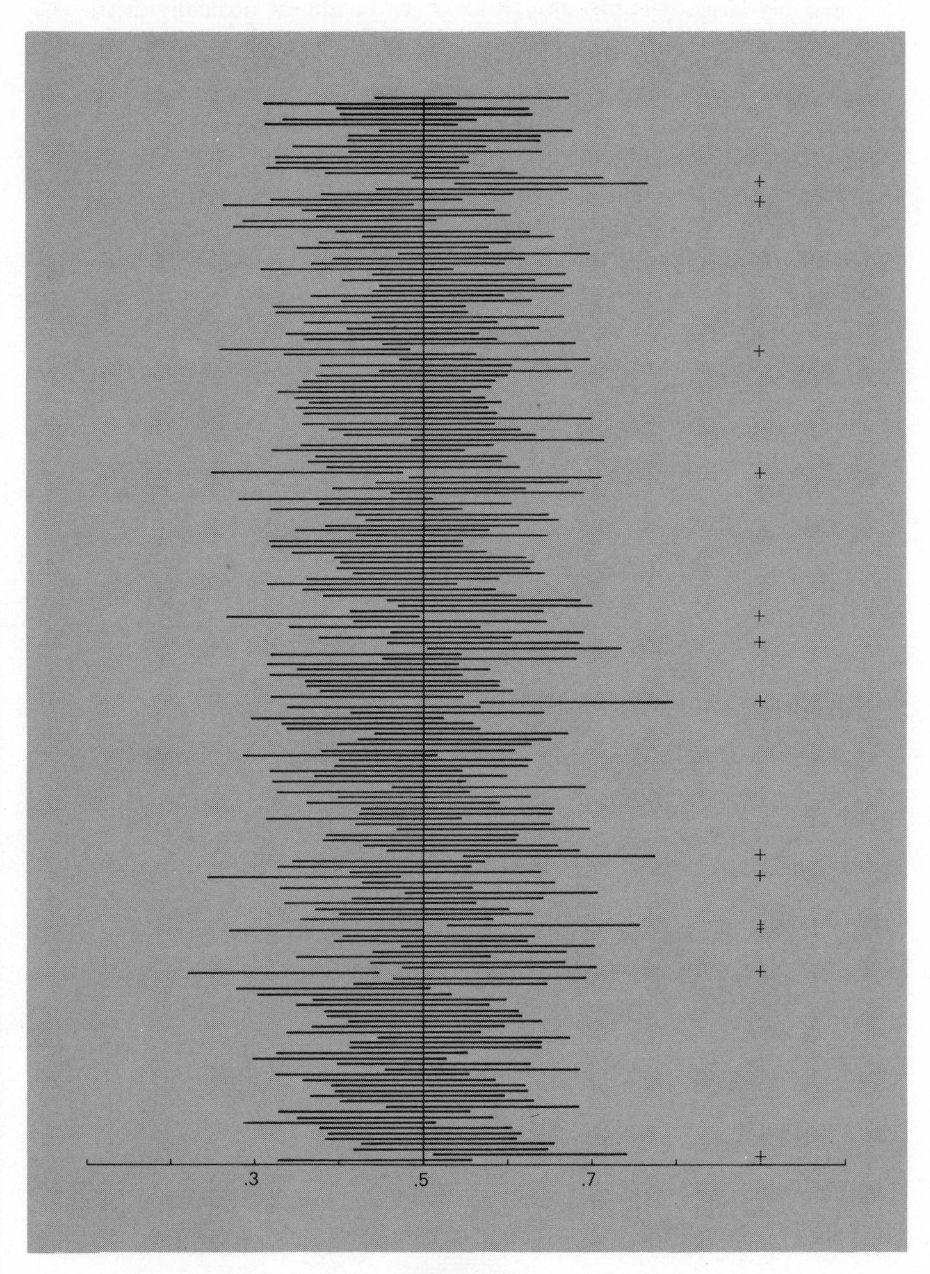

Figure 7.1
Two hundred 95 per cent confidence intervals for the mean of a uniform population ($\mu = .5$), sample size 25.

terribly large, it is big enough for $\bar{X}$ to be almost normally distributed.) A 95 per cent confidence interval for the mean based on a sample of size 25 is given by

$$\bar{X} - \frac{1.96\sigma}{\sqrt{25}} < \mu < \bar{X} + \frac{1.96\sigma}{\sqrt{25}},$$

where the known standard deviation of the population, $\sigma = 1/\sqrt{12}$, can be used rather than the sample standard deviation. This reduces to

$$\bar{X} - .113 < \mu < \bar{X} + .113.$$

Thus the interval is .226 unit wide and is centered at $\bar{X}$. Figure 7.1 shows the results. Each interval is plotted on an x-axis, with successive intervals one above the other. One can easily *count* the number of intervals that actually include the mean, $\mu = .5$, since these have been marked at the side with a +. Five per cent of 200 would be 10, and the actual number (13) differs from 10 for the same reason that the frequency of any event, in a *finite* number of trials, generally differs from its expected frequency.

Problems

7.6. The mean length of 1000 fish of a certain species caught commercially in a certain coastal area is 24 in., with a standard deviation of 1 in. Compute a 90 per cent confidence interval for the population mean, and also a 99 per cent confidence interval.

7.7. In a research study of dental products, in a sample of 294 children, the average number of new cavities per child developed over a 30-month period was 10.88, with a standard deviation of 6.36. Construct a 95 per cent confidence interval for the population mean.

7.8. Referring to the large-sample confidence interval for a population mean based on the sample mean, and other things being equal:

(a) Is an interval based on a sample 25 narrower or wider than one based on a sample of 100? (By what factor?)

(b) Is a 98 per cent confidence interval narrower or wider than a 90 per cent confidence interval?

7.9. Suppose that you take a sample of 100 and construct a 95 per cent confidence interval for a population mean that turns out to have width 0.20.

(a) How large a sample would you have needed to obtain a 95 per cent confidence interval of width .05?

(b) Suppose that it is *not* possible to get more data, that you are stuck with $n = 100$. How confident could you be with a confidence interval of width .10?

7.4 Hypothesis Testing: General

Hypotheses about numerical populations are usually hypotheses about parameter values, although Section 7.10 treats a different case. It may be hypothesized that the parameter has a *particular* value, or more generally that its value lies in some range of values.

Tests of hypotheses about parameter values will be based on data in a sample, of course, but more particularly on a statistic closely related to the parameter in question. The obvious choice of statistic for checking out a claim about a population mean is the sample mean. To test a claim about a population variance, the sample variance would often be used.

Whatever the choice of statistic, to interpret an observed value and so to use it properly it will be necessary to know about the sampling distribution of the statistic. In the light of this knowledge of what *can* happen, and with what odds, one can make an intelligent (if not always perfect) decision about the validity of a claim or hypothesis.

EXAMPLE 7.8

Scores in a nationally administered aptitude test are found over long experience to be distributed with mean 60 and standard deviation 12 (on a percentage scale, say). A high school administrator who thinks his particular school system is unusually fine may question the assumption that his school is like the national population. If this assumption or hypothesis were in fact true, the mean of a random sample of 100 students would have a pattern of variation described approximately by the normal model with mean 60 and standard deviation

$$\sigma_{\bar{X}} = \frac{\sigma}{\sqrt{n}} = \frac{12}{\sqrt{100}} = 1.2.$$

This means, for instance, that the probability of obtaining a sample mean farther from 60 than $3 \times 1.2 = 3.6$ (per cent) is only .0026. Suppose he finds that $\bar{X} = 65$, which *is* farther from 60 than 3.6. Should this be interpreted as a freak result, possible but unusual if the hypothesis $\mu = 60$ is true; or should it be taken as evidence that μ is *not* 60, but something different (greater than 60, perhaps)?

That is, does the sample give credence to the hypothesis $\mu = 60$, or should this hypothesis be rejected as being inconsistent with the data?

It is tempting to use the following kind of reasoning: If the result of my experiment has a small probability of having happened when the hypothesis that I am testing is true, then I should not believe, or accept, that hypothesis. The flaw in such reasoning may be seen in the following example.

EXAMPLE 7.9
A measurement is assumed, under a hypothesis to be tested, to have a normal distribution with mean 10 and variance 4. Suppose that the result of a measurement is 10.26. This is quite an unlikely result, having *zero* probability under the given hypothesis. But as a matter of fact, the value 10.26 is really quite close to the middle of the distribution; the mean is 10 and one standard deviation is 2. So it should not be terribly surprising to observe 10.26, even though one would not bet much ahead of time on that precise value.

Perhaps, in an attempt to avoid the kind of thing illustrated above, it occurs to one that what he really wants is not to reject his hypothesis when an unlikely outcome turns up, but to reject it if the outcome is one of a certain set or region of outcomes or values all of which, together, are unlikely. This is a step in the right direction but still not quite right, since there are lots of regions of small probability in most models.

EXAMPLE 7.10
I have noticed that a certain candy bar seems to be smaller than it used to be, although the nominal weight is still 1 oz. The manufacturer's hypothesis is that, although weights of individual bars vary, the mean weight is 1 oz. So, to test this hypothesis, I buy 10 bars and weigh them. If I find that their combined weight is only 9 oz, can I reject the hypothesis that the mean weight is 1 oz—not because 9 oz is unlikely (which it is, having zero probability), but possibly because 9 oz is one of a set, or one in a region of values whose combined probability is small? Indeed, it would have been better to prepare a *rule* of inference before gathering the data—a rule that in this case might be: "Reject the claim that the mean is 1 oz if the total weight of 10 bars taken at random from the production line is less than 9.5 oz." That is, *any value* of the total weight W *in the range* $W < 9.5$ would lead to rejection, as would $W = 9$ in particular. Moreover,

the probability that $W < 9.5$ is apt to be small (if the bar weights are not too variable and the mean really is 1 oz).

On the other hand, if I used the rule: "Reject the hypothesis of $\mu = 10$ if $9.99 < W < 10.01$," I would again find that the probability of rejection is small. Which rule is better? Reflection will show that preference for the first rule is based on the fact that, like the manufacturer, *I also have a hypothesis about* μ, namely, that $\mu < 1$. The real question, then, is whether to accept the manufacturer's claim (or hypothesis) that $\mu = 1$, *or* reject it and accept *my* claim (or hypothesis) that $\mu < 1$. And notice in particular that if, instead, my claim were that $\mu > 1$, a different kind of rule would be better (e.g., reject $\mu = 1$ if $W > 10.5$). Thus a test cannot really be constructed sensibly without some thought as to the alternative to $\mu = 1$.

A problem of testing hypotheses has these ingredients:

1. A hypothesis to be tested, called the *null hypothesis*, H_0.
2. An alternative hypothesis, H_A.
3. Data, and a test statistic computed from the data, as an aid to choosing between H_0 and H_A.
4. A decision rule, a rule that specifies a certain region of values of the test statistic any of whose occurrences would call for rejecting the hypothesis being tested (and for accepting the alternative).

The test is really a rule for choosing between two hypotheses, H_0 and H_A. It may be that one statistician's null hypothesis is another's alternative; the designation has an element of arbitrariness. However, it is common to find one hypothesis to be that a given population parameter has a particular value, and the other hypothesis to be more vague; the specific single value is often taken as the null hypothesis.

The region of values of the test statistic that call for rejecting the null hypothesis is called the *critical region* of the test. (The values themselves might be called *critical values.*) To formulate a test is to select a critical region. But the question is: *How* does one select a critical region?

One consideration is that a critical region should be chosen to make the probability of its occurrence small when the null hypothesis is true. This is along the lines of the reasoning, earlier called fallacious, that when the null hypothesis is really true there should be only a small chance of rejecting it; but what is "fallacious" is the common misconception that this is the *only* consideration. To reject a hypothesis that is true is a mistake, to be sure, but this applies equally well to both the null *and* the alternative hypothesis. That is, it is also a mistake to reject the alternative (by accepting the null hypothesis) when the alternative is true. These two kinds of mistakes are often exhibited as a table; see Table 7.2. A test will be judged according to how likely it is to lead to one type of error or the other.

Table 7.2

	NULL HYPOTHESIS TRUE	NULL HYPOTHESIS FALSE
Accept null hypothesis	(Correct decision)	Type II error
Reject null hypothesis	Type I error	(Correct decision)

When a null hypothesis is sufficiently specific or precise, it will permit the calculation of a probability that a given test will reject it; this is called, variously, the size of the type I error, the size of the critical region, or the significance level of the test. A common notation for it is α.

The term *size* is used in place of "probability" because although α is the probability that the test statistic falls in the critical region, this probability is computed *under the assumption that the null hypothesis is true*, which may or may not be the case.

The term *significance level* is used because of the idea that when the test statistic falls in the critical region, this is a significant event which signifies that the null hypothesis should be in doubt (because that statistic turned out to have a value that is out of the ordinary, in some sense).

EXAMPLE 7.11

Some question has arisen as to the fairness of a certain game of three-handed "hearts" among players of equal skill. If it is fair, which can be taken as the null hypothesis, the three players should get stuck with the queen of spades equally often. Thus the probability that player A gets it in a given hand is $\frac{1}{3}$. As a test of fairness, player A decides that if he gets stuck five times in five hands, he will declare the game unfair and refuse to continue playing. His *test statistic*, T, is the number of times he gets stuck in five hands; his *critical region* is just the single value $T = 5$. The size of his critical region is the probability of getting stuck five times in a row, computed *under the null hypothesis*:

$$\alpha = \frac{1}{3} \cdot \frac{1}{3} \cdot \frac{1}{3} \cdot \frac{1}{3} \cdot \frac{1}{3} = \frac{1}{243}.$$

If the game is really fair, α is the probability that player A will erroneously declare it unfair and walk out. But it should be realized that he may get $T = 4$, say, and so continue playing (according to his rule), even though there is some cheating going on. That is, his test may fail to detect that the null hypothesis is not true, and this is an error of type II. Its size is not well defined, because it is not possible to calculate the probability that the test accepts fairness when there is cheating. (To say that a game is not fair does not specify a well-defined model for calculating probabilities.)

In the event that the alternative hypothesis is not vague and permits computation of probabilities, the probability that a rule or test accepts H_O when the alternative is true is called the size of the type II error and is denoted by β. It will be found that this is an uncommon situation. More often than not β is not uniquely definable, for the probability of accepting H_O depends on which of the many models included in the alternative to H_O is governing.

Problems

7.10. You know that I have standard coins and also a coin that has been altered to have heads on both sides. I toss a coin without showing it to you beforehand. In testing the null hypothesis that the coin I use is fair, you decide to reject this hypothesis if the coin lands heads in a single toss, and to accept it if the coin lands tails. Determine the sizes of the two types of error associated with your test. Is there a test with a smaller type I error?

7.11. Consider the testing problem suggested in the comic strip pictured in Question 6.12 (page 172). To formalize it, let it be considered a problem of testing the null hypothesis (H_O) that odds are even that the bread will fall buttered side down, against the alternative hypothesis (H_A) that the odds are 9 to 1 (probability of .5 against probability .9) and let the sample size be 25. Suppose that a friend who knows a little more probability than you know (only a little more is necessary) calculates the following probabilities:

	NUMBER OF TIMES BUTTER DOWN			
	18 OR MORE	19 OR MORE	20 OR MORE	21 OR MORE
Under H_O (even odds)	.0082	.0026	.0007	.0002
Under H_A (odds 9:1)	.996	.986	.958	.891

(a) If the rule is to reject H_O if and only if the buttered side is down 19 or more times out of 25, compute the two error sizes.

(b) What would be the rule to use if α could be as big as .0082, and what would be the corresponding β?

(c) If a β of about 10 per cent could be tolerated, how small could α be made, and with what rule?

7.12. The hypothesis that the random number generator used in obtaining artificial samples for Example 7.7 generates numbers from a population with mean $\mu = \frac{1}{2}$ can be tested as follows: Generate a sample of 25 and construct a 95 per cent confidence interval for μ; if this interval includes the value $\frac{1}{2}$ on its interior, accept $\mu = \frac{1}{2}$, otherwise reject it.

(a) In the 200 samples of Example 7.7, how many would have led you to reject the null hypothesis?

(b) If, as in practice, you are going to use only *one* sample of 25, what are the chances that the procedure outlined will lead you to reject the null hypothesis erroneously?

7.5 Large-sample Tests for μ

The two situations to be considered are the following, in which H_O denotes the null hypothesis, H_A the alternative hypothesis, and μ_0 is a specified value of the population mean:

One-sided:

$$H_O\colon \quad \mu = \mu_0 \quad (\text{or } \mu \leq \mu_0);$$

$$H_A\colon \quad \mu > \mu_0.$$

Two-sided:

$$H_O\colon \quad \mu = \mu_0;$$

$$H_A\colon \quad \mu \neq \mu_0.$$

Large-sample tests for these two problems can be constructed using the sample mean as the test statistic, based on the fact that in large samples, the sample mean is approximately normal with mean μ and variance σ^2/n (which in turn is approximable, from the sample, by S^2/n).

The intuitive basis for these tests is that $\overline{X}$ tends to fall near the actual population mean μ, in large samples. Thus, if $\overline{X}$ turns out to be a large number, one should behave as though μ is large, and if $\overline{X}$ is small, as though μ is small. In the matter of choosing between μ_0 and something bigger than μ_0, then, the reasonable approach is to consider any value of $\overline{X}$ that is much bigger than μ_0 as a basis for rejecting $\mu = \mu_0$ (and accepting $\mu > \mu_0$).

EXAMPLE 7.12
It is desired to test the hypothesis that a certain group of students (say athletes) have no better school records, as measured by grade-point average, than the population of all students. With the realization that there are, naturally, individual differences, the problem can be put as one of testing hypotheses about the mean. Suppose that the population mean for all students (a very large population) is 2.5 and the standard deviation is .5. The problem then is to test

$$H_O\colon \ \mu = 2.5 \qquad \text{versus} \qquad H_A\colon \ \mu > 2.5,$$

where μ is the mean of the population of athletes. The test is to be based, say, on a random sample of 100 from the athlete population—more specifically, on the sample mean $\overline{X}$. What values of $\overline{X}$ should constitute the critical region, i.e., call for rejection of H_O?

The value of $\overline{X}$ can fall near 2.5 or *far* from 2.5, on either side, *even if H_O is true* ($\mu = 2.5$). The sampling distribution of $\overline{X}$ summarizes the nature of this variability; the distribution is approximately normal with mean

$$\mu_{\overline{X}} = \mu = 2.5 \qquad (\text{under } H_O)$$

and standard deviation

$$\sigma_{\overline{X}} = \frac{\sigma}{\sqrt{n}} = \frac{.5}{\sqrt{100}} = .05.$$

Now if $\overline{X}$ should happen to be as large, say, as

$$2.5 + 2 \times .05 = 2.6,$$

this would mean that a value has been observed (from a normal distribution) which is two standard deviations out from the center. This can happen—but when it does one might wonder whether it happens not because a freak value has occurred, but because μ is really *larger* than 2.5. Such reflections might suggest adopting this rule or test: "Reject $\mu = 2.5$ if $\overline{X}$ exceeds 2.6." How would this rule perform?

One can now calculate α, which is the probability that if H_O is true (μ is really 2.5), following the rule will cause him to reject H_O

erroneously. Thus

$$\begin{aligned}\alpha &= P(\overline{X} > 2.6) \qquad \text{(computed assuming that } H_0 \text{ is true)}\\ &= 1 - P(\overline{X} < 2.6)\\ &= 1 - \Phi\left(\frac{2.6 - 2.5}{.05}\right) = 1 - \Phi(2) \doteq .025.\end{aligned}$$

This evaluation proceeded by recognizing first that $\overline{X}$ is approximately normal, then measuring the deviation of 2.6 from 2.5 in terms of standard deviation units, and then looking up this value (2) in the standard normal (Φ) table. (See Figure 7.2.)

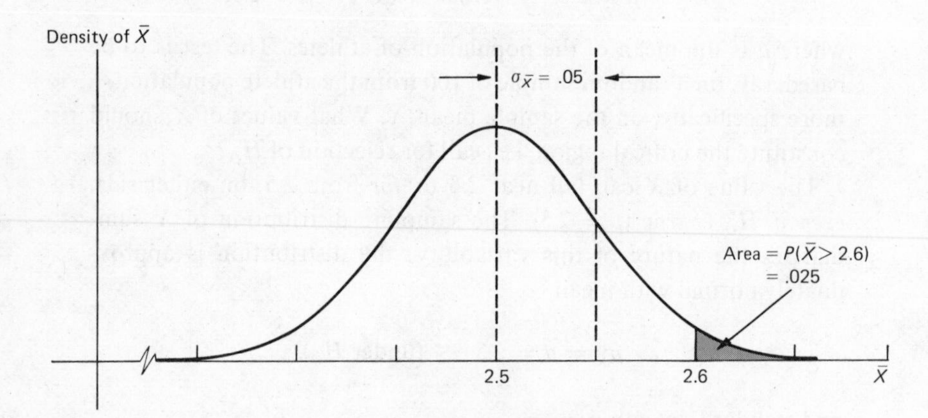

Figure 7.2
Representation of α as an area under the $\overline{X}$-density.

If this α, namely 2.5 per cent, is considered too large, one can devise a test with smaller α, simply by taking a smaller critical region. For example, the critical region $\overline{X} > 2.65$ has an α of about .13 per cent (as the reader can and should verify, following the above computation scheme).

It will be noticed that to permit use of the standard normal table, the critical $\overline{X}$-value had to be converted into a standardized version (called a Z-score by some), by expressing its deviation from the mean being tested (call it μ_0 now) in terms of the unit $\sigma_{\overline{X}}$:

$$Z_{\text{crit}} = \frac{\overline{X}_{\text{crit}} - \mu_0}{\sigma_{\overline{X}}}.$$

This is a simple linear transformation, and observe that if one defines

$$Z = \frac{\overline{X} - \mu_0}{\sigma_{\overline{X}}},$$

then

$$Z > Z_{\text{crit}} \quad \text{if and only if} \quad \overline{X} > \overline{X}_{\text{crit}}.$$

In the above example, then, the test could be expressed in either of two equivalent forms:

$$\text{reject } H_0 \text{ if } \overline{X} > 2.6 = \overline{X}_{\text{crit}}$$

or

$$\text{reject } H_0 \text{ if } Z > 2 = Z_{\text{crit}}.$$

More often than not, it will be the practice here to take Z as the test statistic and give the critical region in terms of Z. Figure 7.3 shows the distribution of Z, an arbitrary critical value K, and a shaded area giving the α for the test with critical region $Z > K$.

Need it be added explicitly that if the alternative to $\mu = \mu_0$ were $\mu < \mu_0$, one would use a critical region of the form $Z < K$?

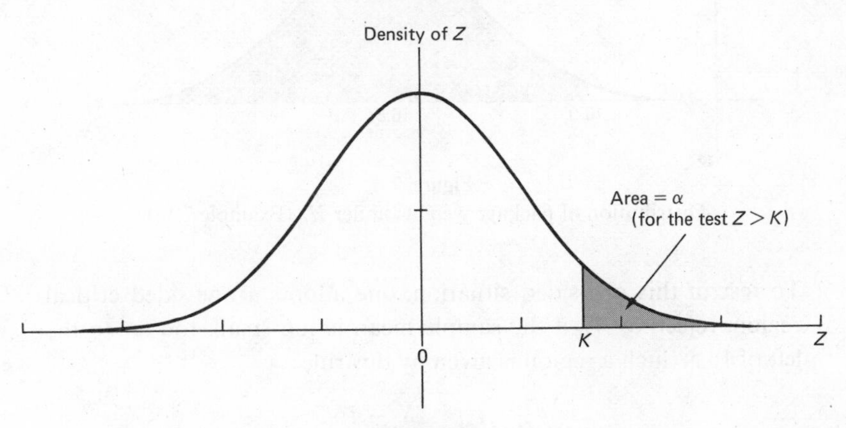

Figure 7.3
Critical region and type I error size.

EXAMPLE 7.13
In Example 2.3 are given the weights of 60 "1-lb" packages of bacon. Presumably, these were weighed to check the claim on the label that the weight of the bacon in the package is indeed 1 lb. The mean

and standard deviation of the 60 weights given in Example 2.3 are

$$\bar{X} = 16.1 \qquad \text{and} \qquad S = .225.$$

The consumer might argue, realizing that weights are bound to vary somewhat, that if the (population) mean weight is 16.5 or more, then over 95 per cent of the packages would weight 1 lb or more. This is because .5 is more than 2 standard deviations of the weight X, assuming that the figure .225 is a reasonable estimate of the actual population standard deviation, and assuming that the weight X is an approximately normal random variable (see Figure 7.4). Thus the consumer's hypothesis might be

$$H_0: \quad \mu \geq 16.5$$

and his alternative (which, if true, would cause him to lodge a complaint),

$$H_A: \quad \mu < 16.5.$$

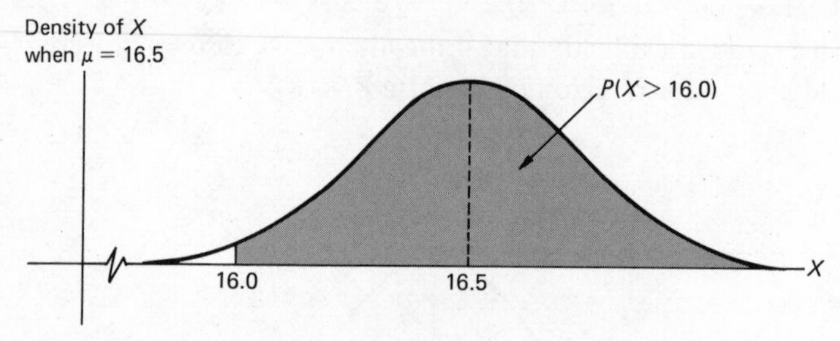

Figure 7.4
Distribution of package weights under H_0 (Example 7.13).

To test in this one-sided situation, one adopts a one-sided critical region, rejecting H_0 if the sample mean is too small, too far to the left of 16.5. Such a region is given by this rule:

$$\text{reject } H_0 \text{ if } Z = \frac{\bar{X} - 16.5}{S/\sqrt{n}} < K,$$

where it is assumed (in using S in place of σ) that n is large. For a rejection probability of .05 when $\mu = 16.5$, the rejection limit would be $K = -1.645$ (see Figure 7.5). For the given data,

$$Z = \frac{16.1 - 16.5}{.225/\sqrt{60}} = -3.4 < -1.645.$$

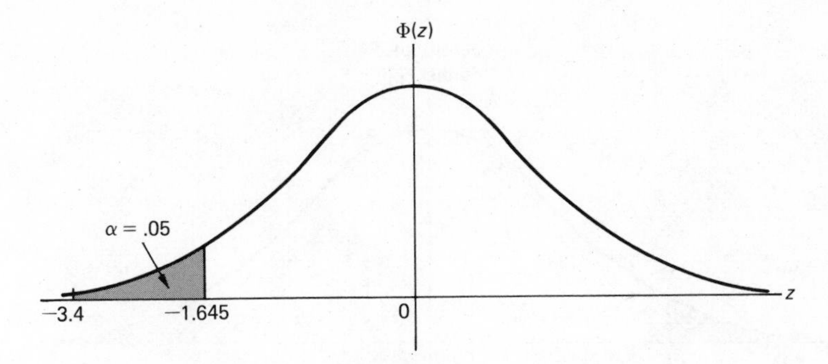

Figure 7.5
Standardized distribution of mean package weight (Example 7.13).

Thus the observed $\overline{X}$ is too far below 16.5 to accept H_O, and so the complaint should be lodged. (The α of .05, strictly speaking, is the probability of falsely rejecting H_O only when $\mu = 16.5$; if μ is actually greater than 16.5, the probability of falsely rejecting H_O would be even less than .05.)

Tests for the two-sided problem,

$$H_O: \quad \mu = \mu_0$$

against

$$H_A: \quad \mu \neq \mu_0,$$

can also be expressed in terms of the standardized statistic

$$Z \equiv \frac{\overline{X} - \mu_0}{\sigma/\sqrt{n}}.$$

As before, the quantity Z measures how far $\overline{X}$ is from μ_0 in terms of standard deviations of $\overline{X}$. The density of Z is as shown in Figure 7.6. In terms of Z, one rejects μ_0 if Z is either too large or too small. That is, the critical region is *two-sided*, like the alternative hypothesis. It is of the form $|Z| > K$, where K is chosen (as shown in Figure 7.6) so that

$$P(|Z| > K) = \alpha \qquad \text{when } H_O \text{ is true.}$$

In these terms, then, one computes the above Z-score corresponding to the observed $\overline{X}$; if this falls *outside* the range $-K < Z < K$, he rejects H_O (otherwise, he accepts H_O). The entries in the following table give the critical value K for the test statistic $|Z|$ corresponding to various commonly used significance levels, for a two-sided test:

α	.20	.15	.10	.05	.01	.005
K	1.28	1.44	1.645	1.96	2.58	2.81

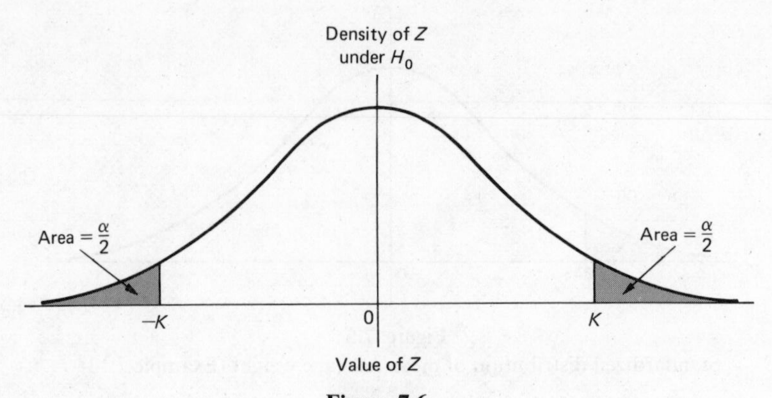

Figure 7.6
Two-sided rejection region for Z.

The entry 1.96 for $\alpha = .05$, for example, is obtained from the standard normal distribution function (Table I) by noting that .025 is the area to the left of -1.96, and .025 the area to the right of $+1.96$.

The statistic Z involves the population σ, and so to use Z one would have to know the value of σ. More often than not σ is not known, but then the estimate S may be used in constructing tests, simply by replacing σ by S in the statistic Z.

In practice, problems of testing the population mean that are genuinely two-sided are not very common.

EXAMPLE 7.14

A sociologist is studying a newly accessible group of jungle people and wants to test the hypothesis that its fertility rate is 2.5, the same as in other similar cultures he knows about. So he questions 70 females who have reached 49 years of age and determines the number of offspring each has had, with these results:

Number of offspring	0	1	2	3	4	5	6	7	8	9
Frequency	6	10	15	20	9	5	2	2	0	1

The mean and standard deviation are

$$\bar{X} = 2.8, \qquad S = 1.78.$$

The test statistic would be (since σ is not known)

$$Z = \frac{\bar{X} - 2.5}{S/\sqrt{n}},$$

and a 5 per cent test would call for rejection of H_0 if Z does not exceed 1.96 in magnitude. Since

$$Z = \frac{2.8 - 2.5}{1.78/\sqrt{70}} = 1.41,$$

it is concluded that the data do not warrant rejection of $\mu = 2.5$ at $\alpha = .05$.

The phraseology at the end of the above example brings up a point that is worth some discussion, since it is a source of contention among statisticians. The question is: If the data do *not* call for *rejecting* H_0, should you *accept* H_0? Some people think that you should not, their reasoning being that there are many values of μ, especially those near μ_0, which could account for the observed $\bar{X}$ just as well as μ_0. In the above example, they would say: Why accept $\mu = 2.5$ when μ could just as well be 2.6 or 2.7? All that the data show, they continue, is that 2.5 cannot be ruled out.

Others hold that if you do not reject a hypothesis, you are necessarily accepting it. And this is no doubt true when a test is conducted for the purpose of providing a rational basis for a decision, for taking action. Presumably, in such a case, there are two possible actions to take, one being better if H_0 is true, and the other if the alternative is true.

It is often the case that a hypothesis of the form $\mu = \mu_0$ is not well put—that for practical purposes μ could be in the *vicinity* of μ_0 and have the same impact as having precisely the value μ_0. It is also often the case that there are considerations other than statistical ones that should help to decide whether really to accept, in the sense of believe, a given null hypothesis.

In examples and problem answers to follow, hypotheses that are not rejected will, for simplicity, be said to be accepted. The interpretations of this will not be argued further.

The tests in this section were presented for the case of large samples, but, as observed previously, how big a sample should be considered to be large is not clear-cut. The largeness has been needed for two reasons: (1) to justify the assumption of the normality of the distribution of the sample mean, and (2) to justify the use of the sample standard deviation in place of the population standard deviation. Generally, the second calls for a larger sample than the first in order that approximations be successful. The size of sample needed to permit approximation of the distribution of the mean by a normal distribution depends on how close to normal in shape the population distribution is. Indeed, if the population itself happens to be normal, then the mean of a sample of *any* size is normal. If the population is unimodal and not too much skewed to one side or the other, a sample size of 15 or 20 may be adequate for the two-place accuracy useful in probability computations. If it has a couple of peaks or is badly skewed, then 50 or more observations may be needed.

Substitution of the sample S for the population σ when the sample size is much below 50 or so can render the results inaccurate; i.e., probabilities are not what they are supposed to be. However, with this grain of salt, the substitution in a small sample can be done and perhaps some information can be gleaned from the result. The next section tells how to get more accurate results in small samples when the population being sampled is normal.

Problems

7.13. Radar speed checks were made on 100 randomly chosen cars passing a certain point on a freeway where the speed limit is 65 mph. The average speed observed was 63 mph, with a standard deviation of 5 mph.
(a) Test the hypothesis that the average speed of motorists at that point is 65 mph.
(b) Construct a 95 per cent confidence interval for the (population) average speed.

7.14. A filling machine is set so that the mean weight of a "5-lb" bag of flour is 5.1 lb. The standard deviation is assumed to be .05 lb.
(a) A sample of 25 bags has a mean weight of 5.08 lb. Should the machine be reset? Assume that at most a 5 per cent chance of stopping production from the resetting when this is unwarranted is to be permitted.
(b) When the mean is 5.1, what is the chance of getting a bag that weighs less than the nominal weight of 5 lb?
(c) What is the probability that the needed adjustment will *not* be made if the mean is really 5.09 and the machine is shut down for resetting only when the mean weight of the sample of 25 bags is less than 5.08? (How serious would a deterioration to 5.09 be—what would then be the chance of getting a bag weighing less than the nominal weight?)

7.15. A poster proclaims that graduates of a technical school will earn an average of $8000 per year five years after graduation. Do you accept this claim, at the 5 per cent level, if a sample of 64 graduates are found to be averaging $7800 (after five years) with a standard deviation of $500? (Decide on the alternative before looking at the data.)

7.16. Power consumption, although random, increases steadily over the years. After adjusting for this trend, a power company estimates that the average domestic use in a certain month will be 490 kWh. To help in the present attempt to reduce power consumption, the company puts on an advertising campaign prior to the given month to try to influence customers to use less power. A sample of 400 bills for the month in question later shows an average consumption of 485 kWh, with a standard deviation of 60 kWh. Was the campaign successful? (Test the hypothesis that the mean consumption for the month is 490 kWh against the alternative that it is less than 490 kWh.)

7.17. Under the impression that the average height of freshman men is 68 in., a health service physician, wondering if dietary and/or genetical factors have led to a change in average height, computes the average height of 324 incoming college freshmen to be 68.5 in., with a standard deviation of 3.6 in. Is this sufficient evidence to reject his preconception that the average height is 68 in.? (Use a 1 per cent significance level.)

7.18. A manufacturing process is adjusted with the intention of bringing a certain dimension of the manufactured piece to 20 cm (on the average). From long experience it is known that the standard deviation of this dimension is .5 cm. A sample of 10 pieces is to be checked, and the process will be halted and adjusted if the average dimension deviates from 20 cm by more than .2 cm.

(a) What assumptions are needed before any probability computations involving the sample mean can be carried out? Make such assumption or assumptions in answering the following questions.

(b) Taking $\mu = 20$ as H_0, what is the size of the type I error?

(c) What rule should be followed if it is desired that this error size should be 5 per cent?

(d) What is the probability that the rule found in (c) leads to a type II error (failing to shut down and adjust) when the mean has actually shifted to $\mu = 20.1$?

7.6 Small-sample Tests for μ

The test for $\mu = \mu_0$ given in Section 7.5 employed the statistic

$$Z \equiv \frac{\overline{X} - \mu_0}{S/\sqrt{n}},$$

which is approximately standard normal (if $\mu = \mu_0$) when the sample size n is large. If n is *not* large, the approximation may be poor, and the method must be modified. Different modifications would be required for different populations, so the best that can be done here is to give the modification for the special case of a *normal* population.

The test for $\mu = \mu_0$ in the case of a small sample from a normal population employs essentially the same statistic as was used for large samples, differing from it* only by a constant factor:

$$T \equiv \frac{\bar{X} - \mu_0}{S/\sqrt{n-1}}.$$

Now, with the same kind of critical region as in the large-sample case, the relation between the critical value and α is obtained from a table of percentiles of what is called the *t-distribution with* $n - 1$ *degrees of freedom*† (Table III of the Appendix). So again the decision to accept or reject $\mu = \mu_0$ is based on the amount of the deviation of the sample mean from μ_0 as measured in units of the standard error of the mean, with the size of deviations deemed significant now determined from the t-table.

The t-distribution, the probability model for the statistic T, is (like the standard normal) symmetric about zero and bell-shaped. Indeed, for the larger number of degrees of freedom, corresponding to larger sample sizes, it is quite close to the standard normal model used in the case of large samples. The t-table entries stop after $n - 1 = 30$, the assumption being that for samples larger than 30 one would use the normal table.

As in using the large-sample normal distribution of Z, so in using the t-table, one uses the 2.5 and 97.5 percentiles as rejection limits for a two-sided test to obtain a significance level of 5 per cent (see Figure 7.7).

EXAMPLE 7.15

To test the claim a certain type of light bulb has a life expectancy of 700 hours, 10 such bulbs were burned until they failed. Their times to failure (in hours) were as follows: 590, 705, 640, 682, 740, 710, 664, 690, 728, and 749. Since $\bar{X} = 689.8$ and $S = 46.14$,

$$T = \frac{689.8 - 700}{46.14/\sqrt{9}} \doteq -.66.$$

This value of T is well within ordinary acceptance limits. For instance, using an α of .10 with a one-sided test, the rejection limit would be

* This assumes that $S^2 = \Sigma(X_i - \bar{X})^2/n$. If the divisor $n - 1$ had been used, the ratio called Z above would be used, rather than T.

† The numbering of "degrees of freedom", at this point, can only be thought of as an indexing parameter of a family of sampling distributions. It might seem to have made more sense here simply to use the sample size as the indexing parameter (and so number the rows of the table according to sample size). However, the t-distribution has other uses in which the relation between sample size and degrees of freedom is different from what it is here.

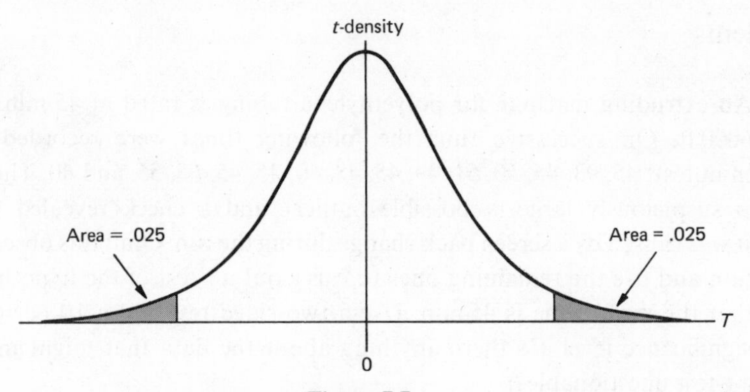

Figure 7.7
Two-sided t-test with $\alpha = .05$.

-1.38 (which is the 10th percentile of the t-distribution with 9 degrees of freedom). It should be pointed out that a normal distribution of times to failure is assumed; otherwise the t-distribution does not describe the null distribution of the statistic T.

The small-sample modification of the Z-test, which replaces the standard normal by the t-distribution as the null distribution of the test statistic, can also be used in constructing a small-sample confidence interval for the population mean. All that is involved is the replacement of constants determined from the normal table by constants from the t-table: The 95 per cent confidence interval for μ in a sample of size 10, for instance, is

$$\bar{X} - 2.26\frac{S}{\sqrt{n-1}} < \mu < \bar{X} + 2.26\frac{S}{\sqrt{n-1}}.$$

EXAMPLE 7.16

For the data on the life of light bulbs used in Example 7.15, with $\bar{X} = 689.8$ and $S = 46.14$, an 80 per cent confidence interval for the expected life μ would be obtained using 1.38, the 90th t-percentile (9 degrees of freedom), in place of 1.28, the 90th normal percentile, as a multiplier:

$$689.8 - 1.38\frac{46.14}{\sqrt{9}} < \mu < 689.8 + 1.38\frac{46.14}{\sqrt{9}}$$

or

$$668.5 < \mu < 711.0.$$

Problems

7.19. An extruding machine for polyethylene tubing is rated at 45 min per 6000 ft. On successive runs the following times were recorded (in minutes): 45, 93, 45, 50, 51, 49, 45, 48, 40, 45, 45, 65, 55, and 40. The 93 is suspiciously large (a possible outlier), and a check revealed that it was caused by a screen pack change during the run. Omit this observation and use the remaining ones to carry out a t-test of the hypothesis that the mean time is 45 min. Use a two-sided test and a 10 per cent significance level. (Is there anything about the data that might make a t-test questionable?)

7.20. Construct a 90 per cent confidence interval for the mean weight of female college students based on the sample of 12 in Problem 7.2, with mean weight 122.58 lb and standard deviation 16.52 lb. (It is necessary to assume a normal population.)

7.21. Use the birth-weight data of Problem 7.5 (mean weight 7.093, standard deviation .97 lb, $n = 30$) to construct a 95 per cent confidence interval for the population mean weight. (Is there any reason to wonder about the assumption of normality, from an inspection of the data? Perhaps there are two populations, normal babies and premature babies, combined to give a bimodal population?) If you were not satisfied with the precision attainable with this sample size ($n = 30$), how large a sample would be needed, approximately, to achieve 95 per cent confidence in an interval extending only .1 lb on either side of $\bar{X}$?

7.22. A sample of 25 airline passengers on a certain nonstop runs shows an average luggage weight of 32 lb with a standard deviation of 5.3 lb. Test the hypothesis that the mean weight of luggage per person on this run is 30. (Use a two-sided test at the 5 per cent level.)

7.23. A sample of 10 rats averages 38 sec to solve a certain maze, with a standard deviation of 6.5 sec. Construct a 90 per cent confidence interval for the mean time required.

7.7 Comparisons

A very common problem is that of determining something about the effectiveness of what is called a *treatment*, which may refer to a *new* treatment as compared to an old or standard one. For this purpose one may conduct an experiment in which the experimental units or objects are subjected to the new treatment, and the results noted. Because the results (in most experimental environments) are best regarded as random, i.e., representable by a model of outcomes and probabilities, the results are merely observations from a population of possible observations. In order to make a comparison of the new treatment with the old, one must consider not only the new population but also the old population of results that could be observed from an application of the old (or standard) treatment. Clearly, one needs a sample of observations from each population, and so the problem of comparison is often called a *two-sample* problem.

Traditional terminology refers to the group of experimental units to which the new treatment is applied as the *treatment group*, and to the group of experimental units to which the old or standard treatment is applied as the *control group*. (The control group may actually be *untreated*. But even if untreated, it is sometimes administered a fake treatment, or placebo, to equalize the psychological effect of being given something.) Having obtained results from both groups, one is faced with a sample $X_1, \ldots, X_m$ from the control population (untreated, or treated with a standard treatment) and a second sample $Y_1, \ldots, Y_n$ from the treatment population (subject to the new treatment, whose relative efficacy is under study). Each sample can be analyzed in terms of measures or statistics introduced earlier, but to make comparisons one must decide what to compare with what.

Just as an appeal to intuition yields useful statistics for studying single populations, so statistics for comparisons are often proposed on an intuitive basis. Indeed, it is generally easier to think of some statistic to compute, given a pair of samples, than it is to understand the significance of what was computed or how to use the result as a tool in inference.

Samples can be compared visually by an inspection of their graphical representations, but such techniques have flaws. First, visual comparisons are subjective, because conclusions and inferences drawn from such inspection depend on the viewer; different people will usually see different things in the data or in the graphs. Second, if the actual difference in populations is small, such difference may be difficult to detect with any assurance by looking at the samples. Indeed, even when there is *no* real treatment difference, there will ordinarily be some difference between samples, owing to the ubiquitous sampling fluctuations. When a difference between samples is noted, it is necessary to decide whether that difference is the result of an actual difference between the populations or is only a manifestation of the phenomenon of sampling variability. This kind of difficulty will plague us in all problems of statistical inference, but it is particularly awkward (if not impossible) to handle adequately when visual judgments are involved.

EXAMPLE 7.17

Table 7.3 is taken from an article in the *Journal of the American Medical Association*.* Given are platelet counts for a sample of 35 normal

Table 7.3

PLATELET COUNT (1000 PER MM^3)	FREQUENCY, CONTROL GROUP	FREQUENCY, CANCER PATIENTS
0–50		2
50–100		0
100–150		2
150–200	8	12
200–250	15	15
250–300	9	18
300–350	3	12
350–400		20
400–450		22
450–500		15
500–550		13
550–600		8
600–650		4
650–700		4
700–750		1
750–800		0
800–850		1
850–900		1
900–950		1
950–1000		1
Over 1000		1
Sum	35	153

males and a sample of 153 patients in a veterans' hospital in whom the diagnosis of lung cancer had been proved histologically. The data are recorded in terms of the number of thousands of platelets per cubic millimeter. Figure 7.8 exhibits the corresponding histograms, with relative frequency on the vertical scale so that the total areas of the histrograms are the same.

It is evident in the frequency tabulations and in the histograms that these samples are almost surely from different populations, and that no sophisticated statistical techniques are needed to point out what is

* S. E. Silvis et al., Vol. 211, No. 11 (1970), pp. 1852–1853.

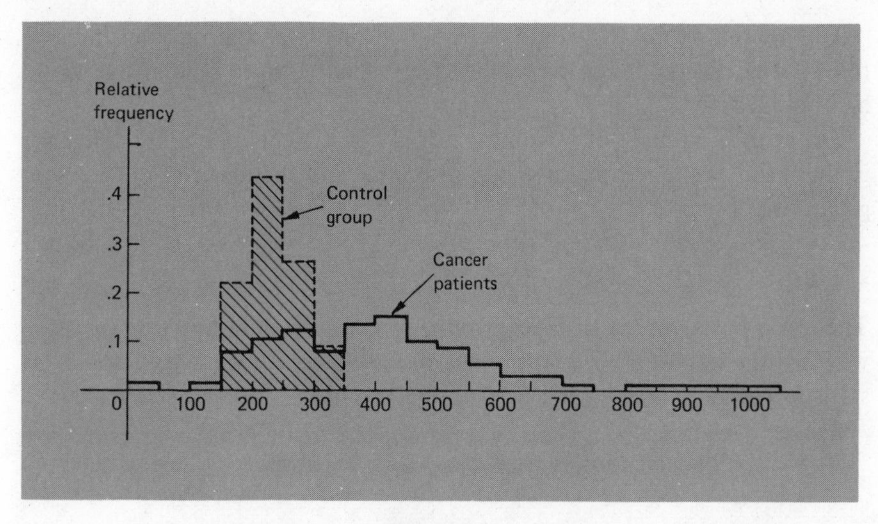

Figure 7.8
Histograms of platelet data of Example 7.17.

obvious. Yet it must be admitted, of course, that even what is obvious might not be true. Data can be freakish and lead to erroneous conclusions.)

To provide better means than tenuous, visual comparisons, a number of statistics, measures of difference between samples, have been proposed, aimed at detecting and estimating particular types of population differences. Thus there may be a shift in location, or a change in dispersion or variability, or a change in some other characteristic from one population to the other, and statistics are tailored to these various possibilities.

7.8 Comparing Means: Large-sample Tests

With the expectation that the (new) treatment may change the average level of response, one tests the null hypothesis

$$H_O: \quad \mu_T = \mu_C$$

(where μ_T is the population mean of the treated subject responses and μ_C of the control). This could also be expressed in terms of $\Delta = \mu_T - \mu_C$:

$$H_O: \quad \Delta = 0 \quad \text{(no difference)},$$

a form that no doubt gives rise to the phrase "null hypothesis."

The alternative to H_0 would be $\Delta > 0$ if one expected only an increase ($\Delta < 0$ if a decrease) in the average level of response. More generally, it would be two-sided: $\Delta \neq 0$.

The sample estimate of Δ would naturally be the corresponding difference in the mean of the treatment-group responses and the mean of the control-group responses:

$$D \equiv \overline{X}_T - \overline{X}_C.$$

The test is based on the distance from D to the null value $\Delta = 0$ in the same way that the test for $\mu = \mu_0$ was based on the distance from $\overline{X}$ to μ_0. The test statistic will be

$$Z = \frac{D - 0}{\sigma_D}$$

if σ_D can be computed, but more often than not σ_D will be estimated from the samples.

The variance of the difference D is the sum of the variances of $\overline{X}_T$ and $\overline{X}_C$ if the treatment sample and control sample are independent random samples:

$$\sigma_D^2 = \text{var } \overline{X}_T + \text{var } \overline{X}_C$$

$$= \frac{\sigma_T^2}{n_T} + \frac{\sigma_C^2}{n_C}.$$

In the case of large samples,

$$\sigma_D^2 \doteq \frac{S_T^2}{n_T} + \frac{S_C^2}{n_C},$$

since sample variances will be good approximations to corresponding population variances. (In the above formulas, as usual, σ^2 is a population variance, S^2 is a sample variance, and n is a sample size; the subscript T or C indicates whether it refers to the treatment group or the control group.)

EXAMPLE 7.18

A metallurgist wishes to compare the Rockwell hardness readings taken on the shank of a shearing pin with those taken on its hexagonal head. He finds that the average and standard deviation of 50 readings of each type are as follows:

$$\text{head:} \quad \overline{X}_{\text{Hd}} = 40.67, \quad S_{\text{Hd}} = 2.0;$$

$$\text{shank:} \quad \overline{X}_{\text{Sh}} = 41.63, \quad S_{\text{Sh}} = 1.8.$$

To test $\mu_{Hd} = \mu_{Sh}$, he computes

$$Z \doteq \frac{41.63 - 40.67}{\sqrt{4/50 + 3.24/50}} = 2.51 .$$

The null hypothesis is to be rejected, with a two-sided critical region, if $|Z| > 1.96$ for a significance level of 5 per cent. So the observed Z calls for rejection at this level.

Although it was convenient to describe the comparison problem in terms of treatment and control, one may wish to compare populations, neither of which is treated in an obvious way, but which exist side by side and cry out for comparison. Thus the population of males and the population of females might be compared with regard to any of a variety of responses or measurement (height, IQ, blood pressure, strength, etc.). More generally, any two species (of animal or plant) could be compared with respect to any measured characteristic. The alternative to the null hypothesis (no difference) is more apt to be two-sided in such cases than in the treatment-control context.

EXAMPLE 7.19 (continuing Example 7.17)
Platelet counts in the blood of 35 patients in a control group (normal healthy males) and in the blood of 153 patients in a treatment group (cancer patients) yielded these statistics:

$$\bar{X} = 235, \quad S_X = 44.4 \quad (m = 35),$$
$$\bar{Y} = 395.9, \quad S_Y = 169.9 \quad (n = 153),$$

where X is a platelet count in the control and Y in the treatment population. Although 35 is perhaps only marginally large, an approximation to σ_D, where $D = \bar{Y} - \bar{X}$, is computed from the above formula:

$$\sigma_D \doteq \sqrt{\frac{S_X^2}{m} + \frac{S_Y^2}{n}}$$

$$= \sqrt{\frac{(44.4)^2}{35} + \frac{(169.9)^2}{153}} = 15.65 .$$

Thus

$$Z = \frac{D - 0}{\sigma_D} \doteq \frac{395.9 - 235}{15.65} = 10.28 .$$

It seems quite clear that at any reasonable significance level one can assert that there is a difference. Actually, this could have been anticipated from an inspection of the histogram in Figure 4.1. One might say that the computation of Z was really not necessary; the conclusion is obvious at the outset.

EXAMPLE 7.20

A large toothpaste manufacturer advertised a report of clinical studies that purportedly showed the superiority of its product. One of the sheets in the folder sent to those who requested the report gives the results shown in Table 7.4. Thus $\overline{X}_1 = 10.88$, $\overline{X}_2 = 13.41$,

Table 7.4

	NUMBER OF CHILDREN	AVERAGE NUMBER OF NEW CAVITIES PER CHILD (30-MONTH PERIOD)
Toothpaste C (with fluoride)	294	10.88
Control toothpaste	284	13.41

and the difference is $\overline{X}_2 - \overline{X}_1 = 2.53$. (Observe that the population variable, "number of cavities per child," is discrete, with values $0, 1, 2, \ldots$.) The consumer is apparently supposed to leap to the conclusion that toothpaste C is better. But is a difference of 2.53 really a significant reduction, or is it explainable by random sampling fluctuations?

This question cannot be answered with the information given, because there is *no indication of population variability*. To determine this, the consumer is forced to consult the dental journal in which the study was reported [*British Dental Journal*, Vol. 124 (1968), pp. 209–213, article by Møller, Holst, and Sørensen]. There one finds sample standard deviations:

$$S_1 = 6.36, \qquad S_2 = 7.20.$$

The standard deviation of the difference of means is then approximated as follows:

$$\sigma_D = \sqrt{\frac{\sigma_T^2}{n_T} + \frac{\sigma_C^2}{n_C}}$$

$$\doteq \sqrt{\frac{6.36^2}{294} + \frac{7.20^2}{284}} = .566,$$

and this provides a measuring stick, $\sigma_D = .566$, in terms of which we can decide whether the observed difference is too large:

$$\frac{D}{\sigma_D} = \frac{2.53}{0.566} = 4.49.$$

Expecting or hoping that the fluoride toothpaste is better, one would no doubt use a one-sided test and reject $H_O(\mu_1 = \mu_2)$ if D/σ_D exceeds 1.645, if a 5 per cent significance level is used. Since $4.49 > 1.645$, the given data would lead to rejection of H_O, or acceptance of the hypothesis that the average number of cavities per child is fewer when he uses the fluoride toothpaste than when he uses the control toothpaste. Before rushing out to buy toothpaste C, however, the reader should realize that this experiment and the results quoted may not be definitive except under the conditions corresponding to those under which the test was conducted. Indeed, the manufacturer's brochure did include results from five other studies under varying conditions (amount of brushing, supervision of brushing, fluoridation of the local water supply, geographical location, etc.). The incorporation of more than one factor into a single experimental design will be introduced briefly in Chapter 11.

Problems

7.24. At the beginning of Section 7.8 it was stated that the variance of a difference is the *sum* of the variances. It is true, generally, that when X and Y are independent, then

$$\text{var}(X - Y) = \text{var}\, X + \text{var}\, Y.$$

Carry out the following experiment to convince yourself of the truth of this relation. Select 50 (or more) pairs of random digits from the table of random numbers (Table VIII of the Appendix); call one of the pair X and the other Y, and calculate $X - Y$ for each pair. Compute the variance of the X's, the variance of the Y's, and the variance of the differences $X - Y$ in this sample of 50 pairs. Observe how close $S_x^2 + S_y^2$ is to S_{x-y}^2, and note also how *far* the *difference* of sample variances is from the variance of the difference. (In two populations with equal variances the difference of the variances would be zero, whereas the variance of differences is certainly not zero.)

7.25. A study of dentifrices [*Journal of Oral Therapeutics and Pharmacology*, Vol. 4 (1968), pp. 443–449, article by Frankl and Alman] showed 19.98

new cavities per child over a three-year period in a sample of 208 children in a treatment group, and 22.39 per child in a sample of 201 children in a control group. The standard deviations are given as 10.61 and 11.96, respectively. Test the hypothesis that the dentrifrice used by the treatment group is no better than that used by the control group. (Use $\alpha = .01$.)

7.26. A sample of 100 male faculty and a sample of 50 female faculty at a large university have mean salaries of $17,800 and $15,100, respectively, with corresponding standard deviation $2900 and $1800. Carry out a test of equality of male and female salaries using the difference of sample means. Is the test valid for anything? (Does it prove discrimination? Discuss possible flaws.)

7.27. Calculus is taught in the traditional way to 600 students and using computers to 250 students. Afterward the students are all given a test designed to test their comprehension of principles (rather than the special techniques used in either method of teaching). The control group (traditional) averaged 70.6 with a standard deviation of 8.30, and the treatment group (computers) averaged 71.5 with a standard deviation of 7.92. Is there a difference?

7.28. A sample of 200 dwelling units in the 13th Ward shows an average of 1.62 bathrooms per unit, and a sample of 200 in the 3rd Ward shows an average of .85 bathroom per unit. Is this a significant difference? (No variances are given, but you can be certain that $\sigma < 3$, say, in each case. So set up the test statistic with $\sigma = 3$ and see how it comes out.)

7.9 Comparing Means: Small-sample Tests

As in the case of the one-sample problem, when the sample sizes are not "large," the distribution of D (the difference between sample means) is generally too complicated and dependent on the population distributions to permit construction of a general procedure. But (again as in the one-sample case) if the populations are normal, a modification of the Z-statistic of Section 7.8 can be used in a t-test, if the population variances can be assumed equal:

$$\sigma_T^2 = \sigma_C^2 = \sigma^2.$$

In this case,

$$\sigma_D^2 = \frac{\sigma^2}{n_T} + \frac{\sigma^2}{n_C} = \sigma^2\left(\frac{1}{n_T} + \frac{1}{n_C}\right).$$

(Again the T and C subscripts refer to the treatment and control groups, respectively.)

The estimator of σ^2 used in constructing a t-statistic is a pooled variance, employing variability information from both samples:

$$S^2 \equiv \frac{n_T S_T^2 + n_C S_C^2}{n_T + n_C - 2},$$

which is an average of squared deviations taken in each sample about the mean of that sample. (The -2 in the denominator is a modification that permits use of existing tables.) Then

$$\sigma_D \doteq S\sqrt{\frac{1}{n_T} + \frac{1}{n_C}},$$

and this is used as a measuring stick to determine whether D is significantly different from 0. Thus one rejects the null hypothesis ($\Delta = 0$) if

$$T = \frac{D}{S\sqrt{1/n_T + 1/n_C}}$$

is far from zero, as determined from the t-distribution with $n_T + n_C - 2$ degrees of freedom.

EXAMPLE 7.21

A worker in a consumer testing laboratory claims that men drivers generally drive more economically (use less gas per mile) than do women. His female associates naturally challenge the claim, and a test of the hypothesis of no difference is conducted as follows: Five men and five women each drive a car (the same car and using the same gasoline) over a specified route through city traffic. Gasoline-consumption figures (in gallons) turn out to be as follows:

MEN (X)	WOMEN (Y)
0.94	1.40
1.20	.98
1.00	1.22
1.06	1.16
1.02	1.34

Assuming normal populations with equal variance (which may be quite an assumption), the t-test can be applied:

$$\overline{X} = 1.044, \qquad S_X^2 = .007584,$$

$$\bar{Y} = 1.220, \qquad S_Y^2 = .0216,$$

$$S^2 = \frac{n_X S_X^2 + n_Y S_Y^2}{n_X + n_Y - 2} = (.135)^2.$$

From these, one has

$$T = \frac{1.044 - 1.220}{(.135)\sqrt{\frac{1}{5} + \frac{1}{5}}} = -2.06.$$

From the problem description it seems that a one-sided alternative ($\Delta < 0$) is appropriate, so a one-sided critical region (reject H_O if $T < k$) is best. For $\alpha = .05$, the critical value (5th percentile of the t-distribution with 8 degrees of freedom) is 1.86. Thus the null hypothesis would be rejected at the 5 per cent level, but it is close: At $\alpha = .025$ one would accept H_O (critical value of 2.31).

7.10 Comparing Locations: Distribution-free Tests

The large-sample test of Section 7.5 is *distribution-free*, i.e., not subject to any restrictions on the nature of the populations whose means are being compared. However, the small-sample test of Section 7.6 is definitely *not* distribution-free; the assumption of normality of the populations is essential. A small-sample test will now be given for the null hypothesis of no population difference, one that uses a statistic whose distribution does not depend on the nature of the populations.

The test to be presented is based on *ranks*. When a set of numbers such as (4, 10, 2, 7) is *ordered*, yielding (2, 4, 7, 10), the rank of each is its position in this ordered sequence. Thus 2 is of first rank, 4 has rank 2, 7 has rank 3, and 10 has rank 4.

Any statistical test is most effective against certain alternatives, and not so effective against others. In constructing a test intuitively, it is helpful (if not essential) to look at the alternatives to the null hypothesis. As in the earlier sections of this chapter, we consider *shift* alternatives—in the case of comparisons, the alternative that one population is shifted (in location) to the right or left, with no other significant difference in population characteristics. Figure 7.9 shows three cases: the null case, that of a shift in Y to the right of X, and that of a shift to the left. Typical observations from X and from Y are marked on the axis in each case as x and y, respectively.

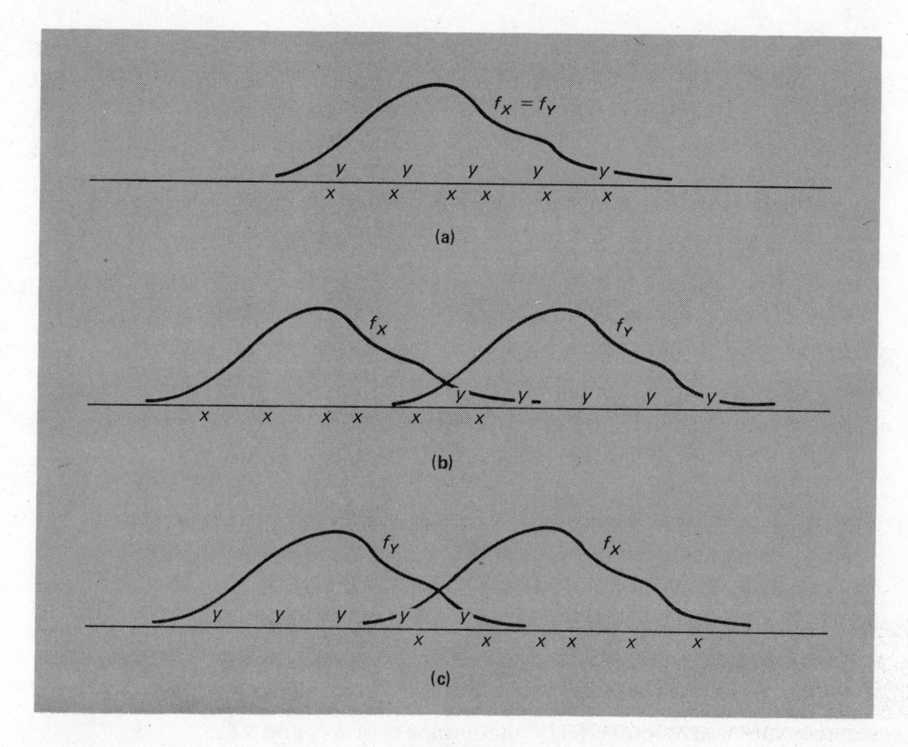

Figure 7.9
Null hypothesis and shift alternatives.

Plotting observations on a scale of values (as the x's and y's are plotted in Figure 7.9) serves to order them, and so to put their *ranks* on display. In particular, if one thinks of the X-observations and the Y-observations as comprising a *single* sample (from the common population of H_O), their ranks in this combined sample are on display. Notice that under H_O, when the X's and Y's are rather well interspersed, there are some small and some large Y-ranks. When Y is really shifted right, the Y-ranks in the combined sample tend to be large, and when shifted left, the Y-ranks are mostly small. To be explicit, for the 11 observations (6 X's and 5 Y's) plotted in Figure 7.9, the Y-ranks are as given in Table 7.5. These observations were plotted as more or

Table 7.5

	Y-RANKS					SUM
Null case	2	4	6	8	10	30
Y shifted right	6	8	9	10	11	44
Y shifted left	1	2	3	4	6	16

less typical, but it is clear that the conclusion is sound. A shift to the right tends to produce a larger rank sum, and a shift to the left a smaller rank sum. No shift tends to yield moderate rank sums, neither large nor small. This then is the *Wilcoxon rank-sum test*: Let R_Y denote the sum of the Y-ranks when the X's and Y's are combined in a single sample, and reject H_O (the null hypothesis of no population difference) if R_Y is either too large or too small, to protect against a two-sided shift alternative. (Clearly, for an alternative of a shift of Y to the right, reject H_O if R_Y is too large.)

The test is practicable because of the fact (not proved here) that under H_O the distribution of the test statistic R_Y depends *only on the sample sizes*. This distribution can be computed exactly, for small sample sizes; and it has been computed, the results being given in Table V (Appendix). For large samples the distribution of R_Y is approximately normal, with mean

$$E(R_Y) = \frac{n}{2}(m + n + 1)$$

and variance

$$\text{var}\,(R_Y) = \frac{mn}{12}(m + n + 1),$$

where m and n are, respectively, the numbers of X's and Y's.

EXAMPLE 7.22

The gasoline-consumption data of Example 7.21 are repeated here:

MEN	WOMEN
0.94	1.40
1.20	0.98
1.00	1.22
1.06	1.16
1.02	1.34

The combined, ordered sample is

$$(\mathit{0.94}, 0.98, \mathit{1.00}, \mathit{1.02}, \mathit{1.06}, 1.16, \mathit{1.20}, 1.22, 1.34, 1.40),$$

in which the X-observations are given in italics. The sum of the X-ranks is clearly

$$R_X = 1 + 3 + 4 + 5 + 7 = 20.$$

Reference to Table V of the Appendix shows that a 5 per cent one-sided test would call for rejecting the null hypothesis (of no difference between men and women in the given driving feat) if $R_X \leq 19$. Since $R_X = 20$, the null hypothesis is (barely) accepted. (Compare Example 7.21, which applied a *t*-test to these data; but notice that in the present case, no assumption of normality is required.)

EXAMPLE 7.23
The composite scores (combining two quizzes and a final) of 14 students in a certain statistics class in the fall of a certain year were as follows:

88, 80, 38, 74.5, 72, 91, 39, 63, 63, 81.5, 64, 70, 66, 64.5.

Two years later the same instructor taught the same course to a class of 32 students, and the following composite scores resulted:

78, 88.5, 66, 72.5, 81, 90.5, 70, 84.5, 56, 75, 56.5

62, 96.5, 86.5, 88.5, 98.5, 87, 62, 82, 90, 39.5, 50

98, 88, 93, 92, 75, 88, 40, 54, 74, 70.

To test the hypothesis that there is no difference between the populations of scores from which these two samples were taken, the instructor computed the X-ranks (X referring to the earlier class) in the combined ordered sequence of 46 scores, and found these ranks to be

1, 2, 11, 12, 13, 14, 15.5, 18, 20, 22, 27, 29, 34.5, 40.

(Although the probability of a tie in a continuous model is zero, the rounding off used here permitted ties. In the cases in which X and Y scores were tied, the instructor assigned the average rank of the tied scores to each of them. This explains the 15.5, for instance.) The Wilcoxon statistic is the sum of the above ranks: $R_X = 249$. Assuming R_X to be an approximately normal variate, with mean

$$E(R_X) = \frac{14}{2}(14 + 32 + 1) = 329$$

and variance

$$\text{var}(R_X) = \frac{14 \cdot 32}{12}(14 + 32 + 1) = (41.9)^2,$$

the Z score of R_X is

$$Z = \frac{R_X - E(R_X)}{\sqrt{\text{var}\, R_X}} = \frac{249 - 329}{41.9} = -1.91.$$

The instructor really did not know what the proper alternative H_1 should be or what significance level to pick; but he noted that the observed value of R_X is nearly two standard deviations away from the mean, and he took this as evidence that there was a real difference. However, because he had not designed an experiment to sort out possible factors, he really could not tell whether the difference was caused by easier exams, or better teaching, or better students in the later class (or some combination of these factors).

Problems

7.29. A sample of 20 incoming freshmen, whose blood pressures were taken by the health service physician, showed an average of 118 with a standard deviation of 10.5. Another sample of 15, whose blood pressures were taken by a nurse, showed an average of 121 with a standard deviation of 11.0. Is there any significant difference in the way the physician and the nurse take blood pressures? (Use a 5 per cent significance level.)

7.30. In a test given to 45 students, the 20 women averaged 60 with a standard deviation of 19, and the 25 men averaged 55 with a standard deviation of 16. Assume that the population variances for men and women are equal and test the hypothesis that the population means are equal against the alternative that women are superior.

7.31. Two fly sprays are tested by successive applications of each to batches of flys. The results, in per cent mortality, are as follows:

Spray A	68, 68, 59, 72, 64, 67, 70, 74
Spray B	60, 67, 61, 62, 67, 63, 56, 58

(a) Carry out the rank-sum test of the hypothesis that there is no difference between the sprays, at the 10 per cent significance level.

(b) Use the t-test for the same problem. (What must be assumed for this to be valid?)

7.32. Two samples of 10 subjects, one sober and one treated with two martinis each, were administered a psychological test involving some reasoning and some motor facility. Scores were as follows (on a scale 0 to 20):

Sober	16, 13, 13, 17, 11, 16, 14, 16, 17, 12
Treated	15, 16, 12, 8, 10, 14, 12, 9, 12, 13

(a) Use a t-test for the null hypothesis of no treatment effect.
(b) Use the Wilcoxon test for the null hypothesis. (In the case of ties assign the average rank.)

7.11 Paired Data for Comparison

Measurements on experimental subjects, obtained for the purpose of determining the effectiveness of a treatment, are usually affected by a host of other factors. An unimaginative approach would be to consider these factors as contributing to the unknown, unpredictable random ingredient of the response. Even so, they would contribute to *variability* in the response; and variability is what makes statistical inferences less than completely reliable.

Variability can often be reduced by designing an experiment so that certain identifiable factors can be taken into account. This is the theme of Chapter 11; here the idea will be developed only for the problem of comparing two populations.

Earlier, in comparing boys and girls, it was assumed that one would take a random sample from the reference population of boys and, independently, a random sample from the reference population of girls. But if age and IQ, say, are likely to be factors in the response or measurement of interest, a process of *pairing* will tend to eliminate the variability due to these factors. Thus one selects pairs of experimental subjects, each pair consisting of a girl and a boy who are of approximately the same age and IQ. The differences in response or measurement within each of n such pairs provide a *single* random sample, and this sample of n differences can be used to test the null hypothesis that the expected difference is zero, that there is no difference between boys and girls with regard to the response being measured. There may seem to be a loss in sensitivity in going from an analysis with $2n$ observations to one with just n observations, but there is usually more gain in sensitivity accomplished through the elimination of extraneous factors as sources of variability.

In testing a new treatment for corrosion resistance of underground pipes, one would treat a sample of pipes with the new method and another sample with the old, and then bury the pipes for a period of time, after which corrosion would be measured. If the soil in which the pipes are buried is a factor in corrosion, i.e., if there is variation in soil composition, one would want the pipes to be buried in such a way that the contribution of this factor would be random, but this might not be easy to do. Instead, pipes could be buried in pairs, one treatment pipe with one control pipe, and (after the allotted time) the single sample of differences in degree of corrosion would provide a basis for testing in which the soil factor is of little influence.

Sometimes data come in a naturally or unavoidably paired form, as in "before" and "after" measurements. Although for n subjects there would seem to be $2n$ observations, only the n differences (the after measurement minus the before measurement in each case) can be used. This is because of the obvious dependence between the two measurements from a given subject. In such problems one might be interested in the amount of the expected difference of before and after measurements, rather than a null hypothesis of no difference, i.e., in estimation rather than in testing.

EXAMPLE 7.24

A nutrient was to be tested for its effect in increasing the weight gain of young pigs. To eliminate the variability that is caused by differences among pig families, pairs of pigs were selected, one pair from each of 10 litters. One pig was chosen at random from each pair to receive the nutrient in addition to the feed used for both pigs in the pair. The weight gains in pounds over a given period of time were recorded as shown in Table 7.6. The average difference and the standard deviation of the differences are

$$\bar{d} = 2.32, \qquad S_d = 1.87.$$

Table 7.6

PAIR NUMBER	REGULAR FEED	WITH FEED NUTRIENT	DIFFERENCE, d
1	35.2	36.0	+.8
2	30.0	32.7	+2.7
3	36.5	39.2	+2.7
4	38.1	37.6	−.5
5	29.4	32.0	+2.6
6	36.0	40.2	+4.2
7	31.3	34.4	+3.1
8	31.6	30.7	−.9
9	34.0	37.2	+3.2
10	31.1	36.4	+5.3

The appropriate t-statistic is then

$$T = \frac{\bar{d} - 0}{S_d/\sqrt{n-1}} = \frac{2.32}{1.87/3} = 3.73.$$

Since a one-sided alternative is clearly in order (the nutrient is expected to be effective, or at least not deleterious), one rejects the hypothesis of no nutrient effect at the 10 per cent level, the observed T-value being greater than 1.38, the 90th percentile of t (9 degrees of freedom). Notice that a *normal* distribution of d has again been assumed.

The point has been made that because of the dependence introduced by pairing between the observations in a pair, one must use the differences as a sample of size n, rather than the individual responses as constituting two samples of size n each. In Example 7.24, for instance, there are 10 weight gains with regular feed and 10 weight gains with the added nutrient; and the average difference $\bar{d}$, computed there as the average of 10 differences, is also equal to the difference of the average of the 10 gains on regular feed and the average of the 10 gains with the nutrient. But the variance of $\bar{d}$, estimated in Example 7.24 by using the variance of the 10 differences, could not be computed in terms of the variances of the two samples of 10, as in the earlier t-test for no population shift, because of the dependence of the observations within the pairs. That is, if X_i and Y_i are dependent, then it is *not* true that the variance of $\bar{X} - \bar{Y}$ is the sum of the variances of $\bar{X}$ and $\bar{Y}$.

The method used in Example 7.24 was essentially a single sample t-test, calling for the assumption of normality of the population differences, although if the sample sizes were larger, the statistic used would be approximately normal even if the population were not. A small-sample method that does not require the assumption of normality of the population is the *Wilcoxon signed-rank test.*

The statistic used in the signed-rank test is the sum of the ranks of the positive differences when all differences are ordered according to magnitude. This sum has a distribution under the null hypothesis that depends only on the number of differences, not on the population distribution, as long as the population is symmetric (see Table VI of the Appendix). The sum of the ranks of the negative differences could be used, equivalently, since the sum of all the ranks is fixed: $1 + 2 + \cdots + n = n(n + 1)/2$.

EXAMPLE 7.25

The weight-gain differences d_i observed in Example 7.24 are repeated here, ordered according to magnitude (ignoring the signs):

$$-.5,\ +.8,\ -.9,\ +2.6,\ +2.7,\ +3.1,\ +3.2,\ +4.2,\ +5.3.$$

The sequence of signs is

$$-\quad +\quad -\quad +\quad +\quad +\quad +\quad +\quad +\quad +,$$

and the ranks of the negative signs add up to $4 = 1 + 3$. This sum is small both because there are few negative differences and because these negative differences are of small magnitude. There is a strong indication that the null hypothesis of no treatment effect is false. A one-sided 5 per cent test, according to Table VI, would call for rejecting the null hypothesis if the sum of the ranks of the negative differences is 10 or smaller. And so with the given data, one rejects H_0 at the 5 per cent level of significance.

Problems

7.33. To test the relative suitability of the soil for growing tomatoes in two garden plots, I plant five matched pairs of plants, one from each pair in each plot. The results are judged by the weight of usable tomatoes from each plant:

	Pair number				
Plot	1	2	3	4	5
A	10.2	12.0	15.7	8.8	13.4
B	13.6	8.9	16.3	10.3	16.1

Test at the 20 per cent level the hypothesis of no differences in the soil
(a) Using the t-test. (What must be assumed?)
(b) Using the signed-rank test.

7.34. To decide whether I want to switch from type C tomato plants to an improved type D, I plant five pairs, one of each type, with the pairs matched according to appearance when I buy them at the nursery and according to where I plant them. The resulting yields are as follows:

	Pair number				
Type	1	2	3	4	5
C	13.0	10.9	12.1	14.0	9.8
D	17.8	14.3	11.9	15.3	11.3

Test the hypothesis of no difference (against what alternative?) at $\alpha = .10$
(a) Using the t-test. (What must be assumed?)
(b) Using the signed-rank test.

7.35. A newspaper report of a study of the effect of smoking on infant size included these data: There were 88 matched pairs of mothers in the study (one smoker and one nonsmoker in each pair), and the mean difference in birth weight was $\frac{1}{2}$ lb (the nonsmokers had heavier babies).

(a) What additional information would be needed to be able to draw a conclusion as to the significance of the $\frac{1}{2}$-lb difference?

(b) The original data would provide an estimate of σ_{X-Y} needed to measure the significance of an observed $\bar{X} - \bar{Y}$. Not having the data we can still learn something: The X and Y are not independent and in this case σ_{X-Y} would tend to be *less* than $\sqrt{\sigma_X^2 + \sigma_Y^2}$, which would apply if we had independence. Using $S_X = .97$ from the birth-weight data of Problem 7.5 as a crude guess for each σ_X and σ_Y, it would follow that σ_{X-Y} would here not exceed $\sqrt{.97^2 + .97^2} = 1.37$. Use this as σ_{X-Y} in judging the possible significance of $\bar{X} - \bar{Y} = .5$.

(c) Discuss with the instructor the factors that might be taken into account when matching.

7.36. Two brands of electric typewriter are being considered for a large-quantity purchase. Eight of each brand are tested by matched pairs of typists, with the following results (in words per minute, adjusted for typing errors):

	PAIR NUMBER							
BRAND	1	2	3	4	5	6	7	8
E	72	62	68	58	70	66	60	70
F	78	60	67	55	75	73	64	64

Test the hypothesis of no difference at $\alpha = .10$

(a) Using the t-test.

(b) Using the signed-rank test.

7.12 Goodness of Fit

It is sometimes possible to formulate a specific model for the given population from a priori or nonstatistical considerations. In such a case it is desirable to test the validity of the model by making observations on the population (i.e., by sampling) to see if these observations can be reasonably explained by the assumed model. The problem is one of testing the *hypothesis* that the proposed model is correct against the alternative that it is not correct. The proposed model is the *null hypothesis*.

One simple and often powerful test is based on the *sample distribution function* and how far this function deviates from the population distribution function that defines the model of the null hypothesis. To be specific, the *Komogorov–Smirnov statistic* D_n is defined to be maximum absolute vertical deviation of the sample distribution function from the population c.d.f. defined by the null hypothesis. Figure 7.10 shows a population c.d.f., a sample d.f., and the corresponding quantity D_n whose value is used as a measure of discrepancy. In order to use the statistic D_n intelligently as the basis of a test, it is necessary to know something about D_n as a random variable. This could be learned potentially either from sampling experiments or from a mathematical study. A mathematical study demonstrates that the distribution of D_n, when the null hypothesis is true, *depends only on n* (the sample size). Moreover, this distribution for each n has been tabulated.

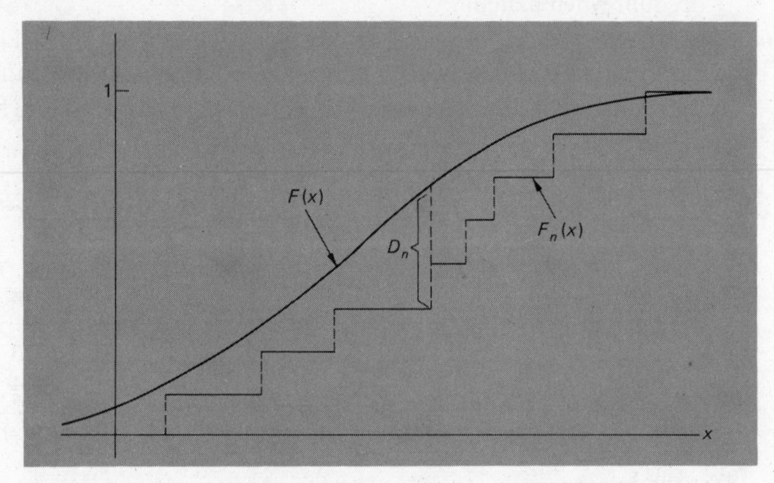

Figure 7.10
The Kolmogorov–Smirnov statistic D_n.

The Kolmogorov–Smirnov test procedure (for the null hypothesis that the assumed distribution is governing the generation of data against the alternative that this is *not* the case) is defined as follows. One chooses a critical value K, and the rule or test is to reject the null hypothesis if and only if the sample yields a value of D_n that exceeds the critical value: $D_n > K$. (This rule is set up on a purely intuitive basis.) Thus large values of D_n are used to infer that the sample results are so far from what is ordinarily obtained when the population does have the distribution of the null hypothesis that some explanation other than the null hypothesis is in order. A given critical value defines a test, and different critical values define different tests. Each test, each critical value, implies a certain significance level, a value of α. This is the probability that if the null hypothesis is correct, the test will erroneously reject it. Table VII of the Appendix gives critical values K corresponding to various commonly used levels of significance and to various sample sizes. Also given are formulas to use for large samples.

EXAMPLE 7.26

In the article giving platelet data quoted in Example 7.17 it is assumed that the platelet count for healthy males is normal with mean 235,000 per mm^3 and standard deviation 44,600 per mm^3. To illustrate the use of the Kolmogorov–Smirnov test of the null hypothesis that lung cancer patients also have platelet counts from this population, the first 25 observations of counts from lung cancer patients are given, arranged in order of magnitude, in the column headed x in Table 7.7,

Table 7.7

x	$F_n(x)$	$N(x)$	x	$F_n(x)$	$N(x)$
173	.04	.082	382	.56	1.000
189	.08	.151	395	.60	1.000
196	.12	.191	399	.64	1.000
207	.16	.265	401	.68	1.000
215	.20	.327	437	.72	1.000
237	.24	.518	480	.76	1.000
275	.28	.815	504	.80	1.000
282	.32	.854	524	.84	1.000
293	.36	.903	634	.88	1.000
300	.40	.927	682	.92	1.000
305	.44	.942	882	.96	1.000
316	.48	.965	999	1.00	1.000
346	.52	.994			

expressed in units of 1000 per mm^3. The column headed $F_n(x)$ gives the step height, the value of the sample distribution function at x. (These go up in steps of $.04 = 1/n = \frac{1}{25}$.) The column headed $N(x)$ gives the value of the normal (cumulative) distribution function, with mean 235 and standard deviation 44.6, at x. For instance,

$$\begin{aligned} N(207) &= P(X \le 207) \\ &= \Phi\left(\frac{207 - 235}{44.6}\right) = .265. \end{aligned}$$

The maximum deviation between $F_n(x)$ and $N(x)$ is obtainable without drawing their graphs when it is realized that this maximum occurs at one of the jump points. However, since $N(x)$ lies above $F_n(x)$, the values to be compared are N at x with F_n at the preceding x (see Figure 7.11). Calculation of such differences shows the maximum to be

$$D_n = .815 - .24 = .575.$$

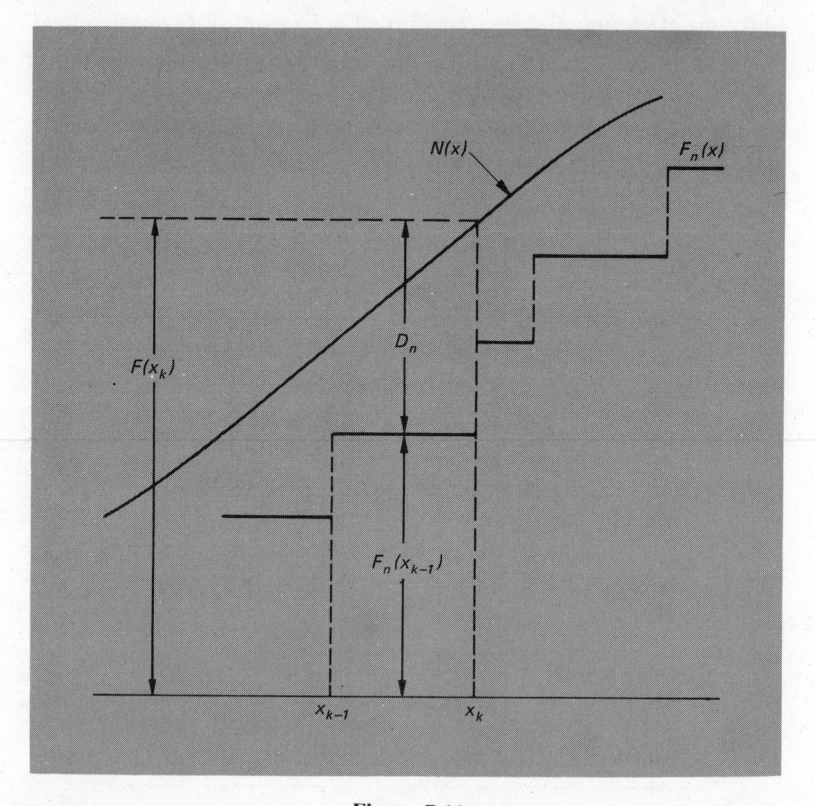

Figure 7.11

This is to be compared with rejection limits from the distribution of D_n for $n = 25$; the following limits are extracted from Table XX:

α	.20	.15	.10	.05	.01
Rejection limit	.21	.22	.24	.264	.32

At any significance level of .01 or greater, the null hypothesis is rejected because the observed $D_n = .575$ exceeds the tabulated rejection limits.

EXAMPLE 7.27
Table VII, which gives critical values of D_n for various sample sizes n, was prepared by means of a mathematical study of the distribution. An empirical approach would be to calculate D_n for many samples, to get an idea of its pattern of variation. Such an empirical procedure was carried out using 250 samples of size 10 from a uniform population (continuous spinning point with scale from 0 to 1), to get an approximation to the distribution of D_{10}. (Because the distribution is the same for any null hypothesis, the choice of the uniform population was simply a convenient one.) The value of D_{10}, of course, was measured as the maximum deviation of the sample distribution function from the c.d.f. of the uniform population. From the 250 values of D_{10} a table of empirical acceptance limits was constructed and is given as Table 7.8, along with the theoretical values (from the mathematical derivation) for comparison:

Table 7.8

α (SIGNIFICANCE LEVEL)	ACCEPTANCE LIMIT (THEORETICAL)	ACCEPTANCE LIMIT (EMPIRICAL)
.20	.322	.329
.15	.342	.342
.10	.368	.363
.05	.409	.398
.01	.486	.453

There is no general method for determining type II errors, or probabilities of accepting the null distribution when it is not the correct model. However, given any specific alternative to the null hypothesis, one can carry out an empirical study like that of Example 7.26 to estimate the chances of accepting H_0 when in fact the actual model is the chosen specific alternative. This is illustrated in the next example.

EXAMPLE 7.28
Consider again the null hypothesis of a uniform distribution on $0 < x < 1$. Two hundred samples of size 10 were obtained from the alternative distribution shown in Figure 7.12 (along with the null distribution), and the value of D_{10} was computed for each sample. For a significance level of, say, .10, the acceptance value is .368; that is, one would reject H_0 if D_{10} exceeds .368. The proportion of the 200 D_{10}-values that actually *did* exceed .368 and so call for rejection of H_0

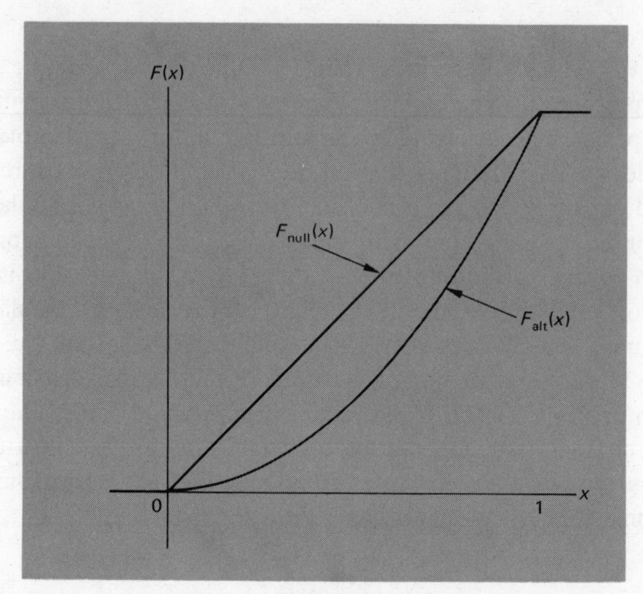

Figure 7.12
Null and alternative distributions for Example 7.28.

turned out to be .58. This proportion is then an estimate of the probability that this 10 per cent test would correctly reject the uniform model when the correct model is the given alternative. (*Note*: The mean under H_A is $\frac{2}{3}$, as compared with $\frac{1}{2}$ under H_O.) The following table gives the results for tests at various α-values:

α (type I error size)	.20	.10	.05	.01
Estimate of β (type II error size for given alternative)	.31	.42	.59	.85

Observe the usual struggle between type I and type II errors: Decreasing one increases the other. For a *given* α, the β could be made smaller only by taking more than 10 observations in a sample.

It is often desired to test the hypothesis that a population is of a certain type, without specifying the values of parameters of the distribution. For example, one is often interested in knowing whether a population can be considered normal. In such a case the Kolmogorov–Smirnov statistic D_n cannot be computed, because the probabilities needed under H_O cannot be computed; the mean and variance must be specified in addition to the assumption of normality. However, if one stretches a point and uses *sample* estimates $\overline{X}$ and S^2, respectively, he can compute a value of D_n and apply the test as before. This device will have the effect of making it harder to reject H_0, since the data were used in choosing the model to fit the data. But this means that the actual α is less than what would be obtained from Table VII; the test would be conservative with regard to α. If observed data call for rejecting normality with this procedure, the evidence is even stronger than suggested by the α used to obtain the critical value of D_n.

EXAMPLE 7.29
Consider testing the hypothesis that the following (artificial) data came from a normal population:

x_i	5	6	12	14	16
f_i	1	5	1	2	1

The mean is 9.1 and the standard deviation is 4.1. The maximum deviation D_n will occur at a point where the sample distribution function jumps, i.e., at one of the observed values. The normal c.d.f. with mean 9.1 and standard deviation 4.1 is evaluated at these points as shown in Table 7.9. The value of D_{10} is seen (Figure 7.13) to be

$$D_{10} = .6 - .224 = .376.$$

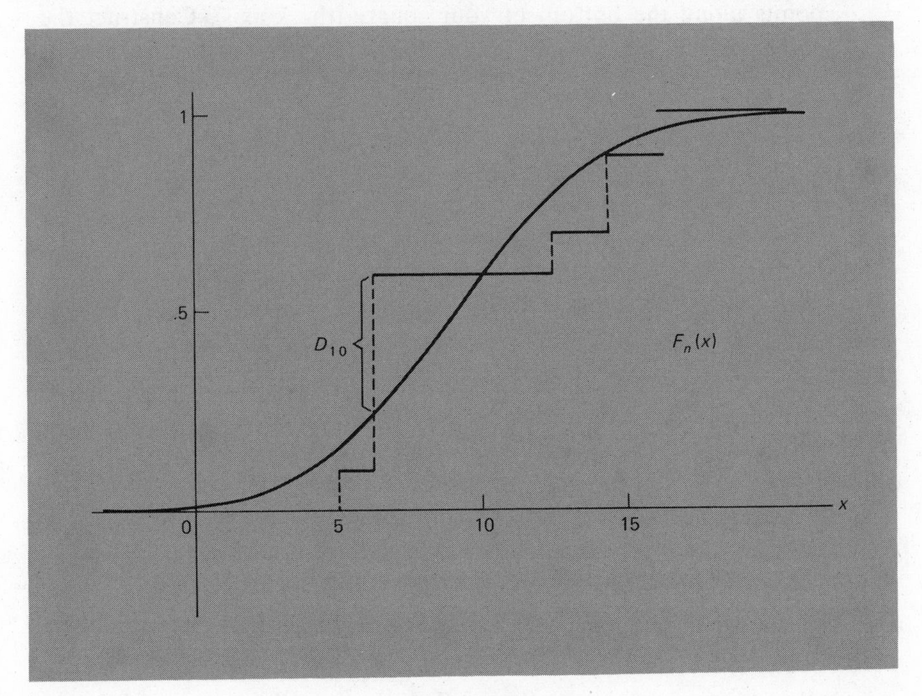

Figure 7.13

Table 7.9

x_i	$\dfrac{x_i - 9.1}{4.1} = z$	$\Phi(z_i)$	$F_n(x_i)$
5	-1.0	.159	.1
6	$-.76$	.224	.6
12	$+.71$	.761	.7
14	1.2	.885	.9
16	1.68	.954	1.0

The 10 per cent critical value of D_{10} is .368 (Table VII), so H_O is rejected.

Problems

7.37. Draw a large square (6–8 in.) with a scale from 0 to 1 marked off along the bottom and left side. Draw in the c.d.f. of the spinning pointer, a "ramp" from (0, 0) to (1, 1). Obtain 20 observations on the spinning pointer from the random number table (Table VIII) and plot them as points along the bottom of your square (the x-axis). Construct the corresponding sample distribution function, a step function starting at (0, 0) and jumping up $\frac{1}{20} = .05$ at each point where an observation is marked. Determine the value of D_{20} by inspection of your graph and carry out the test of goodness of fit of the random numbers to the ramp model at $\alpha = .05$.

7.38. (a) Obtain a sample of 10 normal random numbers from Table IX and plot the sample distribution function. Plot a standard normal c.d.f. on the same axis and determine D_{10}, using this value to test the goodness of fit at $\alpha = .10$.

(b) Have the instructor collect values of D_{10} from all students in the class. These constitute a sample of D_{10}-values that provide an empirical distribution of D_{10}. In particular, roughly 10 per cent of them in a large class should exceed the tabulated critical values. Check this and some other percentiles.

7.39. In Problem 7.19 it was assumed that the observations

45, 45, 50, 51, 49, 45, 48, 40, 45, 45, 65, 55, 40

were from a normal population. Test this hypothesis at $\alpha = .10$.

7.40. Problem 7.21 assumed that the birth-weight data of Problem 7.5 were taken from a normal population. Test this hypothesis at $\alpha = .10$.

8 Inference for Categorical Populations

The model for a categorical population consists of a list of the distinct, possible categories, together with a probability for each one—these probabilities necessarily adding up to 1. Data from such a population is collected by enumerating the number of times each category turns up in a sample of given size. These frequencies are random—they will vary from sample to sample, and the frequencies summarizing the results of a given sample represent just one of the many possible sets of frequencies that could have occurred (and would occur if the sampling process were repeated). Because the true probabilities of the various possible categories are long-run limits of relative frequencies of the categories, inferences are usually based on the particular set of frequencies that summarize the data.

8.1 Estimating a Proportion

An important, simple, special case of a categorical population is that of a dichotomous population, one in which there are just *two* categories. The toss of a coin is a common example of such an experiment, but there are countless others:

A missile shot succeeds or fails in its mission.
An inoculation takes effect or does not.
A manufactured article is defective or good.
A person picked at random is male or female.
A person picked at random favors candidate A or favors candidate B.
A man's height is too small for him to join the army or it is not.
The American League team wins the world series or the National League team does.
A student either drops out of a college program or completes the program.

The reader could add many more examples to this list.

When there are but two outcomes or categories, the complete probability model for the population is defined by the probability of one of them, since the two probabilities involved add up to 1. The categories are often coded 1 and 0, a coding that serves to count (in collecting data) the occurrences of the outcome coded 1. The probability of the outcome coded 1 is usually called p and the other one $q = 1 - p$. Probabilities p and q are thought of as *population proportions*, being the long-run limits of the corresponding proportions in a sequence of trials or observations. Moreover, in the case of drawing an object or person from an actual finite population of objects or people, the probability of observing category 1 is just the proportion of objects or people in the population that fall in that category—it *is* a population proportion.

The coding of the two categories with numerical values converts the population to a numerical population, thereby opening the way for the application of powerful methods developed for numerical populations. In particular, means and variances can now be computed, both for populations and for samples. On the other hand, the numerical structure thus imposed on the population is not so complicated as to be restrictive in its artificiality. That is, a numerical space with only *two* elements (0 and 1) does not even have a meaningful order relation, so there is no artificial ordering of categories.

The population mean and variance are especially easy to compute for a population variable with just two values, 1, and 0. The probability table and necessary products and sums are as given in Table 8.1. Thus the expected value or population mean is

$$E(X) = 1 \cdot p + 0 \cdot q = p,$$

Table 8.1

OBSERVATION	PROBABILITY	(OBS.) × (PROB.)	(OBS.)^2 × (PROB.)
1	p	p	p
0	q	0	0
Sum	1	p	p

and the expected square is

$$E(X^2) = 1^2 \cdot p + 0^2 \cdot q = p,$$

where X has been used to denote the observation (1 or 0) resulting from a single performance of the experiment, or single draw from the population. The population variance is then the difference between the average square and the square of the average:

$$\begin{aligned} \operatorname{var} X &= E(X^2) - [E(X)]^2 \\ &= p - p^2 = pq. \end{aligned}$$

Computations of the mean and variance of a sample are equally simple and parallel closely the population computations. If f denotes the frequency of 1's in a sample of n observations, the data and computations can be summarized as in Table 8.2. The sample mean is then

$$\overline{X} = \frac{1}{n} \sum (\text{Obs.}) \times (\text{Freq.}) = \frac{f}{n},$$

Table 8.2

OBSERVATION	FREQUENCY	(OBS.) × (FREQ.)	(OBS.)^2 × (FREQ.)
1	f	f	f
0	$n - f$	0	0
Sum	n	f	f

and the sample average square is

$$\overline{X^2} = \frac{1}{n} \sum (\text{Obs.})^2 \times (\text{Freq.}) = \frac{f}{n}.$$

The sample variance is then the average square minus the square of the average:

$$S^2 = \overline{X^2} - (\overline{X})^2 = \frac{f}{n} - \left(\frac{f}{n}\right)^2$$
$$= \frac{f}{n}\left(1 - \frac{f}{n}\right).$$

Observe that these sample moment formulas are just like the population formulas, with the population parameter p replaced by the sample relative frequency.

The reason for giving these computations at this point is to show that the relative frequency of 1's, which is a natural statistic to use in making inferences about the population proportion p, is just the sample mean. This implies that the general facts about sample means given earlier would apply here to sample relative frequencies. In particular, the following facts are useful:

1. As a random variable, the relative frequency f/n has a distribution centered at the population mean, p:

$$E\left(\frac{f}{n}\right) = p.$$

2. The variance of the relative frequency f/n is the population variance divided by the sample size:

$$\text{var}\left(\frac{f}{n}\right) = \frac{pq}{n}.$$

3. The mean-squared error (mse) involved in estimating p by f/n is just the variance of f/n:

$$\text{mse} = \frac{p(1 - p)}{n}.$$

4. The standard error of estimate, in estimating p by f/n, is the square root of the mse with p replaced by f/n:

$$\text{standard error} = \sqrt{\frac{(f/n)(1 - f/n)}{n}},$$

which, incidentally, is the sample standard deviation divided by $\sqrt{n}$.

EXAMPLE 8.1

In 1972, based on interviews with 1159 registered voters during the period June 16 to June 19, the Gallup Poll found that 53 per cent would vote for President Nixon against Senator George McGovern (with 37 per cent for McGovern and 10 per cent undecided). How accurate is the 53 per cent figure?

The accuracy can be described in terms of the standard error of estimate. The variance of the sample proportion is

$$\frac{pq}{n} = \frac{p(1-p)}{1159},$$

where p, of course, is unknown. Using the sample mean .53 in place of p to obtain an approximate value of this variance, one has

$$\frac{pq}{n} \doteq \frac{.53 \times .47}{1159} = .0002149,$$

and the square root of this is the standard error of estimate:

$$\text{standard error} = \sqrt{\frac{.53 \times .47}{1159}} = .01466.$$

Thus the observed 53 per cent involves an "average" (rms) error* of about 1.5 percentage points.

8.2 Confidence Interval for a Proportion

Because a population proportion is the population mean (when the outcomes are coded 1 and 0, as in the preceding section), the large-sample technique for obtaining a confidence interval for a population mean in general can be applied to obtain a confidence interval for a proportion (see Section 7.3). In general, a sample mean $\bar{X}$ is approximately normally distributed, in large samples, with mean μ (the population mean) and variance σ^2/n (the population variance divided by the sample size). In the special case of a 0–1 population, then, the sample relative frequency f/n (i.e., relative frequency of 1's) is approximately normally distributed with mean p (the population mean) and variance pq/n (the population variance divided by the sample size).

A normal distribution has the property that about 95 per cent of the probability is within 2 standard deviations of the mean. Hence (for large n)

$$P\left(\left|\frac{f}{n} - p\right| < 2\sqrt{\frac{pq}{n}}\right) \doteq .95.$$

* The computation here assumes that the 1159 voters constitute a random sample from the population of all voters. In fact, the Gallup Poll and other polling organizations use more sophisticated methods of sampling designed to reduce the error of estimation; so the reported 53 per cent may actually be somewhat more reliable than is suggested by the standard error of .0147. The news reports of such polls seldom include any discussion of possible errors.

(That is, the probability is about 95 per cent that the mean f/n is within 2 standard deviations of the center of its distribution.) The inequality inside the parentheses can be rewritten, by transposing various terms, with p in the middle:

$$\frac{f}{n} - 2\sqrt{\frac{pq}{n}} < p < \frac{f}{n} + 2\sqrt{\frac{pq}{n}},$$

an inequality that would also have probability .95. This begins to look like what is needed to define confidence limits, but the extremes involve p, the unknown parameter; the simple tranpositions of terms did not really solve the inequality for p. However, in large samples (and the sample size is assumed large just to use the normal approximation) the quantity pq can be replaced by its sample version $(f/n)(1 - f/n)$, without appreciably disrupting things; thus the probability .95 also applies approximately to the inequality

$$\frac{f}{n} - 2\sqrt{\frac{(f/n)(1 - f/n)}{n}} < p < \frac{f}{n} + 2\sqrt{\frac{(f/n)(1 - f/n)}{n}}.$$

The extremes in this inequality, computable from the sample, define a random interval that has the probability .95 (approximately) of covering p. It is a 95 per cent confidence interval for p. The confidence coefficient .95 corresponds to the multiplier 2 in front of the standard deviation of f/n; other confidence coefficients, of course, would require correspondingly different multipliers:

Confidence coefficient	.99	.95	.90	.80
Multiplier	2.58	1.96	1.64	1.28

EXAMPLE 8.2

A newspaper report (October 1972) of an article in the medical journal *Pediatrics* by Dyment and Botan says that "sneakers are sufficient." The question asked of 500 randomly selected pediatricians was this: "Do you think that rubber-sole sneakers (tennis shoes) are adequate for infants with normal feet?" Seventy-six per cent of those answering said yes, but only 279 of the 500 responded.

The population is the population of all pediatricians, and the characteristic of interest is a pediatrician's opinion about sneakers. The parameter p is the probability that a pediatrician picked at random from the population will say that sneakers are adequate. If the 279 responders are considered to constitute a random sample from the population of pediatricians, then the sample proportion of 76 per cent leads to the following 95 per cent confidence interval for the population proportion p:

$$.76 - 2\sqrt{\frac{.76 \times .24}{279}} < p < .76 + 2\sqrt{\frac{.76 \times .24}{279}},$$

or approximately

$$.734 < p < .786.$$

(The assumption that the 279 replies constitute a proper sample might be challenged with some justification; for, if there happened to be any relation between having a given opinion on sneakers and a willingness to return a questionnaire about sneakers, then the sample used in constructing the confidence interval may be a sample from the wrong population.)

Problems

8.1. A multiple-choice exam contains 75 questions, each with four choices (one correct). A student who has neglected to study picks an answer at random for each question. What is the probability that his score exceeds 40 per cent?

8.2. A random sample of 100 voters is taken from the voters of a large city to determine the voter sentiment in a race for mayor. Determine approximately the probability that fewer than 75 of the 100 voters favor the incumbent mayor if, in fact, the population proportion favoring him is .80?

8.3. A random sample of 400 voters in a city included 180 Democrats. Give an estimate of the population proportion of Democrats, together with the standard error of estimate.

8.4. A survey using a random sample of 1600 households in a city shows 1280 with television sets. Determine a 95 per cent confidence interval for p, the proportion that have television sets among all households in the city.

8.5. If 15 of 100 college students smoke cigarettes, how accurate an estimate of the population proportion does this provide? Construct a 90 per cent confidence interval for that proportion. Answer the same question if the result were 150 of 1000.

8.6. A college newspaper published the results of a poll of 60 students from the student population of 33,000. Thirty-five per cent thought the administration was doing a fair job, or better, in protecting students' rights of privacy concerning access to student files by outside agencies. How reliable is the result?

8.3 Z-test for $p = p_0$

The large-sample approximate normality of the relative frequency f/n of an event can be used to test a null hypothesis of the form $p = p_0$, where p denotes the probability of the event. The test is based on the closeness of f/n to p_0, as measured in units of the standard deviation of f/n, this being a special case of testing $\mu = \mu_0$ where μ is a population mean. (The relative frequency is the sample mean in this case, and $\mu = p$.) The statistic used, then, is this:

$$Z = \frac{f/n - p_0}{\sqrt{[p_0(1 - p_0)]/n}},$$

whose large-sample distribution is approximately standard normal (mean 0, variance 1) if $p = p_0$.

If the alternative to $p = p_0$ is two-sided, $p \neq p_0$, then both large positive and large negative values of Z call for rejection of H_0:

$$\text{reject } H_0 \text{ if } \quad |Z| > K,$$

where, for a given significance level, α, K is chosen so that

$$P_{H_0}(|Z| > K) = \alpha.$$

Figure 8.1 shows the density of Z under H_0 and the areas that total α. (It should be a familiar picture by now.)

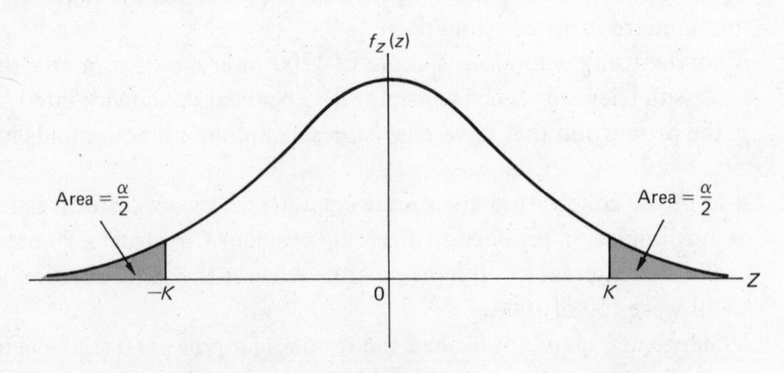

Figure 8.1
Rejection probability under H_0 (two-sided test).

If the alternative is one-sided, $p > p_0$ (or $p < p_0$), the values of Z calling for rejection of p_0 would lie on one side, $Z > K$ (or $Z < K$), and the probability α would be a tail area on one side.

EXAMPLE 8.3
Suppose that in June 1972, President Nixon's campaign staff decided to test the hypothesis

$$H_0: \quad p \leq .5$$

against the alternative

$$H_1: \quad p > .5,$$

using the data gathered by the Gallup organization, as given in Example 8.1. Here p denotes the probability that a person drawn at random from the population of all voters would favor Nixon over McGovern; it is the population proportion of voters favoring Nixon.

The poll showed that 53 per cent of the 1159 persons were in favor of Nixon over McGovern (as of the middle of June). The relevant statistic is

$$Z = \frac{.53 - .50}{\sqrt{(.50 \times .50)/1159}} \doteq 2.05.$$

That is, 53 per cent is a little over 2 standard deviations greater than $p = .5$. The test for H_0 against the one-sided alternative $p > .5$ would be to reject H_0 if Z is too large, i.e., if $Z > K$; and the significance level α would then be the probability that $Z > K$ when actually $p = .5$. This probability is a tail area, shown in Figure 8.2, and can be found in

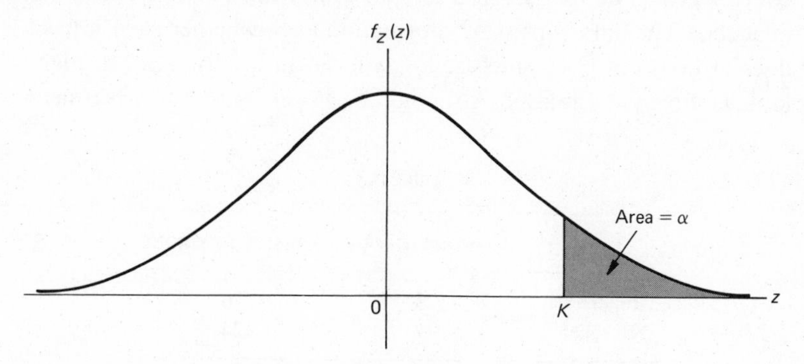

Figure 8.2
Rejection probability under H_0 (one-sided test).

Table I of the Appendix by entering the table at K. For $\alpha = .05$, K is 1.645, a value that is exceeded by the observed Z. The hypothesis would be rejected at $\alpha = .05$; the observed 53 per cent is said to be *significantly* greater than 50 per cent at the 5 per cent level of significance.

8.4 Comparing Proportions

Again using the fact that population proportions are population means (with the usual 0–1 coding), one can devise a large-sample test for the equality of two population proportions by specializing the test given earlier (Section 7.5) for comparing two population means.

Although two populations often exist side by side and so naturally demand comparison in regard to some characteristic (e.g., males and females), it is also frequently the case that one population is treated in some way. The other population is untreated (the control), and the purpose of comparing such populations is to study the effect of the treatment.

EXAMPLE 8.4

It is asserted that girls' hair is more often blond than boys' hair. To challenge or confirm such an assertion, one would want to test the hypothesis that the proportion of girls with blond hair (population 1) is greater than the proportion of boys with blond hair (population 2). A sample from each population would be needed.

EXAMPLE 8.5

To test the effectiveness of a hay fever remedy, 300 hay fever sufferers are divided randomly into a group of 200, given the remedy, and a control group of 100, given a similar-appearing inactive substance (placebo). The data consist of information as to whether each subject does or does not have relief from hay fever during the year. It might look as shown in Table 8.3. Although the 38 per cent for the treatment

Table 8.3

	CONTROL GROUP	TREATMENT GROUP
Relief	23	76
No relief	77	124
Sum	100	200

group is surely greater than the 23 per cent for the control group, one can ask whether it is a "significant" difference, large enough to justify the conclusion that the remedy is effective, or perhaps only typical of the kind of variability attributable to randomness (i.e., to sampling fluctuations).

The large-sample test for the equality of two population proportions is based on a comparison of the corresponding relative frequencies. Let the population proportions be denoted p_1 and p_2, corresponding to frequencies f_1 and f_2, based on random samples of sizes n_1 and n_2. The *null hypothesis* is

$$H_O: \; p_1 = p_2$$

and the alternative may be two-sided or one-sided:

$$H_1: \; p_1 \neq p_2, \qquad \text{or} \qquad H_1: \; p_1 > p_2 \quad \text{(say)}.$$

The sample relative frequency f_1/n_1 is approximately normal, with mean p_1 and variance $p_1(1 - p_1)/n_1$, and f_2/n_2 is approximately normal with mean p_2 and variance $p_2(1 - p_2)/n_2$. Under H_O, the two proportions are equal; let p denote their common value:

$$p = p_1 = p_2 \qquad \text{(under } H_O\text{)},$$

in terms of which the variances of f_1/n_1 and f_2/n_2 are

$$\operatorname{var} \frac{f_i}{n_i} = \frac{p(1 - p)}{n_i}, \qquad (i = 1, 2).$$

The difference in sample proportions,

$$D \equiv \frac{f_1}{n_1} - \frac{f_2}{n_2},$$

is then approximately normal with mean

$$E(D) = E\left(\frac{f_1}{n_1}\right) - E\left(\frac{f_2}{n_1}\right) = p_1 - p_2$$

and variance

$$\begin{aligned}
\operatorname{var}\left(\frac{f_1}{n_1} - \frac{f_2}{n_2}\right) &= \operatorname{var}\left(\frac{f_1}{n_1}\right) + \operatorname{var}\left(\frac{f_2}{n_2}\right) \\
&= \frac{p(1 - p)}{n_1} + \frac{p(1 - p)}{n_2} \\
&= p(1 - p)\left(\frac{1}{n_1} + \frac{1}{n_2}\right).
\end{aligned}$$

Of course, p is not known, so it is approximated by the best available estimate, namely, the proportion of 1's in the *combined* sample:

$$p \doteq \frac{f_1 + f_2}{n_1 + n_2}.$$

The test statistic used is the standardized difference, a quantity that is approximately standard normal under the null hypothesis,

$$Z \equiv \frac{D}{\sqrt{\text{var } D}},$$

where it is understood that var D is approximated using the combined sample proportion. If this is too large, either positively or negatively, one rejects H_O against the two-sided H_1. If it is too large positively, one rejects H_O against a one-sided alternative. That is, the tests are as follows:

$$p_1 = p_2 \quad \text{vs.} \quad p_1 \neq p_2: \quad \text{reject } p_1 = p_2 \quad \text{if} \quad |Z| > K_1$$

and

$$p_1 = p_2 \quad \text{vs.} \quad p_1 > p_2: \quad \text{reject } p_1 = p_2 \quad \text{if} \quad Z > K_2,$$

where K_1 and K_2 are chosen to meet a specified criterion for the type I error size, α.

EXAMPLE 8.6

Suppose that 20 in a sample of 100 boys have blond hair and 13 of 50 girls are blondes. On the basis of this evidence, could it be concluded that the proportion of blond persons is different for girls than for boys?

The difference between the sample proportions is

$$D = \tfrac{20}{100} - \tfrac{13}{50} = -.06,$$

but this is an observation on a random variable (D) whose mean is the difference between population or "true" proportions, and whose variance is

$$\sigma_D^2 = \frac{p_B(1 - p_B)}{n_B} + \frac{p_G(1 - p_G)}{n_G}$$

$$\doteq \frac{.22 \times .28}{100} + \frac{.22 \times .28}{50} = (.072)^2.$$

[Here, .22 is the proportion of blonds in the combined sample, namely, $(20 + 13)/(100 + 50) = \frac{33}{150}$.] Hence the difference, $-.06$, measured in units of σ_D is

$$Z \doteq \frac{-.06}{.072} = -.83,$$

which says that for the given data D is not quite 1 standard deviation to the left of its expected value (zero). If a significance level of 5 per cent is used, it takes an observed D-value which is at least 2 standard deviations removed from the expected zero. So $-.06$ is not significant at the 5 per cent level of significance; the evidence does not warrant (at this level) the inference that the proportion of blonds is different among boys than among girls.

Problems

8.7. The President claims that a majority of the people favor his policy. A public opinion poll based on 1500 persons shows 49 per cent in favor of his policy. Could he be right?

8.8. In a road test by 90 drivers, 57 preferred medium-priced car M to a $16,000 limousine for overall ride. Is this grounds, at the 5 per cent level of significance, to reject the hypothesis that car M rides no better than the limousine? In judging steadiness, 42 of 77 preferred car M; can it be concluded that M is better in steadiness?

8.9. For a number of years the proportion of students entering an engineering school who survive the first year was observed to be about 65 per cent. The first year following concerted efforts at improving this record (by more diligent screening, counseling, etc.), the survivorship rate turned out to be 70 per cent in a class of 800. Is that a significant improvement?

8.10. A study of the Opinion Research Corporation reports responses to the question, "Where would you be most likely to find out all there is about some news you are very much interested in?" Out of 749 responses from persons with incomes of $10,000 and over, 55 per cent replied "Newspapers"; of 1234 responses from persons with incomes of less than $10,000, 50 per cent replied "Newspapers." Test the hypothesis that this preference for newspapers as a source is the same for both income groups.

8.11. Suppose that two public opinion polls, each based on 1000 interviews, show 48 and 51 per cent, respectively, in favor of a certain proposition. Is there reason to believe the populations being sampled are not the same?

8.12. In a study of the effectiveness of a drug for relieving pain, 35 of 50 reported relief after taking the drug, while 28 of 50 reported relief after taking a placebo that looked the same as the drug. (The 100 subjects did not know that any placebo was being substituted.) Is the drug more effective than the placebo?

8.5 Goodness of Fit

Sometimes a model for a discrete population can be derived or postulated from theoretical or other a priori considerations. In such a case it is essential to check the compatibility of the postulated model with the results of actual experiment, to see if the data "fit" the model (or vice versa).

EXAMPLE 8.7
Considerations of geometrical symmetry suggest that the probabilities for the six sides of a *die* are all equal. Thus, if the model is as shown in Table 8.4, then the hypothesis to be tested, suggested by a priori

Table 8.4

OUTCOME	PROBABILITY
1	p_1
2	p_2
3	p_3
4	p_4
5	p_5
6	p_6
	1

reasoning, is this:

$$H_O:\ p_1 = p_2 = p_3 = p_4 = p_5 = p_6 = \tfrac{1}{6}.$$

The general alternative hypothesis would be that this is not correct.

EXAMPLE 8.8
A classical example of the application of the goodness-of-fit test is in the study of genetic models. In crossing two types of corn, four distinct types appear in the offspring: green, golden, green-striped, and golden-green-striped. In a simple Mendelian model, these occur in the proportion 9:3:3:1. This means that probabilities for the four categories under the null hypothesis (that this model is correct) are $\frac{9}{16}, \frac{3}{16}, \frac{3}{16}, \frac{1}{16}$ (where, of course, $16 = 9 + 3 + 3 + 1$); see Table 8.5.

The classical *chi-square goodness-of-fit test* was devised about 1900 by K. Pearson. It is based on a comparison between observed frequencies of categories and corresponding expected frequencies under the hypothesis to

Table 8.5

CATEGORY	PROBABILITY (H_O)
Green	$\frac{9}{16}$
Golden	$\frac{3}{16}$
Green-striped	$\frac{3}{16}$
Golden-green-striped	$\frac{1}{16}$
Total	1

be tested. That is, if the probability of a certain category in the model of the null hypothesis is p_0, the expected value of the frequency of observations falling in that category is np_0, in a sample of n observations. Let the notation be as follows:

O_i = observed frequency of category i,

e_i = expected frequency of category i under the null hypothesis,

k = number of categories.

The statistic used to measure fit is

$$\chi^2 \equiv \sum_{i=1}^{k} \frac{(O_i - e_i)^2}{e_i}.$$

If there is a good fit, then the observed and expected frequencies will not differ by much, and χ^2 will be small; if the fit is poor, χ^2 will be large. The null distribution is therefore rejected if $\chi^2 > K$, where K is a constant chosen so that the test has a prescribed significance level, α.

Once again, before the test statistic is usable in inference, its distribution or pattern of variation from sample to sample must be known. This distribution is actually quite complicated, but it is a redeeming feature of the statistic that in the case of large samples its distribution under H_O depends (approximately) only on k, the number of categories. It can be shown, indeed, that the large-sample distribution of χ^2 is approximately what is called a *chi-square distribution with $k - 1$ degrees of freedom.* These distributions, for different numbers of *degrees of freedom* (a parameter that indexes this family of sampling distribution), are given by percentiles in Table II (Appendix). The significance level of this test is the probability that the statistic falls in the critical region (and so causes rejection of H_O) when H_O is true.

EXAMPLE 8.9

When a die is tossed 120 times, the expected frequency of each side is $120 \times \frac{1}{6} = 20$, under the hypothesis that the die is "straight" (not loaded). Therefore, to test this hypothesis, actual frequencies are compared with 20. Suppose that the results are as follows:

Side	1	2	3	4	5	6
Frequency	14	17	26	23	23	17

The test statistic is

$$\chi^2 = \frac{(14-20)^2}{20} + \frac{(17-20)^2}{20} + \frac{(26-20)^2}{20} + \frac{(23-20)^2}{20}$$

$$+ \frac{(23-20)^2}{20} + \frac{(17-20)^2}{20}$$

$$= 5.4.$$

This number is to be considered, as usual, with respect to what could happen when the die is straight, i.e., with the sampling distribution of the statistic χ^2. Under H_O, this is chi-square with 5 degrees of freedom; the percentiles are given in Table II, and the density function is sketched in Figure 8.3. In the figure, the experimental result $\chi^2 = 5.4$ is marked, as is the value $K = 9.24$, which corresponds to $\alpha = .10$. Since the observed 5.4 does not exceed the 90th percentile (9.24), the null hypothesis (that the die is straight) is accepted.

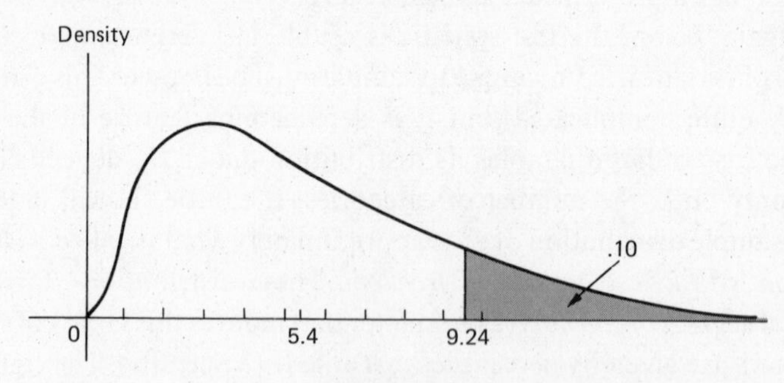

Figure 8.3
Chi-square density, 5 degrees of freedom.

It might be mentioned that the term *expected frequency* should not be taken at face value. Any "expected" value is not really expected in the ordinary sense of the word, at least, not in a single trial. Rather, it refers to the center of the distribution and is expected only in connection with a great many trials, as a long-run average. The number 20 in Example 8.9 is the center of the distribution of each category's frequency; even with a straight die, some faces will occur fewer and others more than 20 times in 100 trials. As a matter of fact, to observe exactly 20 in each category would be unusual indeed. More generally, a value of χ^2 near zero is unusual, as seen from an inspection of the density in Figure 8.3. (On the other hand, although large values of χ^2 are rare, and taken as reason to reject H_0, it would be unreasonable to put the rare small values into a critical region, since they correspond to a very good fit. This is another illustration of the fact that improbability alone is not sufficient as a basis for rejection.)

The chi-square test for goodness of fit can be adapted to a variety of problems in which the theoretical probabilities of the categories to be tested as H_0 are not given numbers but given functions of one or more population parameters. In such a case, it is necessary to use the sample to estimate the unknown parameters so that "expected frequencies" e_i can be calculated for use in the statistic χ^2. This done, the statistic χ^2 still has a chi-square distribution but with the $k - 1$ (degrees of freedom) reduced by the number of estimated parameters. This modification will be used in the next section.

8.6 Comparing Categorical Populations

When two populations are not numerical, it does not make sense to compare such things as "locations" or "variabilities"; these are undefined. About all that one can do is to compare corresponding probabilities of categories. (It would be assumed that to be considered for comparison in the first place, the same set of categories would apply to each population.)

A simple, special case is that in which there are only two categories, a case handled earlier under the heading "Comparing Population Proportions." There, to be sure, the probabilities of the categories were compared, but the comparison was accomplished by interpreting the probability of success as a population mean, with a suitable coding of categories (1 for success, 0 for failure). When there are more than two categories, this approach breaks down.

The approach to be considered here applies for any number of categories but will be described first for the case of two. Suppose that data are collected as shown in Table 8.6. If the population proportions (probabilities) were the same, the samples could be logically combined to yield the estimate $\frac{11}{14}$ for the probability of success. If that *were* the common p, the expected frequencies in nine trials and five trials would be as shown in Table 8.7. Thinking

Table 8.6

	SAMPLE FROM POPULATION 1 (FREQUENCY)	SAMPLE FROM POPULATION 2 (FREQUENCY)	COMBINED SAMPLE
Success	7	4	11
Failure	2	1	3
Total	9	5	14

Table 8.7

	EXPECTED FREQUENCY (SAMPLE 1)	EXPECTED FREQUENCY (SAMPLE 2)
Success	$9 \times \frac{11}{14}$	$5 \times \frac{11}{14}$
Failure	$9 \times \frac{3}{14}$	$5 \times \frac{3}{14}$
Total	9	5

of these as cell frequencies as in a goodness-of-fit test, one might construct this statistic (weighted sum of squares of observed minus expected):

$$\chi^2 = \frac{(7 - 9 \times \frac{11}{14})^2}{9 \times \frac{11}{14}} + \frac{(2 - 9 \times \frac{3}{14})^2}{9 \times \frac{3}{14}} + \frac{(4 - 5 \times \frac{11}{14})^2}{5 \times \frac{11}{14}} + \frac{(1 - 5 \times \frac{3}{14})^2}{5 \times \frac{3}{14}}$$

$$= \frac{1}{9 \times 14}\left(\frac{1}{11} + \frac{1}{3}\right) + \frac{1}{5 \times 14}\left(\frac{1}{11} + \frac{1}{3}\right) = \frac{14}{33 \times 45}.$$

But this is precisely the value of the statistic Z^2 used in Section 8.4 based on the difference between sample proportions:

$$Z^2 = \frac{(\frac{7}{9} - \frac{4}{5})^2}{\frac{11}{14} \times \frac{3}{14}(\frac{1}{9} + \frac{1}{5})} = \frac{14}{33 \times 45}.$$

The point of this little exercise in arithmetic is first to illustrate what can be proved by simple algebra—that for two categories the χ^2-statistic will always turn out to be Z^2. Then, it is observed that the χ^2-statistic used is easily generalized to the case of more than two categories. The resulting statistic, for k categories, will have approximately (as it turns out) a χ^2 distribution with $k - 1$ degrees of freedom under the null hypothesis, when the sample size is large.

EXAMPLE 8.10

Suppose that 100 persons are selected from each of two wards and queried about their political leanings, with the results as in Table 8.8.

Table 8.8

	WARD 1	WARD 2	COMBINED SAMPLE
Republican	48	44	92
Democrat	45	39	84
Independent	5	11	16
Miscellaneous	2	4	6

Do these results support (or reject) the claim that the two wards are essentially the same in political composition? The estimates of proportions from the combined sample are .46, .42, .08, and .03 (combined sample frequencies divided by 200), and the expected frequencies in any sample of size 100 are 46, 42, 8, and 3, respectively. Then

$$\chi^2 = \frac{(48-46)^2}{46} + \frac{(45-42)^2}{42} + \frac{(5-8)^2}{8} + \frac{(2-3)^2}{3} + \frac{(44-46)^2}{46}$$

$$+ \frac{(39-42)^2}{42} + \frac{(11-8)^2}{8} + \frac{(4-3)^2}{3}$$

$$= 3.52.$$

Under H_0 this statistic would have approximately a chi-square distribution with 3 degrees of freedom (number of categories less 1), whose 95th percentile is 7.81. Since $7.81 > 3.52$, the observed χ^2 is not "significant" at the 5 per cent level, so the null hypothesis is accepted.

Problems

8.13. (a) Use the chi-square test for goodness of fit of a sample of 100 random digits (Table VIII) to the model with equal probabilities assigned to the 10 digits.

(b) Have the instructor collect values of χ^2, such as the one you computed in (a), one from each member of the class, and from these construct an empirical sampling distribution for χ^2 (which should resemble the chi-square distribution with 9 degrees of freedom).

8.14. Obtain an ordinary die and, using 100 tosses as your data, test the hypothesis that it is a fair die. [If you cannot locate a die, you can simulate die tossing by using random digits (ignoring 0, 7, 8, and 9)].

8.15. In 30 settings of gill nets in Lake Minnewaska, the following counts of northern pike were obtained:

Number of fish, x_i	0	1	2	3	4
Frequency of sets containing x_i fish	12	10	6	1	1

Test the hypothesis that the model for the number of pike per set is as follows:

Number of fish	0	1	2	3	4 or more
Probability	.368	.368	.184	.061	.019

8.16. Toss a coin 100 times and record the frequencies of heads and tails.

(a) Compute the chi-square statistic for testing the hypothesis that $p_H = p_T = \frac{1}{2}$.

(b) Compute the Z-statistic (as in Section 8.3) for testing the hypothesis that $p_H = .5$. Is there any relation between this Z and the chi-square value you formed in (a)? (Try squaring Z.)

8.17. A project in art education involved two kindergarten classes of 24 children each, an AM control group, and a PM experimental group given a new form of art instruction. Both classes took a pretest and posttest, consisting of drawing a truck, the drawings being judged on a scale from 0 to 10. The gains in score following the period of experiment were summarized as shown in Table 8.9.

(a) Use the chi-square test for the hypothesis that these are samples from the same discrete population. (*Note:* You will find several expected frequencies less than 1, perhaps too many for a valid test. Some say that these should all be 5 or more, but studies have shown that *some* may be as small as 1 without jeopardizing the test.)

(b) Use the two-sample t-test for the hypothesis that the population means are the same. (But, can you really assume equal population variances?)

(c) (For discussion.) Which of (a) or (b) is the more appropriate tests? (Some things to consider: What are the alternative hypotheses? Is the gain really a discrete variable? Is the scale meaningful; is a 4 twice as good as a 2? Is the population of gains normal?) Notice

Table 8.9

AM CLASS		PM CLASS	
GAIN	FREQUENCY	GAIN	FREQUENCY
−1	1	0	1
0	8	1	6
1	9	2	4
2	4	3	4
3	1	4	6
4	1	5	1
		6	1
		7	1
	24		24

that you *can* carry out the test in each case, whether or not it is appropriate, and this is often done. (Is anything proved if the conditions for a test are not met? Or if they are met?)

8.18. Repeat one of Problems 8.10 through 8.12 using the chi-square statistic of Section 8.6 for the test. (Compare the value of χ^2 found with Z^2, where Z was the statistic used when you did the problem earlier.)

8.19. Two samples, one of Democrats and one of Republicans, were obtained to determine the popularity of the President, with results as given in Table 8.10. Are the corresponding populations (of Republicans and of Democrats) the same?

Table 8.10

	DEMOCRATS	REPUBLICANS
Think he is doing a good job	30	48
Marginal or noncommittal	20	46
Think he is doing a poor job	50	6
	100	100

Correlation and Independence

Perhaps the most interesting and important statistical problems are those involving relationships among several variables. The simplest cases are those involving just two variables, and these are of two types: (1) those in which both variables are random variables, best represented by the mathematical structure of a probability or stochastic model; and (2) those in which one variable is an adjustment or a dial setting (a *controlled* variable that is to be set and fixed at several levels in the course of collecting data) and the other variable is a *random* variable (a response, the model for whose randomness is presumably determined by the setting of the controlled variable). Chapter 10 deals with the second type of problem, and the present chapter takes up the first, the problem of inference for a bivariate population.

A *bivariate population*, or *bivariate distribution*, is the model for an experiment of chance in which *two* measurements are made on each experimental unit. Examples are numerous:

1. For each individual drawn at random from a population, two measurements are made, height and weight. (Two numerical variables)
2. A student achieves a certain score on a college entrance examination and goes on to attain a final grade-point average upon graduation. (Two numerical variables)
3. Each person in a population can be categorized according to two schemes, eye color and sex. (Two categorical variables)
4. Each person in a population can be "measured" according to IQ and according to hair color. (One numerical and one categorical variable)

Problems of inference for bivariate populations usually are concerned with the *existence* of a relationship between the two variables and with the *strength* of the relationship (if there is one). If there is a relationship between the two variables, one might expect to be able to use information about one of the variables to *predict* the value of the other. And indeed, that is possible to some extent; but the methods are identical with methods used when the predicting variable is a controlled variable, and these will be taken up in Chapter 10.

Independence is the word used to indicate a complete lack of relationship between two variables, whereby information about the value of one of the variables does not affect the probability model used for the other variable. If one has observed the first variable in the bivariate pair (X, Y) and found the value of X to be (say) 3, he knows nothing more and nothing different about Y than he knew before learning that $X = 3$. Or, in general, any event that has to do only with the value of X is independent of any event that has to do only with the value of Y. A common example of independent random variables is that of the toss of two dice; ideally, at least, the number of points showing on one die is independent of the number showing on the other die, in the sense that if someone announces the number of points on one die, this does not change the fact that the appropriate model for the other postulates equal likelihood for its six faces.

When there *is* a relationship between two random variables, it can be simple or complex. The simplest kind of relationship is that in which the two variables tend to vary together (or to vary reciprocally). It may be that a large value of one implies a tendency for the other to be large also, and a small value of one implies a small value of the other. (Or, a large value of one may imply a tendency toward a small value for the other, and conversely.) In such instances one says that there exists a *correlation* between the variables.

When a correlation exists, the variables are not, cannot be, independent. But the *lack* of this type of relationship does, not, alas, imply complete independence, for there may be a dependency of another type. For example, suppose that you toss a dart at a circular target and count only tosses in which the dart hits the target. The amount of vertical miss need not be positive (a hit above the center line) just because the horizontal miss is positive (a hit to the right of center); but knowing the amount of the vertical miss does restrict the range of values of the horizontal miss (since darts that miss the circle altogether are not counted). The vertical miss and the horizontal miss are not independent, even though they may not be related in the sense of correlation—varying together or reciprocally.

9.1 Bivariate Distributions: Categorical

To see how to set up a model for a bivariate population, it is helpful to examine some data and how they are recorded. Observations (X, Y), in which both X and Y are *categorical*, can be tabulated in a *two-way table*, with the categories of X listed in (say) the horizontal margin, and the categories of Y in the vertical margin. When a particular combination of categories occurs, a mark is made in the corresponding cell of the table; and when the marks are counted, a frequency is obtained for each cell. The resulting table of frequencies is called a *contingency table*.

EXAMPLE 9.1
The students in a class of 39 students were classified according to eye color and hair color. A table with the categories red, blond, brown along one margin, and the categories blue, brown, hazel, green along the other margin, was prepared and filled in according to each student's characteristics. The counts observed were as shown in Table 9.1. Notice that in addition to the cell frequencies,

Table 9.1

	RED	BLOND	BROWN	
Blue	1	7	13	21
Brown	1	0	8	9
Hazel	0	2	5	7
Green	0	2	0	2
	2	11	26	39

there have been given sums of the frequencies in rows and in columns. The last column, for instance, gives the frequencies of students classified according to eye color only. And the 39 in the lower right corner is the sum of the row sums as well as the sum of the column sums. Each student falls into one of 4 categories when he is classified according to eye color, into one of 3 categories when classified according to hair color, and into one of 12 categories when classified according to the combination of hair and eye colors.

It is instructive also to divide each frequency in a contingency table by the number of observations, a process that yields *relative* frequencies; see Table 9.2. Notice that the row sums and column sums are themselves relative frequencies and that they each add up to 1.

Table 9.2

	RED	BLOND	BROWN	
Blue	$\frac{1}{39}$	$\frac{7}{39}$	$\frac{13}{39}$	$\frac{21}{39}$
Brown	$\frac{1}{39}$	0	$\frac{8}{39}$	$\frac{9}{39}$
Hazel	0	$\frac{2}{39}$	$\frac{5}{39}$	$\frac{7}{79}$
Green	0	$\frac{2}{39}$	0	$\frac{2}{39}$
	$\frac{2}{39}$	$\frac{11}{39}$	$\frac{26}{39}$	1

Just as the relative frequencies of a category in a single classification scheme (such as that of either margin in a two-way table) exhibit a long-run stability as more and more observations are taken, so the relative frequencies of the cells in a two-way classification have such long-run stability; for the *cells are simply categories in a more complicated classification*. The long-run limits of the cell relative frequencies are then *cell probabilities*. Of course, the marginal totals of cell relative frequencies are themselves relative frequencies (for the categories in the marginal classification schemes) and approach probabilities as long-run limits, which probabilities are sums of cell probabilities. Thus the *model* for the two-way classification is defined by a two-way table of probabilities.

EXAMPLE 9.2

When a student is classed according to both eye color and hair color, and if the student is picked at random from a certain population of students, the mathematical model describing what can happen when a single observation is made is given by a two-way table, Table 9.3. Here p_{ij} denotes the probability of eye color i together with hair

Table 9.3

	RED (1)	BLOND (2)	BROWN (3)	
Blue (1)	p_{11}	p_{12}	p_{13}	$p_{1\cdot}$
Brown (2)	p_{21}	p_{22}	p_{23}	$p_{2\cdot}$
Hazel (3)	p_{31}	p_{32}	p_{33}	$p_{3\cdot}$
Green (4)	p_{41}	p_{42}	p_{43}	$p_{4\cdot}$
	$p_{\cdot 1}$	$p_{\cdot 2}$	$p_{\cdot 3}$	1

color j, where the colors have been given numerical codes to avoid spelling them out or using the first letters (b could mean blue, blond, or brown). The marginal entries are totals and are themselves probabilities. For example,

$$p_{1\cdot} = p_{11} + p_{12} + p_{13} = P(\text{eye color } 1)$$

for blue-eyed students are either blue-eyed and red-headed, blue-eyed and blond, or blue-eyed and brown-haired, according to the scheme used in this situation. The reason for the dot in the subscript is that if one used simply p_1, it would not be clear whether this is the probability of eye color 1 or of hair color 1. The dot indicates which subscript has been summed out. The symbols p_{ij} are used in this table instead of numbers because, even though it is assumed that there *is* a probability of the combination eye color i with hair color j, the actual value of that probability is not known. Given the data in Example 9.1, one might hazard the guess (or estimate) that the probability of a blue-eyed blond is $\frac{7}{39}$, since this is the relative frequency of blue-eyed blonds in the sample of 39 students. But such a guess is fraught with the usual possibilities of error of estimation encountered in using a sample of only 39 observations.

In summary, the model for the two-way classification is a two-way table of probabilities, in which the marginal totals are themselves probabilities. And for each set of marginal probabilities, the sum is 1, which is also the sum of all the cell probabilities.

9.2 Independence of Classifications

One may notice in the two-way classification in the above examples that information is given about the proportion of the blonds who have blue eyes, the proportion of brown-eyed students who have brown hair, and so on. These proportions, in their ideal versions (thinking of a great number of observations, or of the model for a single observation) define what are called *conditional probabilities*. That is, if one looks only at blue-eyed students, or, in a sequence of observations of students generally, discards all results but those for blue-eyed students, the long-run relative frequency of blonds defines the conditional probability of blonds in the reduced population of blue-eyed students:

$$P(\text{blond} \mid \text{blue-eyed}) = \text{probability that a student has blond hair, } \textit{given} \text{ that he has blue eyes.}$$

(The vertical bar will always be used in this sense and is read "given.") Examination of the table in the examples shows that the conditional relative frequency of blonds among the blue-eyed students in the sample is

$$\frac{7}{21} = \frac{\frac{7}{39}}{\frac{21}{39}} = \frac{\text{rel. freq. of blue-eyed blonds}}{\text{rel. freq. of blue-eyed students}}.$$

Because of this kind of relationship, one defines conditional probabilities in the ideal model in similar fashion:

$$P(A|B) = \frac{P(A \text{ and } B)}{P(B)}.$$

It can happen that the conditional probability of category A (say) in one classification scheme, given that a certain category in the other scheme is observed, is the same as the probability of category A *without* any information about the other scheme: $P(A|B) = P(A)$. If events A and B stand in this relation, they are said to be *independent events*, and if this relation is true for every category A in one classification scheme and every category B in the other, the two classifications are said to be *independent*. Thus, if

$$P(\text{hair color } i \mid \text{eye color } j) = P(\text{hair color } i)$$

for every i and j, then hair color is said to be independent of eye color. To check such independence in a model, of course, one would have to know the actual probabilities, the long-run or ideal relative frequencies. On the other hand, even a finite sample can give some guidance.

EXAMPLE 9.3

To return to hair color versus eye color, Table 9.1 shows that $\frac{13}{21}$ of the blue-eyed students have brown hair, whereas $\frac{8}{9}$ of the brown-eyed students have brown hair and $\frac{26}{39}$ of all the students (in the sample) have brown hair. One might estimate the probabilities as follows:

$$P(\text{brown hair} \mid \text{blue eyes}) = \tfrac{13}{21},$$

$$P(\text{brown hair} \mid \text{brown eyes}) = \tfrac{8}{9},$$

$$P(\text{brown hair}) = \tfrac{26}{39}.$$

The relative frequency of brown-haired students among the blue-eyed students in the class is not very different from the relative frequency of brown-haired students in the class (.62 compared with .67). So possibly the events brown hair and blue eyes in the model could be independent. But $\frac{8}{9}$ is so different from $\frac{26}{39}$ that it would be indeed surprising if the event brown hair and the event brown eyes turned out to be independent in the model. And if they are not, the classification of students by hair color and the classification by eye color are not independent classification schemes.

The condition defining independence can be recast by using the definition of conditional probability. For events A and B to be independent, it must be that

$$P(A) = P(B|A) = \frac{P(A \text{ and } B)}{P(B)},$$

or

$$P(A \text{ and } B) = P(A)P(B).$$

In the case of independent classifications, then, in which every category of the one is independent of every category of the other, the structure of probabilities is such that every cell probability is a product of marginal probabilities. For classifications to be independent, and using the notation developed above,

$$p_{ij} = p_{i\cdot}p_{\cdot j} \qquad \text{for all } i, j.$$

That is, the probability that an observation is both of type i in one category *and* of type j in the other category (and so falls in the ij-cell) is the product of the (marginal) probability that it is of type i and the (marginal) probability that it is of type j. The whole two-way probability table, then, consists of entries that are products of the corresponding marginal entries.

EXAMPLE 9.4

Suppose that an experiment consists of the simultaneous toss of a coin and an ordinary die (cube). Identifying the sides in the usual fashion, and assuming equal likelihood for the possible outcomes for the coin and equal likelihood for the outcomes of the die, one has the marginal probabilities in the ideal or mathematical case shown in

Table 9.4

	1	2	3	4	5	6	
H							$\frac{1}{2}$
T							$\frac{1}{2}$
	$\frac{1}{6}$	$\frac{1}{6}$	$\frac{1}{6}$	$\frac{1}{6}$	$\frac{1}{6}$	$\frac{1}{6}$	

Table 9.4. To fill in the body of the table (the cells) with appropriate probabilities, one would have to *assume* something about the relationship of the toss of the die and the toss of the coin. If it is assumed that they are unrelated in the complete sense of *independence*, so that the equal likelihood of the six faces persists even if it becomes known how the coin falls, then the probabilities of the cells are obtained by *multiplying* the $\frac{1}{2}$'s in the right margin by the $\frac{1}{6}$'s in the bottom margin; see Table 9.5. (Observe, incidentally, that the 12 cells are equally likely in this model.)

Table 9.5

	DIE						
COIN	1	2	3	4	5	6	
H	$\frac{1}{12}$	$\frac{1}{12}$	$\frac{1}{12}$	$\frac{1}{12}$	$\frac{1}{12}$	$\frac{1}{12}$	$\frac{1}{2}$
T	$\frac{1}{12}$	$\frac{1}{12}$	$\frac{1}{12}$	$\frac{1}{12}$	$\frac{1}{12}$	$\frac{1}{12}$	$\frac{1}{2}$
	$\frac{1}{6}$	$\frac{1}{6}$	$\frac{1}{6}$	$\frac{1}{6}$	$\frac{1}{6}$	$\frac{1}{6}$	1

EXAMPLE 9.5

An experiment consists of drawing two chips at random, one at a time without replacement, from a bowl containing four chips numbered 1, 2, 3, and 4. The *pair* of numbers (X, Y), where X is the first number drawn and Y is the second, has a bivariate distribution. To learn about this distribution, the experiment was carried out 80 times, with results as given in Table 9.6. Except for the entry for cell (2, 1), it seems reasonable to suspect that the off-diagonal cells may be equally likely. [The diagonal entries (i, i) have zero frequencies because they cannot occur—for when a chip is drawn, it cannot be drawn again.] This suspected probability distribution would be defined by Table 9.7. Notice that the marginal distributions say that

Table 9.6

	1	2	3	4
1	0	8	5	6
2	3	0	6	6
3	6	6	0	9
4	10	7	8	0

Table 9.7

	1	2	3	4	
1	0	$\frac{1}{12}$	$\frac{1}{12}$	$\frac{1}{12}$	$\frac{1}{4}$
2	$\frac{1}{12}$	0	$\frac{1}{12}$	$\frac{1}{12}$	$\frac{1}{4}$
3	$\frac{1}{12}$	$\frac{1}{12}$	0	$\frac{1}{12}$	$\frac{1}{4}$
4	$\frac{1}{12}$	$\frac{1}{12}$	$\frac{1}{12}$	0	$\frac{1}{4}$
	$\frac{1}{4}$	$\frac{1}{4}$	$\frac{1}{4}$	$\frac{1}{4}$	1

the four values are equally likely to appear as the first number drawn, and also equally likely to appear as the second number drawn (not knowing what the first number is). Notice also that the numbers X and Y are *not independent* in this model, for the cell probabilities would all have to be $\frac{1}{4} \times \frac{1}{4} = \frac{1}{16}$. Indeed, a zero could not appear in a cell unless one of the marginal probabilities were also zero.

The hypothesis that the classification schemes in a two-way contingency table are independent can be tested using a goodness-of-fit test, and an appropriate chi-square statistic—if the sample size is large. The idea will be explained by means of an example.

EXAMPLE 9.6
It is desired to test the hypothesis that hair color and eye color are independent. The data given earlier from 39 students are repeated here as Table 9.8. The experiment of selecting a student and observing hair color and eye color can be considered as having a discrete model with 12 categories (the 12 cells of the table). The model to be tested is that of independence, which is described by saying that

$$p_{ij} = p_{i\cdot}p_{\cdot j},$$

where p_{ij} is the probability of the ij-cell, and the $p_{i\cdot}$ and $p_{\cdot j}$ are row

Table 9.8

	RED	BLOND	BROWN
Blue	1	7	13
Brown	1	0	8
Hazel	0	2	5
Green	0	2	0

and column sums of cell probabilities. Thus the model of the null hypothesis is defined by the seven marginal probabilities, but these are unknown. To estimate them for the purpose of constructing the chi-square statistic, one uses corresponding *relative frequencies*. These are as shown in Table 9.9.

Table 9.9

	RED	BLOND	BROWN	
Blue				$\frac{21}{39}$
Brown				$\frac{9}{39}$
Hazel				$\frac{7}{39}$
Green				$\frac{2}{39}$
	$\frac{2}{39}$	$\frac{11}{39}$	$\frac{26}{39}$	

Under the hypothesis of independence, the probability of a blue-eyed, brown-haired student would be

$$\tfrac{21}{39} \times \tfrac{26}{39},$$

and the expected cell frequency (in 39 observations) would be 39 times this:

$$39 \times \tfrac{21}{39} \times \tfrac{26}{39} = 14.$$

The observed number in this cell is 13, not too far from the expected 14; however, one must consider *all* the cells in the same fashion. For the blue-eyed redheads, the comparison is

$$39 \times \tfrac{2}{39} \times \tfrac{21}{39} = 1.07 \text{ versus } 1.$$

The chi-square statistic is computed as follows:

$$\chi^2 = \frac{(13 - 14)^2}{14} + \frac{(1 - 1.07)^2}{1.07} + \cdots = 9.58,$$

where the sum is carried out over similar calculations for each of the 12 cells.

The question is whether the observed value of χ^2 is too large to permit acceptance of the hypothesis of independence. Under *that* hypothesis, the statistic χ^2 has approximately a chi-square distribution with $12 - 1 - 5 = 6$ degrees of freedom for large samples. The density of this distribution is as shown in Figure 9.1. If large values of χ^2 are regarded as evidence of the falsity of the hypothesis of independence, then anything larger than 12.6 would call for rejection of this hypothesis at the 5 per cent level of significance. (The sample size of 39 is perhaps a little small for a really good approximation; this means that the 5 per cent may not be very accurate as a measure of type I error.)

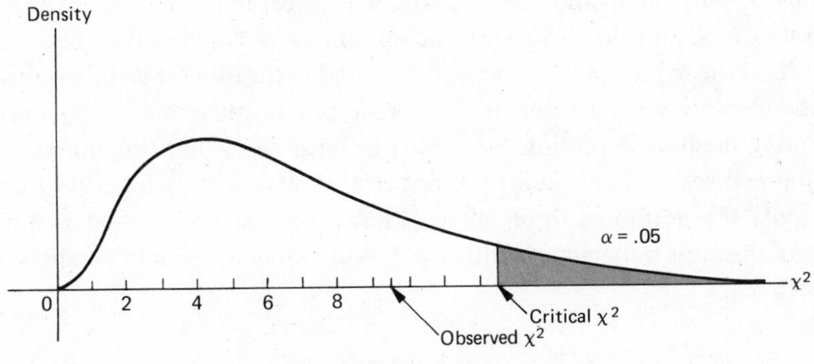

Figure 9.1
Chi-square density, 6 degrees of freedom.

The generalization to r by s two-way tables, with r categories in one classification and s categories in the other, is fairly obvious. Perhaps the only thing that should be stated more explicitly is that the number of degrees of freedom in the r by s case is

$$rs - 1 - (r - 1) - (s - 1) = (r - 1)(s - 1).$$

For, there are $r - 1$ free parameters to estimate on one margin and $s - 1$ on the other. (In the above example, $r - 1 = 3$, $s - 1 = 2$.)

The astute reader may have noticed that the calculation of expected frequency for a cell in the above computation of χ^2 turns out to be exactly the same as in the earlier computation of χ^2 for comparing categorical populations (Section 8.6). This means that the test for independence of two classification schemes is exactly the same as the test for the identicalness of two (or more) populations. The difference is in the model and the interpretation of the test result in the context of that model.

Thus, if one wants to know whether sex is related to hair color, he can use one of two approaches. In one, he considers males and females as two populations each with the categorical variable "hair color" defined on it. He takes samples of given size from each population and bases his comparison of the populations on the results in the two samples. In the other approach, he considers the single population of all people with the bivariate classification scheme (sex, hair color) defined on it. He takes a sample from that population and classifies each member of the sample according to sex and hair color. The tabulations of results will in both approaches be a $2 \times k$ (if there are k hair-color categories) table of frequencies. In the first approach one set of marginal totals will be the given sample sizes, and in the other approach all marginal sums are random.

Just looking at the table of frequencies will not tell one which approach was used; only the method of sampling will reveal that. The test is the same in both cases, and the decision to accept H_O or H_1 is the same, but in one case H_O is the hypothesis of independence, and in the other it is the hypothesis of no difference among populations. Yet, despite the differences in model and sampling method, accepting H_O means in both cases that (for the example mentioned) sex and hair color are not related. It is simply a matter of how one puts the notion of "not related" into a mathematical model. And in view of this, it is only common justice that the same test would be applicable in both cases.

Problems

9.1. An observer at an intersection recorded, for each of 200 cars passing through the intersection, whether or not the driver stopped for the stop sign and whether the driver was male or female. Test for independence of sex and obedience, given his results as follows:

	MALE	FEMALE
Stopped	30	50
Kept moving	100	20

9.2. A graduate student in education brought in these data, summarizing reactions to the statement: "The PTA has representation at local school faculty meetings":

	PTA PRESIDENTS	PRINCIPALS
Strongly approve	14	3
Approve	13	18
Uncertain	13	7
Disapprove	5	13
Strongly disapprove	2	3

Is the feeling about PTA representation related to whether a person is a PTA president or a principal? (Carry out a chi-square test. Is this a test for independence or for the identicalness of two populations?)

9.3. A random sample of 100 students was surveyed as to political preference of the student and of his parent: Republican, Democrat, or independent, with the following results:

		PARENT		
		D	R	I
STUDENT	D	25	8	7
	R	0	9	1
	I	10	8	32

Test for independence at the 1 per cent level.

9.4. (Calculator desirable.) An opinion research group asked this question of 2023 persons: "Do you read newspapers less, more, or about the same as you did a few years ago?" The results, together with information on the level of education, are given in the table below. Test for independence of response to the question and level of education at $\alpha = .01$.

	SOME COLLEGE	HIGH SCHOOL	LESS
Read more	209	298	227
Read less	87	131	202
Same	180	284	348
Don't know	10	15	32

9.5. The ace, king, queen, jack, and 10 of each of the four suits in a bridge deck are called *honor* cards. (There are also eight nonhonor cards in each suit.) Determine the following probabilities, for events relating to the single cut (random draw) of one card.

(a) $P(\text{honor}|\text{heart})$ (b) $P(\text{honor})$
(c) $P(\text{black}|\text{honor})$ (d) $P(\text{black})$

What conclusions can be drawn (in terms of independence)?

9.6. Calculations of the type of those in the preceding problem show that the classification of a card drawn at random as to suit or color is independent of its classification according to honor or nonhonor.

(a) Cut a shuffled deck at random 100 times and record the result in a contingency table with the four suits on one margin and honor or nonhonor on the other margin. Carry out the chi-square test for independence.

(b) Remove the spade honors and the red 2's, 3's, and 4's (11 cards in all) and cut the doctored deck 100 times, as in (a). Test the results for independence.

9.3 Bivariate Distributions: Continuous

When the two observations made on each experimental unit are continuous random variables, a continuous bivariate model is required. An examination of the scheme used for collecting data will lead to the structure of the appropriate model.

Although the recording of an observation that is a value of a continuous variable is complicated by the continuity, its numerical nature permits the device of a graphical plot, and such visual representation is of great help in appreciating the significance of data. The natural way to plot a pair of observations (X, Y) is as a point in a rectangular coordinate scheme. Each such pair is then seen as a dot, and a sample of n pairs appears as a collection of n dots or points on a rectangular coordinate system. Such a plot is often called a *scatter diagram.*

EXAMPLE 9.7

The weight (in pounds) and height (in inches) of each of 38 students were requested and recorded, with the results shown in Table 9.10. (Actually these were the same students whose eye and hair color were recorded for Example 9.1. There is one less observation pair here because one student declined to give her weight.) These pairs (W, H) are plotted in Figure 9.2. Notice that there are some repetitions, two students with the pair (175, 72), for example. In the figure, these repeated values are indicated by two dots close together, but it is clear that there would be problems if too many such repetitions were

Table 9.10

W	H	W	H	W	H	W	H
175	72	195	75	175	70	185	78
133	65	112	63	175	74	130	61
125	68	175	69	155	70	190	74
155	71	110	64	160	69	130	64
150	70	175	72	185	72	120	61
155	68	105	62	165	71	154	72
115	65	175	72	135	64	115	66
125	67	140	68	170	71	134	66
105	62	145	68	210	78		
170	71	165	67	190	71		

encountered. Of course, the repeated values arise because the values have been rounded off to the nearest pound and the nearest inch, customary in measuring weight and height for most purposes. Indeed, it would appear that students who weigh over 135 lbs all rounded off their weights to the nearest 5 lb. And this may not be unreasonable in view of hour-to-hour and day-to-day variations in a person's weight.

It is possible, from a picture such as that in Figure 9.2, to glean some notion of the underlying distribution (what regions are more likely than others) and some idea as to the existence of a relationship between height and weight. But to get at this ideal, thought of as what would be embodied in infinitely many observations, the use of such pictures would become a little unwieldy.

What is done, usually, is to use the given round-off scheme (i.e., the round-off inherent in the process of reading measurements), or to effect an even coarser round-off by regrouping, in a manner analogous to that leading to a histogram in the univariate case. If the axis for each variable (let them be denoted by X and Y in the general discussion) is divided into class intervals in the usual way, there is induced a partition of the plane of (X, Y) values into rectangular cells whose sides are essentially the univariate class intervals. With data entered in one of the (finitely many) cells, the situation is like that of categorical data, and the results of sampling can be expressed in terms of cell *frequencies* (or *relative* frequencies, upon division by the sample size). As in the univariate case, where class intervals are ordered, so in the bivariate case the cells have order relationships that are significant.

In the *univariate* case, a visual presentation of a sample distribution in the form of a histogram is possible, using the second dimension to represent frequency or relative frequency. In the *bi*variate case, the two dimensions of

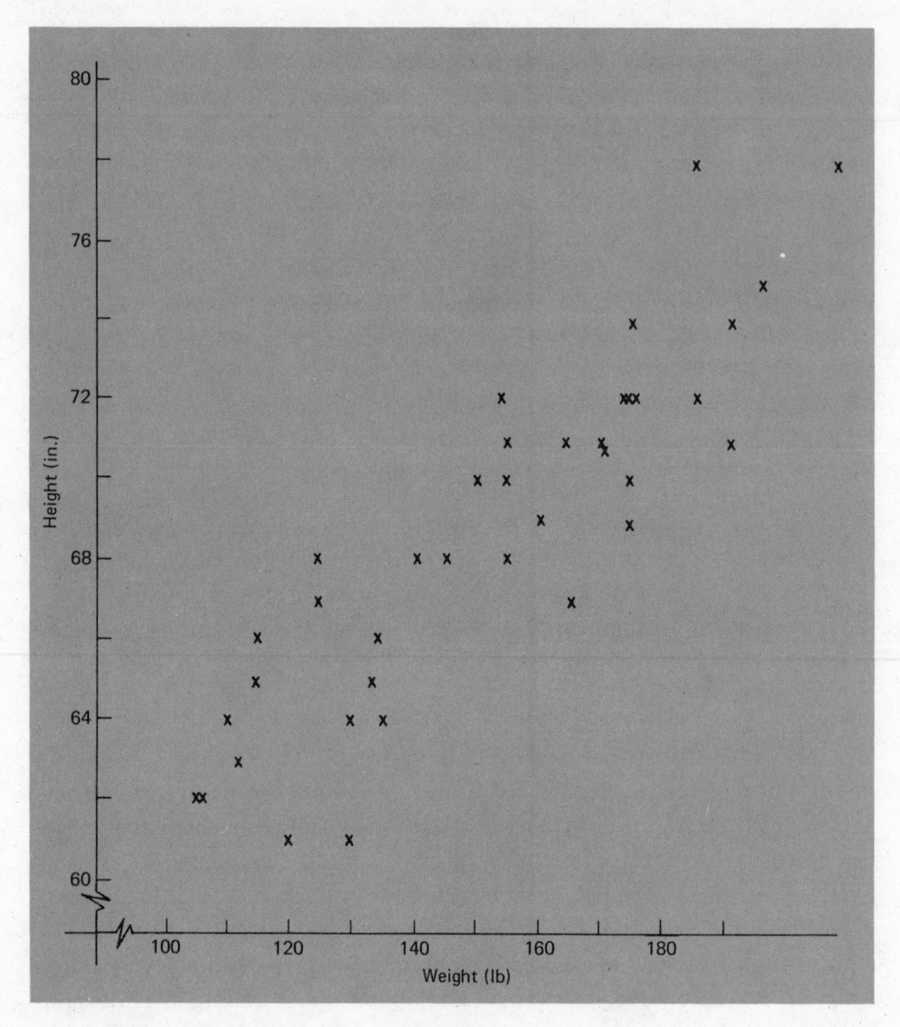

Figure 9.2
Height versus weight in Example 9.7.

the plane are used for the two variates. A third dimension would be required for a graphical representation, a kind of histogram. Since such three-dimensional pictures are not easy to draw, the device is not often used as a practical aid. Nevertheless, a histogram *could* be constructed using three-dimensional blocks of some sort for a given set of bivariate observations.

As in the case of a univariate histogram, in which a continuous density curve is approached as the sample size is increased and the class interval size is reduced, so in the bivariate case it can be observed that a (three-dimensional) histogram converges to a limiting configuration, a *surface* that defines a *bivariate density function*.

Probability is defined by means of a bivariate density function as *volume* under the surface that represents it, this volume being the quantity analogous to area under a density function's graph in the univariate case. Calculations based on this definition involve either the calculus or numerical analysis and so will not be considered here. Suffice it to say that a probability model for a continuous bivariate distribution is assumed to exist, and it is helpful to think of this model in terms of a density surface whose height at any point (x, y) is proportional to the concentration of probability in the neighborhood of that point. The model describes what *can* happen at a single observation of (X, Y), but it is also conceived of as the idealization that describes the distribution of a great many such observations.

EXAMPLE 9.8

The data of the preceding example, giving height versus weight for each of 38 students, has been regrouped and plotted in a three-dimensional histogram in Figure 9.3. Some of the difficulties encountered in trying to represent a three-dimensional figure on a flat, two-dimensional page are evident. In particular, some of the bars are hidden by others.

With data rounded off or regrouped into a finite number of cells, the test for independence, presented in Section 9.2 for categorical phenomena, can be applied to the present case. It provides a large sample test for independence of the marginal variates X and Y. The test is based, however, on categories, without regard to the numerical ordering inherent in the fact that X and Y are numbers. A concept of relationship that depends essentially on ordering of the values of the variates will be considered in the next section.

9.4 Linear Correlation

Granted that there *is* a model for a continuous bivariate experiment, defined by the idealization of the three-dimensional histogram called the density surface of the distribution, it remains to describe and characterize this kind of model, much as one-dimensional models were described by the mean, variance, distribution function, etc. As in the case of a single variate, in which sample statistics were idealized to population parameters, so in the case of a bivariate distribution, it will be illuminating to look first at some sample statistics.

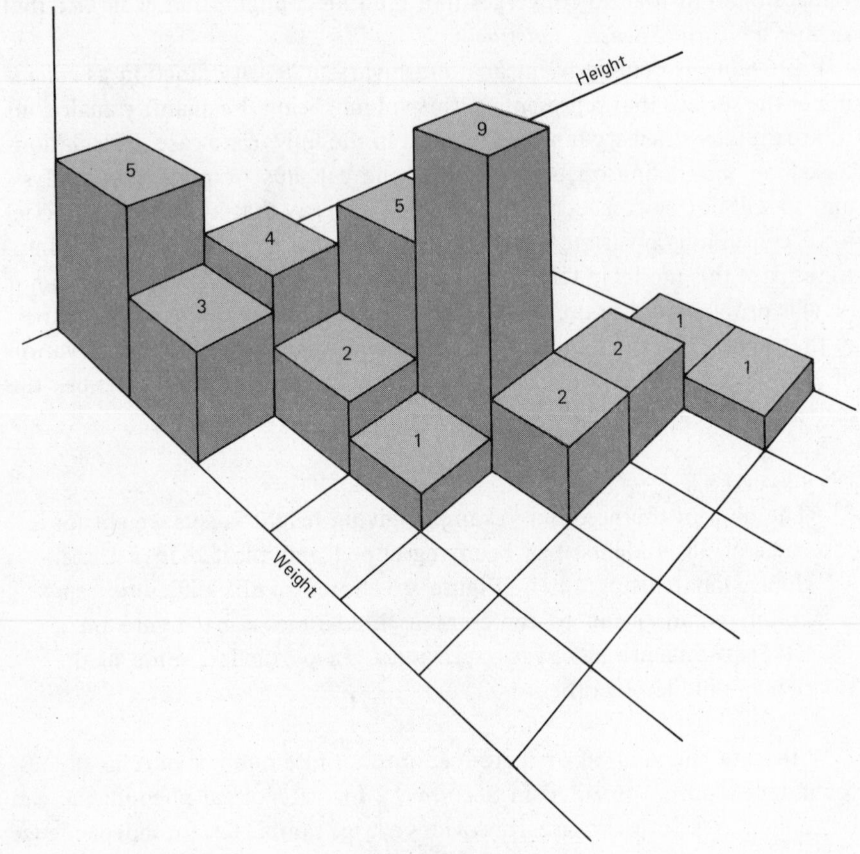

Figure 9.3
Bivariate histogram (Example 9.8).

A sample, it will be recalled, is a set of pairs: $(X_1, Y_1), \ldots, (X_n, Y_n)$, and these pairs can be represented graphically in a scatter diagram. The mean and variance for the X's and the Y's can be computed separately, of course, but these quantities tell nothing about the *joint* variation of X with Y. A useful quantity related to their joint variation is obtained by computing average products:

$$\overline{XY} = \frac{1}{n} \sum X_i Y_i.$$

This is an example of what is called a *mixed moment* or a *product moment*. A more useful quantity is obtained by applying this averaging of products to the *deviations* about the means, the deviations of the X's about their mean and the deviation of the Y's about their mean. The sample *covariance* is defined to be the average of the products of the corresponding deviations:

$$S_{X,Y} = \frac{1}{n}\sum(X_i - \overline{X})(Y_i - \overline{Y}) = \overline{XY} - \overline{X}\cdot\overline{Y}.$$

[This quantity reduces to the variance if computed for the pairs (X_i, X_i).] Because the covariance is dependent on the scale of measurement of the X_i and the Y_i, a more useful measure of covariation is obtained by dividing the covariance by the standard deviation of each coordinate in the sample:

$$r_{X,Y} = \frac{S_{X,Y}}{S_X S_Y}.$$

This is the *coefficient of linear correlation*, or, more simply, the *correlation coefficient*. The following examples indicate the sense in which this measures covariation.

EXAMPLE 9.9

Consider the very artificial and simple set of four observations: (1, 1), (1, −1), (−1, 1), and (−1, −1). The calculations for X and Y separately are

$$X:\ 1, 1, -1, -1; \qquad \overline{X} = 0, \qquad S_X^2 = \tfrac{1}{4}\sum(X_i - \overline{X})^2 = 1.$$

$$Y:\ 1, -1, 1, -1; \qquad \overline{Y} = 0, \qquad S_Y^2 = \tfrac{1}{4}\sum(Y_i - \overline{Y})^2 = 1.$$

Thus $S_X = S_Y = 1$. And the average of the products of the deviations is

$$S_{X,Y} = \tfrac{1}{4}[(1)(1) + (1)(-1) + (-1)(1) + (-1)(-1)] = 0.$$

The correlation coefficient is therefore zero:

$$r_{X,Y} = \frac{0}{(1)(1)} = 0.$$

Notice that the product of two deviations of the same sign is positive, and the product of two deviations of the opposite sign is negative. In the present case there are as many positive products as negative products, and they exactly cancel each other out. There is no tendency of a value of X to one side of its mean to be associated with or to imply a value of Y to the same side of its mean [see Figure 9.4(a)].

Another simple set of points with zero correlation is this: (−2, 2), (2, 2), (−3, 0), (3, 0), and (0, 1). The average product is 0, and the average of the x-values is 0; therefore, the covariance $S_{X,Y}$ is 0 [see Figure 9.4(b)].

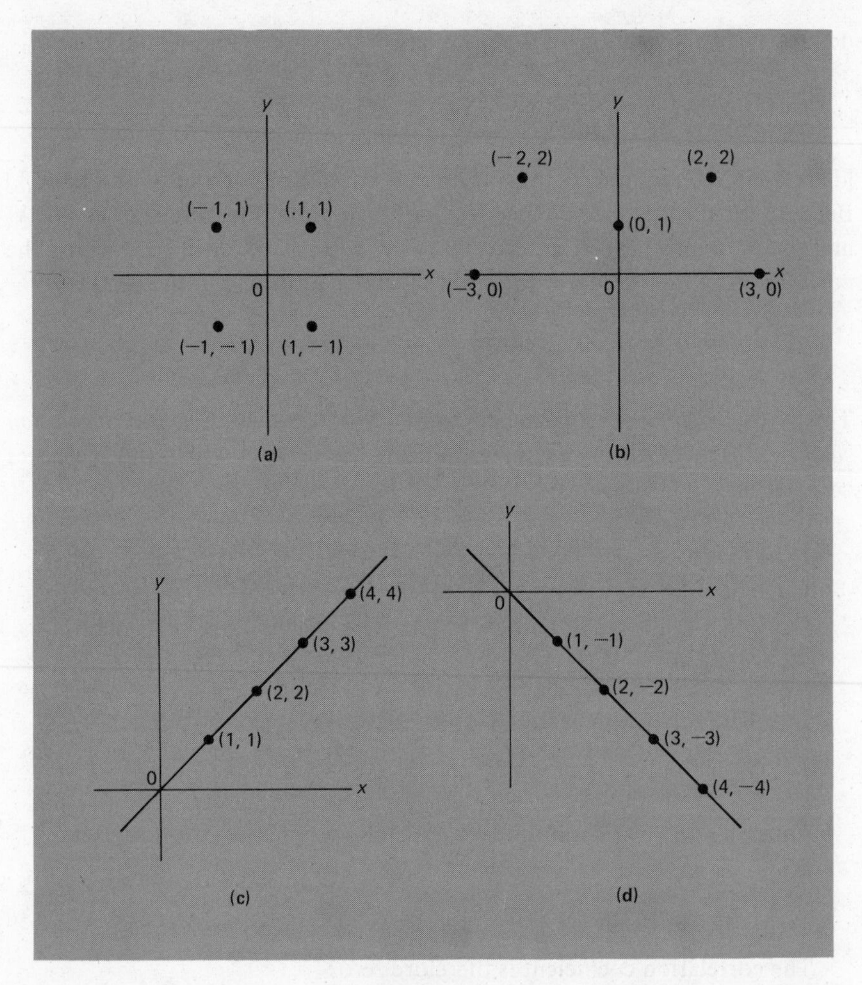

Figure 9.4
(a) $r = 0$; (b) $r = 0$; (c) $r = 1$; and (d) $r = -1$.

Consider next the data consisting of these four points: (1, 1), (2, 2), (3, 3), and (4, 4). The mean of the X's is 2.5, as is the mean of the Y's. The deviations, in each case, are $(-1.5, -1.5)$, $(-.5, -.5)$, $(.5, .5)$, and $(1.5, 1.5)$. The average product is precisely the same as the average squares:

$$S_{X,Y} = S_X^2 = S_Y^2 = 1.5^2 + .5^2 + .5^2 + 1.5^2 = 5,$$

so the correlation coefficient is

$$r_{X,Y} = \frac{S_{X,Y}}{S_X S_Y} = \frac{5}{\sqrt{5}\sqrt{5}} = 1.$$

Here there is total reinforcement, no cancellation of products. The X-deviations are positive when the Y-deviations are positive and negative when the Y-deviations are negative. The values of X and the values of Y *vary together* [see Figure 9.4(c)].

Finally, consider the points $(1, -1)$, $(2, -2)$, $(3, -3)$, and $(4, -4)$. It is easy to carry out the same kind of calculation, with similar results *except* that all the products will be negative, and the correlation coefficient is -1 [see Figure 9.4(d)].

These four examples illustrate two extremes and the case in the middle. It can be shown that $r_{X,Y}$ can never be larger than 1 nor smaller than -1, and that if it is equal to either of these extremes the points lie on a straight line. Hence the name *linear* correlation; the quantity $r_{X,Y}$ measures, in a sense, how close the points come to falling on a straight line. In the first case considered, the points were corners of a square—about as far as possible from falling on a line, and the value or the correlation coefficient was zero.

Figure 9.5 gives the scatter diagrams for several artificially generated samples of size 100. In each case the sample correlation coefficient is shown. The scales are not indicated, since the correlation coefficient does not depend on the scales. The figures should be studied as to the relationship between the hint of linearity and the value of the correlation coefficient.

The reason for computing a correlation coefficient for the points in a given sample, as in the case of all sampling, is to give some information about the underlying mechanisms, about the relationship between the variables X and Y in the ideal or model for the experiment. As usual, the sample values X_i and Y_i are random variables, and any statistic computed from them, such as $r_{X,Y}$, is a random variable. It varies from sample to sample and has a certain sampling distribution. The problem of statistical inference, as usual, is to draw inferences about the population on the basis of a *single* sample.

The *population*, or the model for the experiment that produces the pairs (X_i, Y_i), also has a coefficient of linear correlation, which can be thought of as the long-run limit of the sample correlation coefficient values as the sample size increases without limit. This population parameter is usually designated by the lowercase Greek letter r, rho: $\rho_{X,Y}$, or, more simply, by ρ. It measures the degree to which the value of X and the value of Y fall on a line *in the population distribution*.

The significance of the presence of a degree of covariation, as measured by ρ, is that if ρ is near 1 or -1, one ought to be able to predict the value of X from the value of Y, or the value of Y from the value of X. This point will be discussed at greater length in Section 9.6.

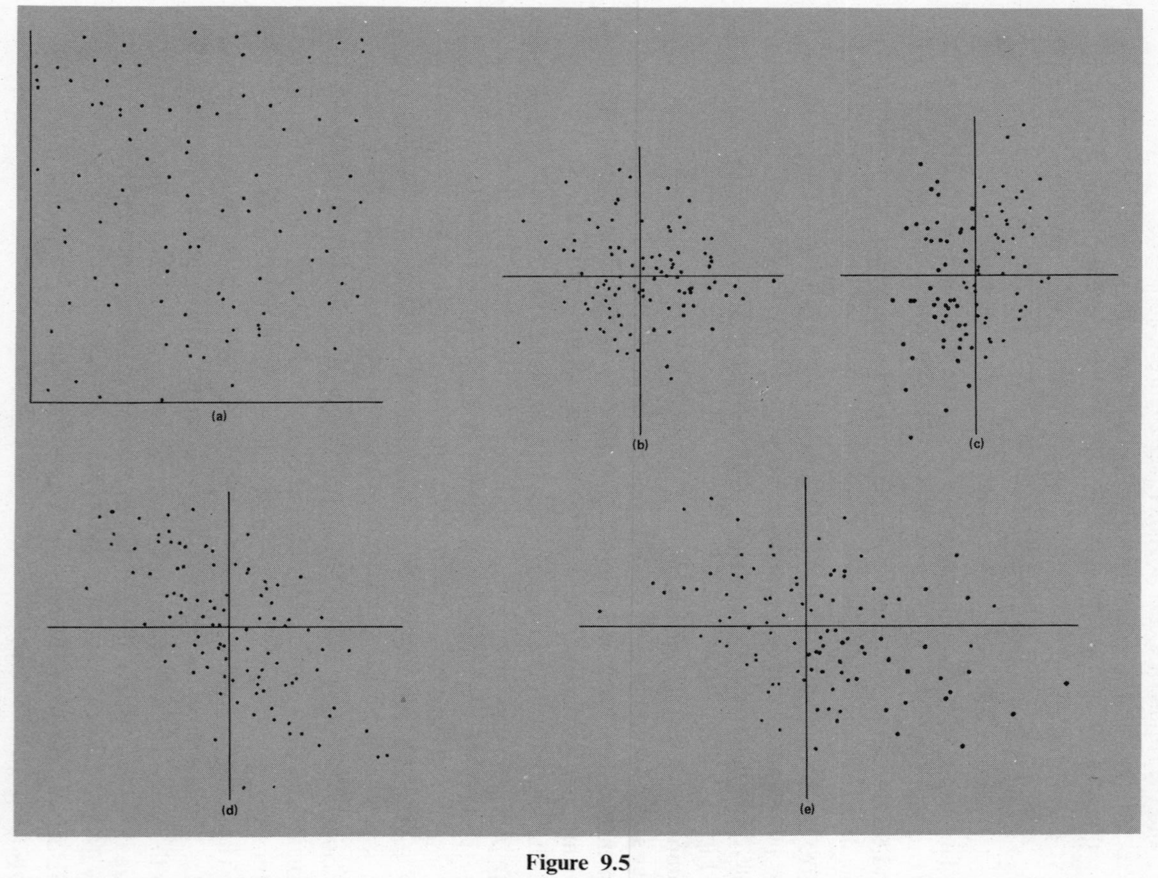

Figure 9.5
(a) $r = -.0476$; $\rho = 0$. (b) $r = -.03$; $\rho = 0$. (c) $r = .137$; $\rho = .1414$. (d) r $= -.348$; $\rho = -.316$.
(e) $r = -.657$; $\rho = -.60$.

It should be pointed out that a high degree of association, as indicated by a large value of r^2 (or, in the population, ρ^2), is *not* necessarily indicative of a *causal relation* between the variables X and Y. That is, a large X may not be *caused* by a large Y, even though they tend to be large together and small together. Nevertheless, the presence of an association can be useful and exploited in prediction. Moreover, when two variables are related in the sense of being linearly correlated (having a nonzero coefficient of correlation), this may be sufficient reason to think about and look for a possible causal relationship.

It is hoped that questions of inference such as those addressed in earlier contexts will have occurred to the reader. Thus r is a statistic, computed from a sample of pairs, and so has a sampling distribution determined by the parent population. How is its value related to ρ? Is it successful as an estimator, and can it be used to test hypotheses about ρ? More specifically, since the sample correlation coefficient need not be 0 when $\rho = 0$, how far from zero can r get before one should reject $\rho = 0$ and accept $\rho \neq 0$ (or some one-sided alternative)? These questions are not taken up here, but answers are available in more advanced texts.

Problems

9.7. Two judges of a piano contest turned in ratings as follows:

	PERFORMER							
JUDGE	1	2	3	4	5	6	7	8
1	92	90	85	96	92	88	96	88
2	89	90	88	93	90	85	95	90

Plot the data and compute the correlation coefficient of the ratings of the two judges. (The arithmetic will be much simpler if the ratings are first converted to numbers of points above 90. That is, let $U_i = X_i - 90$, $V_i = Y_i - 90$, and use the fact that the correlation of the U's and V's is the same as that of the X's and Y's.)

9.8. Ten couples comparing notes gathered the following data on the numbers of years of college attended:

Husband	4	1	7	6	5	4	7	5	4	6
Wife	4	0	2	4	4	0	6	2	2	4

Determine the coefficient of correlation.

9.9. A student fascinated by correlation took a notion to calculate the correlation coefficient between the jersey number and the weight for the 40 players on a certain professional football team. He found that $r = .60$. Is this surprising?

9.10. (a) A survey of 1000 taxpayers showed a correlation of .76 between annual income and years of schooling. Does this show that education tends to increase one's income?

(b) Someone once computed a strong positive correlation between the number of stork nests and the number of babies born (in a given year) in a series of English villages. Would this show that storks bring babies? (Or that babies attract storks?) Are there explanations for the high correlations in these instances?

9.11. After slaughter, 26 turkeys were subjectively rated for tenderness on a scale from 1 to 5 (1 = very tender). After the turkeys were prepared, tenderness was measured objectively according to the amount of force needed to cut the meat. The ratings were as follows:

SUBJECTIVE	OBJECTIVE	SUBJECTIVE	OBJECTIVE	SUBJECTIVE	OBJECTIVE
5	5.85	4	6.70	5	8.58
3	4.95	4	7.18	3	5.70
3	7.82	2	4.45	4	4.70
2	4.78	3	6.78	4	4.22
4	5.58	5	7.52	4	4.88
2	6.50	4	6.92	3	5.22
3	3.98	5	12.54	3	5.18
5	6.75	4	4.95	4	7.00

(*Note:* Two turkeys disappeared during storage, both with a subjective rating of 1.) Compute the correlation coefficient for the two methods of judging.

9.5 Bivariate Normal Distribution

Probability models for a pair of continuous random variables (X, Y) are manifold and multifarious, but there is one in particular that is not only a good approximation to the true model in many practical situations, but is also simple enough to be analyzed theoretically in considerable detail. That model is the generalization (to two variables) of the normal model considered earlier (Sections 3.6 and 4.6).

The normal density in *one* dimension is a bell-shaped, symmetrical function with tails rapidly converging to zero, such as that shown in Figure 9.6. Just as a univariate density is represented by a curve, so in two dimensions a density is represented geometrically by a *surface* (above the XY-plane); and the *bivariate normal* density is a bell-shaped surface, with various symmetries, such as that shown in Figure 9.7. To help give an idea of the shape, various cross sections (which happen to be ellipses) parallel to the XY-plane have been sketched in. The equation of the surface* need not concern us at this level of exposition, but certain facts about these models can be given, and perhaps appreciated, even if not proved here mathematically:

1. The bivariate normal density is completely described in terms of the first and second moments, i.e., in terms of means, variances, and covariance or correlation: $\mu_x, \mu_y, \sigma_x^2, \sigma_y^2$, and $\rho_{x,y}$. The point (μ_x, μ_y) is a point of symmetry, and the density has its maximum there.

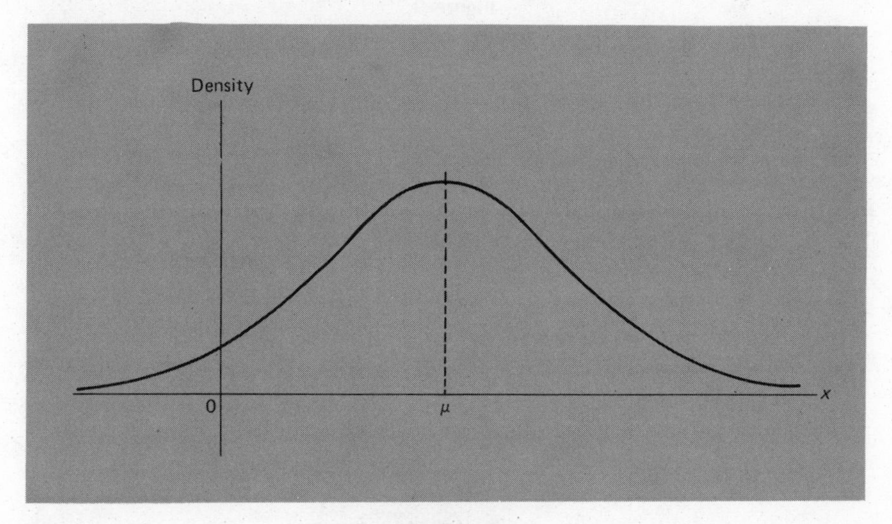

Figure 9.6
Univariate normal density.

2. There are two planes of symmetry; if these planes happen to coincide with the vertical coordinate planes, the correlation is zero, and the variables X and Y are independent.

3. Any cross section parallel to a coordinate plane is essentially a one-dimensional *normal* curve (see Figure 9.8). Such a cross section is obtained from the density by *holding one variable* (x or y) *fixed*; the resulting curve is

* The equation is of the form

$$f(x, y) = K \exp(Ax^2 + Bxy + Cy^2 + Dx + Ey).$$

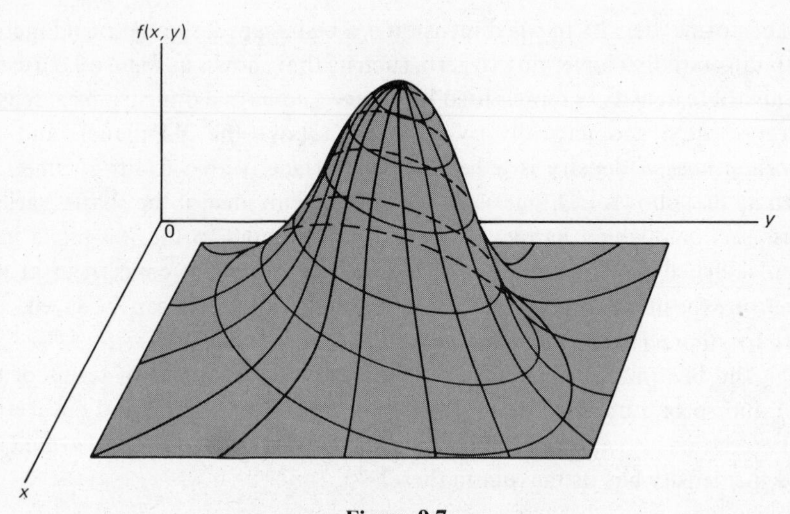

Figure 9.7
Bivariate normal density.

essentially the density of the *conditional* probability distribution of one variable, given the value of the other. (By "essentially" is meant that to be a bona fide density, with area 1, it would have to be multiplied by a constant.) Thus, in any bivariate normal distribution, the conditional distribution of X given $Y = y_0$, or of Y given $X = x_0$, is *normal*.

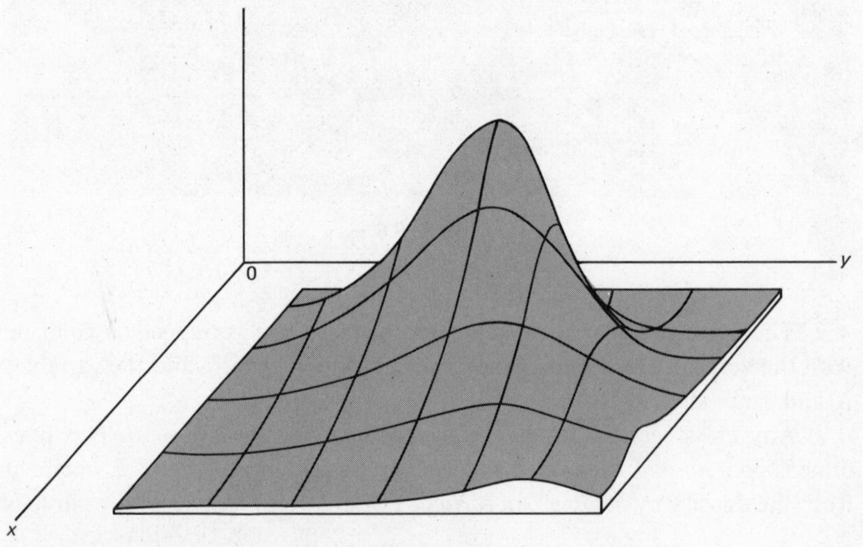

Figure 9.8
Cross sections of bivariate normal density.

4. The variable X, by itself, with no reference to or knowledge of Y, is univariate normal, as is Y by itself. (This is more complicated to interpret geometrically, so the interpretation will not be attempted.)

5. If one marks the *mean*—the center of symmetry for each cross section—on the horizontal axis of that section, it will be found that the locus of such centers is a straight line. The mean so marked is the mean of the *conditional* distribution; it will be denoted $E(X|Y = y)$ or $E(Y|X = x)$ according as the information given is that $Y = y$ or $X = x$, respectively. The locus of conditional means given $Y = y$ defines the *regression function* of X on Y (and given $X = x$, the regression function of Y on X). In this bivariate normal model, the regression functions are *linear*. [Moreover, they intersect at the point (μ_x, μ_y), which is the point over which the surface is centered.]

6. The variances of all cross-sectional (conditional) distributions parallel to one of the coordinate planes are the same. (This fact is not really intuitively evident, or seen in the graph, since the densitities in the cross sections are not fixed up to have area 1.) Thus the distance from the conditional mean out to the inflection point of the cross-section curve is constant. In symbols,

$$\text{var}(Y|X = x) = \text{constant},$$
$$\text{var}(X|Y = y) = \text{constant}.$$

Another fact about these normal densities, not evident in the graph, is a relationship among the conditional and unconditional variances and the correlation coefficient:

$$\text{var}(Y|X = x) = (1 - \rho^2)\sigma_y^2.$$

(This can be established simply from the facts that the regression is linear and the conditional variance is constant.) Notice that if X and Y are independent, in which case $\rho^2 = 0$, the conditional variance is equal to the absolute variance. Notice also that if $\rho^2 = 1$, with X and Y perfectly correlated, the conditional variance is zero; for if X is given, and Y is perfectly linearly related to it, then Y can be computed and has no variability.

9.6 Predicting One Variable from a Value of the Other

Predicting the value of any random variable U to be the mean of its distribution is a prediction that has minimum mean squared error of prediction:

$$E(U - A)^2 \text{ is smallest for } A = EU.$$

(Also, the size of the smallest mean squared error is just the variance of U.)

So if one observes, in a bivariate experiment (X, Y), that $X = 4$, the predicted value of Y that has the smallest mean squared error of prediction would be the *mean* of Y given $X = 4$: $E(Y|X = 4)$. More generally, one predicts Y, given $X = x$, to be $E(Y|X = x)$; in other words, the regression curve gives the best predictions in terms of mean squared error.

In practical situations, of course, the precise model for (X, Y) is not known, and so the value of $E(Y|X = x)$ is not known. But assume, for the moment, that the regression function $E(Y|X = x)$ *is* known. In such a case, and using $E(Y|X = x)$ to predict Y given $X = x$, the mean squared error of prediction would be the *conditional variance* of Y:

$$\text{mean squared error} = \text{var}(Y|X = x).$$

Now, in the bivariate normal case, this is independent of x and can be written

$$\text{mean squared error} = (1 - \rho^2)\sigma_y^2.$$

From this one can see the relation between correlation and the predictability of Y given the value of X. For, if X is *not* given, one would predict Y to be EY, with a mean squared error of σ_y^2. But if X is given, the mean squared error can be reduced by a factor $1 - \rho^2$. Or, on the scale of Y, the *root* mean squared error can be reduced by a factor $\sqrt{1 - \rho^2}$:

$$\text{rms error} = \sqrt{1 - \rho^2}\,\sigma_y.$$

EXAMPLE 9.10

In using such things as high school rank to predict college performance, it is not uncommon to find a correlation between them of about .6. Assuming a normal distribution, and assuming the regression function to be known and used as a predictor, the root mean squared error would be (with X = HSR and Y = college performance score):

$$\sqrt{1 - \rho^2}\,\sigma_y = \sqrt{1 - (.6)^2}\,\sigma_y = \sqrt{.64}\,\sigma_y = .8\sigma_y.$$

That is, knowing high school rank enables one to predict college performance with only .8 of the (rms) error he would make without knowing high school rank.

Next, what if the regression function is *not* known? The statistical approach would be to gather *data* about the experiment (X, Y), which means to perform the experiment say n times and record the results: $(X_1, Y_1), (X_2, Y_2), \ldots, (X_n, Y_n)$. Then from these data the statistician would estimate the regression function and use the estimated function as his predictor. If the regression function is *linear*, as it is when the population (X, Y) is bivariate normal,

$$E(Y|X = x) \quad = \quad \alpha + \beta x,$$

the statistician's job is to estimate α and β. This task will be considered in more detail in Chapter 10. It will be seen, as might be expected, that the uncertainty in the location of the regression line (as estimated using data) will increase the mean squared error of prediction.

Problems

9.12. Draw a graph of the function $\sqrt{1 - \rho^2}$ to represent visually how this varies with ρ, i.e., to show how the factor of reduction in rms prediction error made possible by using the regression function depends on the degree of correlation.

9.13. Obtain 25 successive pairs of random normal numbers from Table IX of the Appendix as a sample of 25 observations (X_i, Y_i) from a bivariate normal distribution with $\mu_X = \mu_Y = 0$, $\sigma_X = \sigma_Y = 1$, and $\rho = 0$. Plot the 25 points on a rectangular coordinate system, and notice the small degree of correlation. (Your *sample* may exhibit *some* correlation even though the *population* correlation is zero.)

9.14. Use 25 pairs (X_i, Y_i) from Table IX (as in Problem 9.13) to generate *new* pairs (U_i, V_i), where $U = X$ and $V = X + Y$. Plot the UV-pairs on a rectangular coordinate system. It can be shown that for the UV-population $\mu_U = \mu_V = 0$, $\mu_U = 1$, $\sigma_V = 2$, $\sigma_{U,V} = 1$, and $\rho_{U,V} = 1/\sqrt{2}$. The theoretical regression line is $V = U$; sketch this on your plot. If you have a calculator, calculate the sample correlation $r_{U,V}$.

9.15. The data given are pairs (X, Y) from a bivariate normal population with $\mu_X = 10$, $\mu_Y = 5$, $\sigma_X = 1$, $\sigma_Y = \sqrt{5}$, and $\sigma_{X,Y} = 2$. The population regression line is given by the equation $y = 2x - 15$. For each X,

X	Y
10.39	6.30
11.20	8.90
10.72	5.38
10.54	9.58
8.65	.389
9.43	4.55
10.25	5.46
10.43	5.80
8.76	2.56
11.68	8.17

predict the value of Y using the regression line and compute the prediction error:

$$e = \text{predicted } Y - \text{observed } Y.$$

Average the squares of these errors to obtain the mean squared error of prediction, and compare this with the mean squared error that would be made using the Y-mean of 5 as a prediction (which would be a best prediction *if* X were not observed). (The ideal factor relating these mean squared errors is $1 - \rho^2$.) It would be instructive to plot the regression line and the 10 data points, and to mark the various prediction errors.

10 Linear Regression

At the beginning of Chapter 9 two types of bivariate situations were described: (1) those in which both variables are random, and (2) those in which one variable is a design adjustment or a dial setting, a controlled variable whose level determines the nature of the other variable, which in turn is thought of as a *response*. Chapter 9 considered type 1 situations and discussed in particular the problem of predicting one variable when a value of the other is given; this prediction involved the conditional distribution of the one variable given the value of the other. Even in type 1 situations, one is interested in studying the problem as though one variable were known or fixed or constant (at certain levels). Indeed, the analysis of these aspects of type 1 situations is exactly the same as for the type 2 situation, whose setting will now be made more precise.

10.1 The Regression Model

In many fields of application there are pairs of variables, one of which responds linearly (or approximately linearly) to variation in the other. Thus one variable is a *linear function* of the other:

$$y = \alpha + \beta x.$$

(Physics abounds with such instances: Ohm's law, Newton's law, etc. In other fields, not so neat and precise in their theories, there are often indications and suggestions of such functional relationships, and one hopes that simple laws exist, although they may be obscured by inadequacies of definition or measurement.)

In any case, actual data, pairs $(x_1, y_1), \ldots, (x_n, y_n)$, will be found to fall, not *on* a line, but scattered about an actual linear relationship (if, indeed, there is one). The data will be collected by selecting values $x_1, \ldots, x_n$ of the controlled variable, to be set in (by dial, or what have you), and then measuring a value of the response variable corresponding to each x. It will be assumed, for various reasons, that the response at $x = x_i$ is Y_i, a *random variable*, and that it is the random variation in these responses that obscures the functional relationship. In the case of *linear* regression, the response when $x = x_i$ is assumed to be of the form

$$Y_i = \alpha + \beta x_i + \varepsilon_i,$$

where α and β are unknown parameters of the true linear relationship and ε_i is a random error. It is usually assumed that ε_i has mean zero, so that $\alpha + \beta x_i$ is the mean or expected value of the response Y_i at $x = x_i$.

The statistician's job is to make inferences, given data $(x_1, Y_1), \ldots, (x_n, Y_n)$, about the parameters α and β, or possibly to use approximate knowledge of α and β to predict the response Y at still other dial settings (values of the controlled variable x). In making inferences precise, he would have to know the probability distributions of any statistics he might want to use, and this would depend on the model he assumes for the randomness in Y, or for the random component ε in the response. A common model for the errors $\varepsilon_1, \ldots, \varepsilon_n$ makes the following assumptions:

1. $\varepsilon_1, \ldots, \varepsilon_n$ are independent random variables.
2. $E(\varepsilon_1) = \cdots = E(\varepsilon_n) = 0$.
3. $\text{var}\, \varepsilon_1 = \cdots = \text{var}\, \varepsilon_n = \sigma^2$ (the same variability for all dial settings).
4. The errors ε_i have a *normal* distribution.

Whether these assumptions are satisfied in a given application is a matter that warrants careful consideration. But they are not unreasonable assumptions and are at least approximately fulfilled in some applications.

10.2 The Method of Least Squares

The problem of determining an actual linear relation from corrupted data can be attacked simply by asking: How do you pass a line through a set of points that do not lie on a line? Of course, it cannot be done, except approximately. One obvious method is to do it by eye, a method that has obvious drawbacks. More systematically, one could lay down a line in an arbitrary position, decide on a measure of how well it fits the data points in that position, and then move the line around so as to maximize this measure of fit.

Any measure of fit should take into account how close *each* point comes to lying on the line. Because the random variation in the response Y is a vertical variation (assuming the response Y to be plotted vertically as usual, with x horizontal), it is natural to measure the vertical distance from each point to the line being considered. Some combination of these vertical discrepancies, then, is natural to use in assessing overall success in the task of fitting a line to the data. The method of *least squares* uses the average squared vertical discrepancy to measure how far a proposed line is from fitting well, and one adjusts the line until this average squared discrepancy is as *small* as possible.

The vertical distance from a data point (x_i, Y_i) to a line $y = a + bx$ is the difference between its ordinate and the ordinate on the line at $x = x_i$:

$$d_i \equiv Y_i - (a + bx_i)$$

(see Figure 10.1). The average square of these is the measure that will be used.

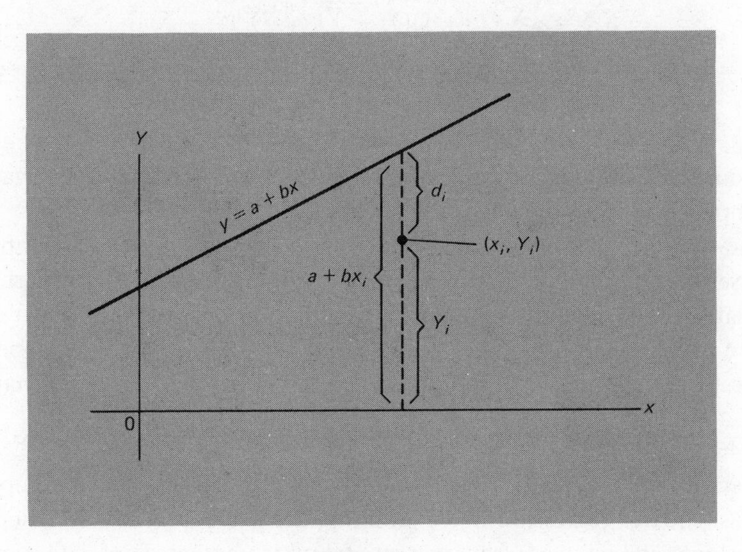

Figure 10.1
Residual about a regression line.

$$D = \frac{1}{n}\sum_{1}^{n}[Y_i - (a + bx_i)]^2.$$

This measure is dependent on the intercept a and slope b of the trial line $y = a + bx$. Varying a and b will vary the value of D, so it is simply a matter of choosing the a and b that minimize D.

Because D is just a quadratic function of a and b, it is possible to determine the minimizing values by methods of algebra, but calculus techniques will also work. No matter how the minimization is carried out, the result is the same, and the following formulas give the values of a and b that do the job; these values will be called, respectively, $\hat{\alpha}$ and $\hat{\beta}$:

$$\hat{\beta} = \frac{S_{xY}}{S_x^2},$$

$$\hat{\alpha} = \bar{Y} - \hat{\beta}\bar{x}.$$

In these formulas $\bar{x}$ and $\bar{Y}$ denote the arithmetic means, respectively, of $(x_1, \ldots, x_n)$ and $(Y_1, \ldots, Y_n)$. The quantity S_x^2 denotes the variance of the numbers $(x_1, \ldots, x_n)$:

$$S_x^2 = \frac{1}{n}\sum (x_i - \bar{x})^2 = \frac{1}{n}\sum x_i^2 - \bar{x}^2,$$

and S_{xY} denotes the covariance of the n pairs (x_i, Y_i):

$$S_{xY} = \frac{1}{n}\sum (x_i - \bar{x})(Y_i - \bar{Y}) = \frac{1}{n}\sum (x_i - \bar{x})Y_i$$

$$= \frac{1}{n}\sum x_i Y_i - \bar{x}\bar{Y}.$$

(All sums extend over the range of indices $i = 1$ to $i = n$, that is, over all the n sample points.)

The variance S_x^2 is not a statistic in the usual sense; it is determined by the dial settings $x_1, \ldots, x_n$, which are part of the design of the experiment. Similarly, the covariance S_{xY} involves randomness only in the Y's that enter.

The linear relationship that is the end result of adjusting the slope and intercept so that the sum of squared deviations about it is minimized, namely,

$$y = \hat{\alpha} + \hat{\beta}x,$$

is referred to as the *empirical regression function*, and its graph as the empirical or least-squares regression line. Naturally, the position of this line depends on the data and will vary from one set of data to another set of data from the same experiment.

The minimum *value* of the sum of squared residuals has a significance; divided by n, it provides an estimate of the *error variance*, σ^2:

$$\hat{\sigma}^2 = \frac{1}{n}\sum(Y_i - \hat{\alpha} - \hat{\beta}x_i)^2.$$

This estimate can be calculated directly from the formula, or indirectly by means of the relation

$$\hat{\sigma}^2 = S_Y^2(1 - r^2),$$

where r is the coefficient of linear correlation between the x's and Y's in the sample; it can be computed as follows:

$$r = \frac{S_{xY}}{S_x S_Y}.$$

(The second formula for $\hat{\sigma}^2$ is often easier to work with than the first one, which averages the squared residuals about the least-squares line.)

EXAMPLE 10.1

To illustrate the computations in fitting a line by the method of least squares, consider the simple, artificial problem of fitting to these three data points: (1, 1), (2, 0), and (3, 3) (Figure 10.2). The calculations

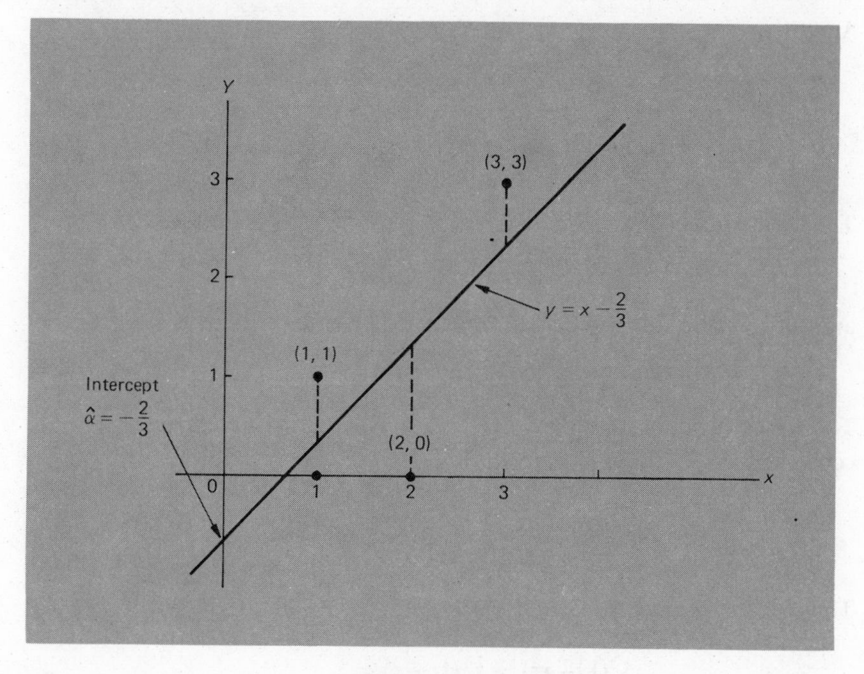

Figure 10.2
Empirical regression line.

are facilitated by a tabular presentation; see Table 10.1. From this are obtained the various moments.

Table 10.1

	x_i	Y_i	x_i^2	Y_i^2	x_iY_i
	1	1	1	1	1
	2	0	4	0	0
	3	3	9	9	9
Sum	6	4	14	10	10

$$\bar{x} = \frac{6}{3} = 2, \qquad \bar{Y} = \frac{4}{3},$$

$$S_x^2 = \frac{14}{3} - \bar{x}^2 = \frac{2}{3}, \qquad S_Y^2 = \frac{10}{3} - \bar{Y}^2 = \frac{14}{9},$$

$$S_{xY} = \frac{1}{n} \sum x_i Y_i - \bar{x}\bar{Y} = \frac{10}{3} - 2 \times \frac{4}{3} = \frac{2}{3}.$$

And then

$$\hat{\beta} = \frac{S_{xY}}{S_x^2} = \frac{\frac{2}{3}}{\frac{2}{3}} = 1,$$

$$\hat{\alpha} = \bar{Y} - \hat{\beta}\bar{x} = \frac{4}{3} - 1 \times 2 = -\frac{2}{3}.$$

The least-squares line is

$$y = -\tfrac{2}{3} + x$$

and the average squared deviation of the data points about it is

$$\tfrac{1}{3}[(1 - \tfrac{1}{3})^2 + (0 - \tfrac{4}{3})^2 + (3 - \tfrac{7}{3})^2] = \tfrac{8}{9}.$$

The easier formula for this is $S_Y^2(1 - r^2)$, where r is the correlation coefficient:

$$r = \frac{S_{xy}}{S_x S_Y} = \frac{\frac{2}{3}}{\sqrt{\frac{2}{3} \times \frac{14}{9}}} = \sqrt{\frac{3}{7}} \doteq .65.$$

Thus

$$S_Y^2(1 - r^2) = \tfrac{14}{9}(1 - \tfrac{3}{7}) = \tfrac{8}{9},$$

which agrees with the above estimate of error variance.

EXAMPLE 10.2

An experiment was designed to study the effect of dissolved sulfur on the surface tension of liquid copper. The decrease in surface tension was assumed to be a linear function of the logarithm of the percentage of sulfur. Data were obtained as follows:*

x = log per cent sulfur	Y = decrease in surface tension (deg/cm)
−3.38	308
−2.38	426
−1.20	590
−.92	624
−.49	649
−.19	727

Calculations of slope and intercept according to the formulas for $\hat{\beta}$ and $\hat{\alpha}$, respectively, yielded the following empirical regression function:

$$y = 736 + 127.6x.$$

The correlation coefficient is $r = .996$, so $1 - r^2 = .008$ and

$$\hat{\sigma}^2 = S_Y^2(1 - r^2) = 81.37 = (9.02)^2.$$

The data and the empirical regression line are plotted in Figure 10.3.

Problems

10.1. To get acquainted with the arithmetic of least squares, work with the following artificial data, consisting of just three pairs (x, y):

x	y	x^2	y^2	xy	$a + bx_i$	$y_i - (a + bx_i)$
−1	1					
1	2					
2	0					

(a) Plot the given points on a rectangular coordinate system, and make a guess as to what you think would be a good-fitting line.

* Modified from data of Baer and Kellogg, *Journal of Metals*, Vol. 5 (1953), pp. 643–648.

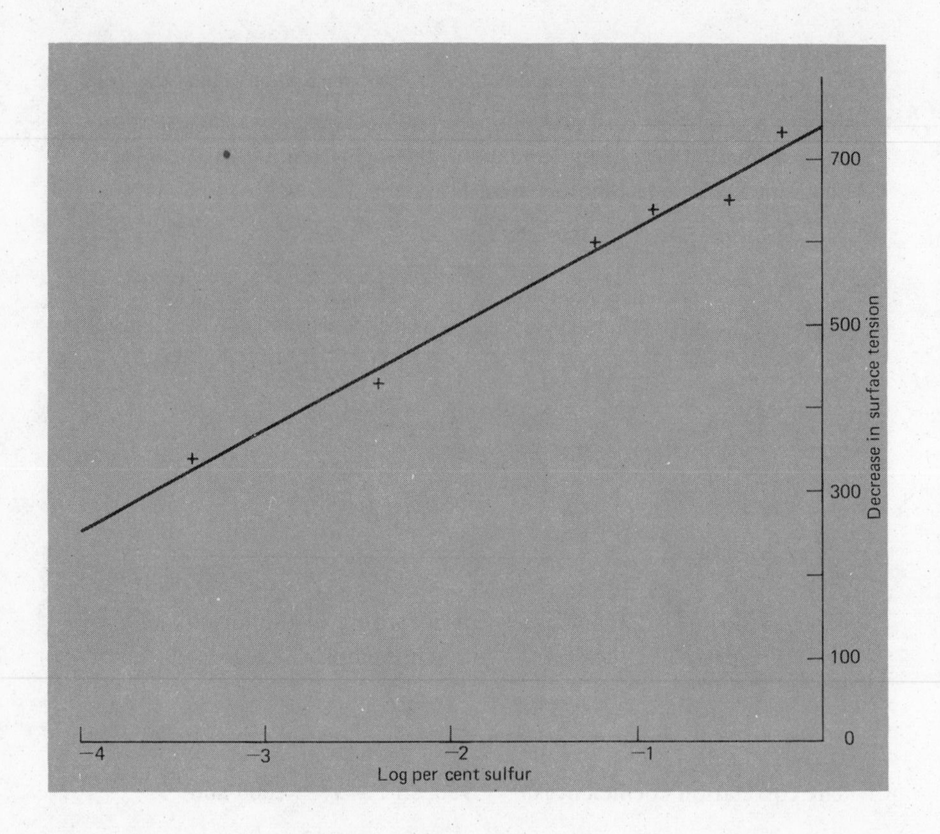

Figure 10.3
Regression line for Example 10.2.

(b) Determine the least-squares line $y = a + bx$ and draw it on your plot.

(c) Calculate the average squared residual about the least-squares line, $y = a + bx$:

$$\frac{1}{n}\sum [y_i - (a + bx_i)]^2.$$

[This is the minimum *value* of the quantity minimized by the least-squares line (among all possible lines). It is a good estimate of σ^2.]

(d) Calculate $S_y^2(1 - r^2)$, which should agree with your answer to (c). (Is the relationship strong enough so that knowing x permits a much better prediction of Y?)

10.2. The manager of a chain of stores, one in each of five similar towns, studied the effect of advertising by spending various amounts in the different communities at the time of a special sale. The data consist of the amounts spent (in hundreds of dollars) and the resulting sales. The sales figure given in each case represents a factor of increase over ordinary sales (when there is no sale) for that store.

Amount	1	3	5	7	9
Sales	1.4	1.2	2.0	2.2	2.2

(a) Plot the data and determine the least-squares line.
(b) What sales-increase factor would you anticipate for an advertising budget of $1500? (Give the naïve answer obtained by using the regression line, but then ask yourself whether this is realistic. Can you assume that the regression function is linear out to $1500?)

10.3. The data on which the ice went out of Lake Minnetonka is given for every tenth year for 80 years:

YEAR	DATE
1892	April 11
1902	April 1
1912	April 13
1922	April 17
1932	April 14
1942	April 4
1952	April 22
1962	April 25
1972	April 26

(a) Determine the least-squares line for these data and plot this line and the data. (If you have no calculator, use these coded variables:

X = number of decades after 1932,

Y = number of days after April 14.

It is a simple matter to put in the original values at the end.) Do statistics show that the climate is getting colder?
(b) Data are actually available for each year, and the successive 10-year average dates are as follows:

YEARS	AVERAGE DATE
1887–1896	April 14
1897–1906	April 16
1907–1916	April 16
1917–1926	April 15
1927–1936	April 12
1937–1946	April 13
1947–1956	April 18
1957–1966	April 16
1967–1972	April 15

Now what do you say* about (a)?

10.4. The following data† give the effects of 15 daily doses of thyroxine, in various amounts, on the kidney transamidinase activities of thyroidectomized rats. The quantity X is the logarithm of the daily dose per 100 g of body weight, and Y is a measure whose units are μmoles GA formed/g kidney/hr (where GA is guanidinoacetic acid).

X	−2	−1.7	−1.4	−1.1	−.8	−.5
Y	18.6	21.8	27.0	35.7	41.1	46.9

Obtain the least-squares regression line.

10.3 Inferences in Regression Problems

There are a number of problems of inference that might be posed in the context of a regression model, problems of estimating, testing, and predicting.

The least-squares technique has provided *estimates* of the parameter of the regression model, of α, β, and σ^2. Thus the slope $\hat{\beta}$ of the least-squares line is an estimate of the slope β of the "true" regression line, and the intercept $\hat{\alpha}$ of

* Compare these two quotes:

"It would be impossible to manipulate statistics. It's all done by computers." (Remark by a nominee to the post of chief of the principal statistics gathering agency of the federal government.)

"Ah, les statistiques! Your Secretary of Defense loves statistics. We Vietnamese can give him all he wants. If you want them to go up, they will go up. If you want them to go down, they will go down." (Quoted from a Vietnamese general in *The New York Times*, June 9, 1967.) In view of Problem 10.3, which quote is nearer right?

† Adapted from an article in *Endocrinology*, Vol. 87 (1970), pp. 1237–1244, by Van Pilsum et al. The Y-values are actually means of values for 10 rats each.

the least-squares line is an estimate of the intercept α of the true regression line. The logical question to raise is: How good are they?

It can be shown that $\hat{\beta}$ is normally distributed (assuming that the responses Y are normal) with mean β and variance σ^2/ns_x^2. So the smaller the measurement error σ^2, the better the estimate; and the larger the value of

$$ns_x^2 = \sum (x_i - \bar{x})^2,$$

the better the estimate. To make this latter quantity larger, it suffices to use as many observations as possible and to take them at as widely separated x-values as possible. The controlled variable x ordinarily has practical limits, and for a good estimate of slope one should take half his observations at as large an x-value as is practical and the other half at as small an x-value as is practical.

An interval estimate for β can be constructed from the fact that the variable

$$T = \sqrt{(n - 2)S_x^2}\ \frac{\hat{\beta} - \beta}{\hat{\sigma}}$$

has a t-distribution with $n - 2$ degrees of freedom (Table III of the Appendix). It is as follows:

$$\hat{\beta} - k\frac{\hat{\sigma}}{\sqrt{(n - 2)S_x^2}} < \beta < \hat{\beta} + k\frac{\hat{\sigma}}{\sqrt{(n - 2)S_x^2}},$$

where k is a percentile of the t-distribution corresponding to the desired confidence coefficient. For 90 per cent confidence, k is the 95th percentile; for 95 per cent, it is the 97.5th percentile; and so on. [The quantity $\hat{\sigma}/\sqrt{(n - 2)S_x^2}$ is just an estimate of the standard deviation of $\hat{\beta}$. It is the *standard error* of estimate, so the confidence interval for β is formed in essentially the same fashion as was the confidence interval for a mean.]

EXAMPLE 10.3

In Example 10.2, dealing with the apparently linear relation between the amount of dissolved sulfur and the surface tension of liquid copper, the estimate of slope was found to be $\hat{\beta} = 127.6$ (sulfur in units of log of per cent sulfur and tension in degrees per centimeter). Further calculations yield (with $n - 2 = 4$)

$$S_x = 1.135, \qquad \sqrt{(n - 2)S_x^2} = 2.227,$$

and

$$\sigma_{\hat{\beta}} \doteq \frac{\hat{\sigma}}{\sqrt{(n - 2)S_x^2}} = \frac{9.02}{2.227} = 4.05.$$

The 95 per cent confidence limits are 2.78 (the 97.5th percentile of t with 4 degrees of freedom) times this standard error on either side of $\hat{\beta}$. That is, the 95 per cent confidence interval is

$$127.6 - 2.78 \times 4.05 < \beta < 127.6 + 2.78 \times 4.05$$

$$\text{or} \qquad 116.34 \quad \text{to} \quad 138.86.$$

A test for the hypothesis that β has a specified value β_0 can be constructed by measuring how far $\hat{\beta}$ is from this hypothesized value β_0 in terms of σ-units—the σ, i.e., of the estimate $\hat{\beta}$. The statistic used is

$$T = \frac{\hat{\beta} - \beta_0}{\hat{\sigma}}\sqrt{(n-2)S_x^2},$$

which has a t-distribution with $n - 2$ degrees of freedom *if* β is really equal to β_0. One value of β that might be tested more than others is $\beta_0 = 0$, corresponding to a *horizontal* regression line, which would say that the controlled variable setting is irrelevant and that Y does not really depend on x.

EXAMPLE 10.4

The data of Problem 10.3(b), giving the average date of ice breakup for nine successive decades, yielded a slope estimate of $\hat{\beta} = .1$ (y in days after April 14, and x in decades after 1927–1936). With $n = 9$ and $S_x^2 = \frac{60}{9}$, one obtains (to test $\beta = 0$)

$$T = \frac{\hat{\beta} - 0}{\hat{\sigma}}\sqrt{(n-2)S_x^2} = \frac{.1}{1.64}\sqrt{7 \times \frac{60}{9}} = .42.$$

This is between -2.36 and $+2.36$, so one accepts $\beta = 0$ at the 5 per cent level of significance in a two-sided test. (The 97.5th percentile of t with 7 degrees of freedom is 2.36.)

There is usually little interest in estimating α, but it is important to know the value of $\alpha + \beta x$ for any given x, since this linear function of x provides a good *prediction* of the response y at the setting x. Not knowing α or β, the next best thing to use in predicting a response at $x = x_0$, say, would be the value of the least-squares regression function, $\hat{\alpha} + \hat{\beta}x_0$. This has $\alpha + \beta x_0$ as its expected value, and the mean-squared error of prediction is

$$\begin{aligned}\text{mean squared error} &= E[(\hat{\alpha} + \hat{\beta}x_0) - (\alpha + \beta x_0)]^2 \\ &= \sigma^2\left[1 + \frac{1}{n} + \frac{(x_0 - \bar{x})^2}{nS_x^2}\right],\end{aligned}$$

which can be approximated by using $\hat{\sigma}^2$ in place of σ^2. Observe that because of the term $(x_0 - \bar{x})^2$, the prediction deteriorates when the value x_0 is taken far from the center of the x-values used in obtaining the least-squares line.

EXAMPLE 10.5

Suppose, referring to Example 10.2, that it was desired to predict the decrease in surface tension corresponding to a sulfur concentration given by $x_0 = -4$ (log percent sulfur). The predicted value would be

$$Y = \hat{\alpha} + (-4)\hat{\beta} = 736 + (-4) \times 127.6 = 225.6.$$

The rms prediction error would be approximately

$$\hat{\sigma}\sqrt{1 + \frac{1}{n} + \frac{(x_0 - \bar{x})^2}{nS_x^2}} = 9.02\sqrt{1 + \frac{1}{6} + \frac{2.573}{7.44}} \doteq 11.$$

Another kind of inference problem, encountered in the field of *bioassay*, is that of estimating the value of x that produces an observed response. In such problems x is a *dose* (of some drug), and it is often necessary to estimate the amount of an unknown dose from the response which it produces and from a linear regression relationship estimated from other data. This problem will not be considered further, except to note that an obvious estimate of x is

$$\hat{x} = \frac{Y - \hat{\alpha}}{\hat{\beta}},$$

obtained by solving the regression equation for x.

10.4 Prediction in the Bivariate Normal Model

At the end of Chapter 9 it was pointed out that to predict the value Y of a bivariate normal pair (X, Y) given the value of X, it was necessary to estimate the parameters α and β in the regression function,

$$E(Y|X = x) = \alpha + \beta x,$$

from a random sample $(X_1, Y_1), (X_2, Y_2), \ldots, (X_n, Y_n)$. Reference to the properties of conditional distributions of Y given X listed there shows that Y_1 given $X_1 = x_1$, Y_2 given $X_2 = x_2, \ldots, Y_n$ given $X_n = x_n$ are precisely the same as assumed in the regression model in this chapter, in which the X-value is a controlled variable. For this reason the least-squares estimator of α and β

developed in this chapter are used to estimate $\alpha + \beta x$ in the bivariate normal model, i.e., used in predicting the value of Y given $X = x$.

It was seen in Chapter 9 that if α and β are *known*, the prediction $\alpha + \beta x$ has a mean squared error equal to

$$\text{var}(Y|X = x) = \sigma_Y^2(1 - \rho^2).$$

Now, this *conditional* variance is the σ^2 of the present chapter (the variance of Y at a given X-value, which is a constant). If, not knowing α and β, one uses $\hat{\alpha} + \hat{\beta}x$ as the predicted value of Y at $X = x$, there is an additional error of estimation involved; the formula given at the end of Section 10.3 was

$$\text{mean squared error} = \sigma^2\left[1 + \frac{1}{n} + \frac{(x - \bar{x})^2}{nS_x^2}\right].$$

Using this with σ^2 replaced by $\text{var}(Y|X = x)$, one has

$$\text{mean squared error} = (1 - \rho^2)\sigma_Y^2\left[1 + \frac{1}{n} + \frac{(x - \bar{x})^2}{nS_x^2}\right],$$

as the measure of success in prediction, *given* $X_1 = x_1, \ldots,$ and $X_n = x_n$. (If one takes into account the variability in these X's, the formula is slightly modified, a point that will not be discussed further here.)

EXAMPLE 10.6

Table 10.2 gives brown bear counts (by aerial survey) and average wind velocity on 20 days in the Chignik–Black Lakes area of the

Table 10.2

WIND VELOCITY, X (mph)	NUMBER OF BROWN BEARS, Y	WIND VELOCITY, X (mph)	NUMBER OF BROWN BEARS, Y
2.1	99	10.5	79
16.7	60	18.6	57
21.1	30	20.3	54
15.9	63	11.9	69
4.9	82	6.9	87
11.8	76	20.6	47
23.6	43	13.5	73
4.0	89	14.0	72
21.5	49	6.9	84
24.4	36	27.2	23

Alaska Peninsula (reported at a 1963 wildlife and resources conference). The results of appropriate calculations are as follows:

$$\bar{X} = 14.82, \quad S_X^2 = 51.504, \quad S_{XY} = -141.547,$$

$$\bar{Y} = 63.6, \quad S_Y^2 = 419.04, \quad r = -.9635,$$

$$\hat{\alpha} = 104.33, \quad \hat{\beta} = -2.75.$$

The predicted number of bears where the wind velocity is 30 mph would be

$$\hat{\alpha} + \hat{\beta}(30) = 104.33 - 82.45 \doteq 22,$$

with rms prediction error (given these wind velocities)

$$\left\{(1 - .9284)(419.04)\left[1 + \frac{1}{20} + \frac{(30 - 14.82)^2}{1030}\right]\right\}^{1/2} \doteq 6.2.$$

Although these computations have served to illustrate the various concepts in regression, some comments are in order. For one thing, it may be unlikely that one would want to *predict* the number of bears for a given wind velocity. It is perhaps more to the point to conclude simply that wind velocity is a factor in the movement of bears.

As to the linear model used, the scatter diagram (Figure 10.4) actually suggests more of a curved regression function than a linear one. Along with the least-squares regression line, an empirical quadratic regression function ($\alpha + \beta x + \gamma x^2$), determined by the method of least squares, is shown in Figure 10.4. The mean squared deviations of the data about the regression functions were computed to be

$$\hat{\sigma}^2_{\text{linear}} = S_Y^2(1 - r^2) = 30.0,$$

$$\hat{\sigma}^2_{\text{quad}} = 24.3,$$

which reinforce the intuitive conclusion that the underlying relation is more quadratic than linear. Of course, if the regression is actually quadratic, then the model for (X, Y) is not bivariate normal. For that matter, the scattergram does not really look like a plot of data from a bivariate normal distribution; indeed, the marginal data (X alone and Y alone) do not look much like univariate normal data—there is not much bunching around the middle, as there should be if the bivariate model is normal. (Incidentally, the bear count is necessarily an integer, so strictly speaking the continuous normal model would not apply; however, a continuous model may well serve as a good approximation.)

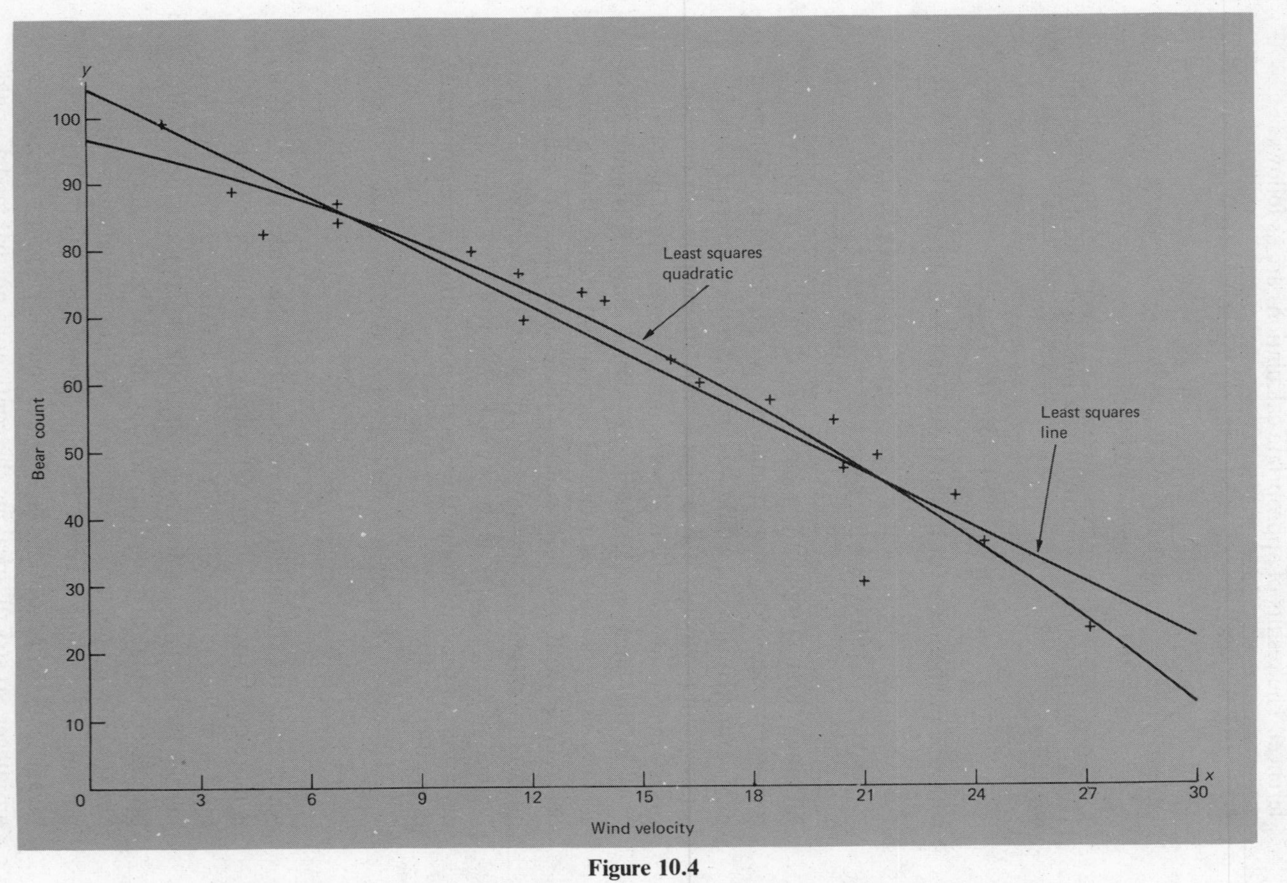

Figure 10.4
Scatter diagram and empirical regression lines, bear count versus wind velocity.

Problems

10.5. Obtain a 90 per cent confidence interval for the slope β of the true regression line in Problem 10.4. (From the computations of that problem, $S_x = .5123$, $S_y = 10.22$, and $\hat{\beta} = 19.82$.)

10.6. From the data of Problem 10.4, predict the value of y corresponding to a log daily dose of -1.5, and give the rms error of prediction. (The least-squares line was $y = 56.6 + 19.82x$.)

10.7. If in Problem 10.4 one observed a response of $Y = 24$, what would you estimate to be the log daily dose that produced this response?

10.8. Given $\bar{x} = 8.2$, $S_x^2 = 10.24$, $\hat{\alpha} = 6.31$, $\hat{\beta} = .092$, $n = 50$, and $\hat{\sigma}^2 = 4.6$, would you accept the hypothesis that $\beta = 0$ against the alternative that $\beta \neq 0$?

10.9. An educator has data on 500 students consisting of a grade-point average (GPA) after two years of college and a preentrance aptitude score for each student. The GPA's are scattered about a mean $\overline{Y} = 2.72$ with a standard deviation $S_Y = .36$, and the aptitude scores have mean $\overline{X} = 70$ and standard deviation $S_X = 10$.

(a) Without using any relationship between X and Y, the predicted GPA for any new student would be $\overline{Y} = 2.72$. What is the rms error involved in this prediction?

(b) The educator computes $r = .60$ and a least-squares regression function $y = .2 + .036x$. Using these, what GPA would he predict for a new student who scores 82 on the aptitude test, and what is the rms error involved in this prediction?

11 Analysis of Variance

The effect of a treatment on a response was studied in Chapters 7 and 8, which gave techniques for comparing the mean of a treated population with that of a control population. In Chapter 10, regression analysis dealt with a regulating variable that, in effect, treats the population or response variable at various levels given by selected values of the regulating variable.

Between these two situations is one in which a regulating factor or treatment is applied in more than one variety, not numerically indexed. For example, an experiment may be designed to determine whether the brand of seed used really makes a difference in corn yield. The experiment would involve measuring the yield from several plots, each planted with a different brand of seed. The analysis of the results of such an experiment will be considered in Section 11.1, while Section 11.2 will take up the more challenging situation in which more than one factor may be affecting a response. These are but glimpses into the extensive area of the design and analysis of experiments.

11.1 Single-factor Analysis

Consider a hypothetical (but not altogether unrealistic) situation in which manufactured parts are produced by a machine with an operator. The question arises: Does the output (say, in parts per hour) depend on the operator factor? To answer this, one might conduct an experiment in which each operator's output is checked for 1 hour's operating time. Perhaps the results are as follows:

Operator	1	2	3	4
Output	16	19	21	17

From this it could be rashly concluded that there is an operator effect, because the outputs are all different, until operator 1 points out the obvious—that his output, and that of any operator, is not the same for every trial of 1 hour's run. There is *variability* in the response, in the output, and another experiment like the one just conducted would probably have different results for each operator. The question then presents itself: Is there any more variability from operator to operator than there is among the several runs of a single operator? If not, it could hardly be concluded that there is an operator effect, and certainly this conclusion could not be reached on the basis of the single trial for each operator.

So a new experiment is proposed: Let each operator try the hour's run several times. (This is referred to as *replication.*) Now, if there really is no operator effect, the results might turn out to look as shown in Table 11.1 if

Table 11.1

	OPERATOR			
	1	2	3	4
Trial 1	17	19	20	16
Trial 2	20	21	17	21
Trial 3	20	20	17	20

each operator has three tries. Upon looking at this array of results, one would be tempted not to bother a statistician but to conclude intuitively that the variation between operators is not inconsistent with the variation in a single operator's performance. If this intuition is justifiable, a statistical analysis should bear it out. Moreover, a statistical analysis should be sufficiently sophisticated to yield information in the more usual case when intuition is a little shaky in its conclusion.

Both intuition and a statistical analysis are based on *assumptions* about the nature of the variation in the outputs. The assumptions would be hard to pin down in the case of intuition, partly because intuition is a personal thing. For a statistical analysis it would be assumed that each operator's output is a random variable, being representable by some (unknown) probability model, i.e., by a probability distribution. In the simplest case it is assumed further that the variance of this distribution is the same in the case of each operator. Differences among the means, on the other hand, would be indicative of an operator effect.

The response, the output of 1 hour's operating time in the present example, will be denoted generally by X_{ij}, the double subscript identifying this as the output of the ith operator's jth trial run. For operator i, then, the results would be referred to (assuming k trials) as

$$X_{i1}, X_{i2}, \ldots, X_{ik},$$

which constitute a set of k observations or a sample from the population of operator i's possible outputs. Let μ_i denote the mean or expected value of this population:

$$\mu_i = E(X_{ij}).$$

With this mean separated out, the response can be written

$$X_{ij} = \mu_i + e_{ij},$$

which expresses it as its mean plus a random component or deviation about that mean. Clearly, the mean of the random component e_{ij} is zero, and the assumption mentioned earlier about variances means that all the random components have the same variance, call it σ^2:

$$E(e_{ij}) = 0 \quad \text{and} \quad \operatorname{var}(e_{ij}) = \sigma^2.$$

The sample of outputs of a particular operator will provide an estimate of the mean output for that operator; for operator i, the sample mean is

$$\overline{X}_i = \frac{1}{k} \sum_{i=1}^{k} X_{ij}.$$

The same sample will provide an estimate of variability, i.e., of σ^2:

$$S_i^2 = \frac{1}{k}\sum_{i=1}^{k}(X_{ij} - \overline{X}_i)^2.$$

The particular results (from three fictitious trials each) given earlier are repeated here, together with the sample means and variances, in Table 11.2. The average of the sample averages,

$$\overline{X} = \frac{1}{m}\sum_{i=1}^{m}\overline{X}_i = \frac{1}{mk}\sum_{i=1}^{m}\sum_{j=1}^{k}X_{ij},$$

Table 11.2

OPERATOR	OUTPUT, X_{ij}	SAMPLE MEAN, $\overline{X}_i$	$\sum_i (X_{ij} - \overline{X}_i)^2$	SAMPLE VARIANCE, S_i^2
1	17, 20, 20	19	6	2
2	19, 21, 20	20	2	$\frac{2}{3}$
3	20, 17, 17	18	6	2
4	16, 21, 20	19	14	$\frac{14}{3}$
		$\overline{X} = 19$		$S_E^2 = \frac{7}{3}$

is shown, as is the average of the variances,

$$S_E^2 \equiv \frac{1}{m}(S_1^2 + \cdots + S_m^2) = \frac{28}{12} = \frac{1}{mk}\sum_{i=1}^{m}\sum_{j=1}^{k}(X_{ij} - \overline{X}_i)^2.$$

(The number of operators has been given the general name m, and the number of replicates for each operator, the name k.)

Why average the sample variances? Although each sample does indeed provide an estimate of σ^2, the *error variance*, these estimates are all different; moreover, each one is based on a small number of observations. A combination of these estimates can be more reliable.* The combination denoted S_E^2 is called a *pooled variance*, for the information on variability in the various samples is pooled in it. It is also called a *within-samples* estimate of variance, being based on the variation within each sample about that sample mean.

In developing a test for a given null hypothesis, it is sometimes helpful to take a look at data generated when the null hypothesis is false, as they compare

* It should be at least mentioned that the combining operation of a simple average is only appropriate when the number of replications is the same for each sample; more generally, the sample means and variances would need to be weighted according to sample size in averaging them.

with data generated when the null hypothesis is true. With this in mind, consider the data in Table 11.3. Notice that these data were obtained from the

Table 11.3

OPERATOR	OUTPUT, X_{ij}	SAMPLE MEAN, $\overline{X}_i$	$\sum_i (X_{ij} - \overline{X}_i)^2$	SAMPLE VARIANCE, S_i^2
1	17, 20, 20	19	6	2
2	22, 24, 23	23	2	$\frac{2}{3}$
3	16, 13, 13	14	6	2
4	21, 26, 25	24	14	$\frac{14}{3}$
		$\overline{X} = 20$		$S_E^2 = \frac{7}{3}$

first set simply by adding 3 to the outputs of operator 2, subtracting 4 from those of operator 3, and adding 5 to those of operator 4, i.e., by artificially inserting an operator effect. Just as the first set of data seemed to say intuitively that there was no operator effect, so this second set would probably be sufficient to convince anyone that there is an operator effect, even without any kind of statistical analysis.

Notice that the sample variances and the pooled variance are the same as they were for the original set of data (before constant amounts were added to or subtracted from a given operator's outputs); this is because the variances are based on differences. But the sample means now exhibit more variation from sample to sample.

The amount of variation among the sample means will serve as the basis for judging whether or not an operator effect is present. But, as usual, to be able to decide that any statistic is too large, or larger than it should be to be accepted as within ordinary sampling fluctuations, it is necessary to have a standard of comparison. That standard is provided here by the pooled variance, a quantity not sensitive to the presence or absence of an operator effect.

Some variation among the sample means is expected, even under the null hypothesis of no operator effect, because a sample mean is a random variable, with variance σ^2/m (where σ^2 is the population variance and m the sample size). This variance of the sample mean can be estimated from the data; for the four values of $\overline{X}$ in the present example constitute a sample of 4 from a population of $\overline{X}$-values with variance $\sigma^2/3$. The sample variance, which provides this estimate of $\sigma^2/3$, is

$$\frac{1}{4}\sum_1^4 (\overline{X}_i - \overline{X})^2,$$

and the estimate of σ^2 (which, of course, is three times the estimate of $\sigma^2/3$) is

$$\frac{3}{4}\sum_1^4 (\overline{X}_i - \overline{X})^2 \equiv \text{between-samples estimate of } \sigma^2.$$

More generally, for k samples of size m, the estimate would be

$$\frac{m}{k}\sum_1^k (\overline{X}_i - \overline{X})^2.$$

For the first set of data, the between-samples estimate of σ^2 is

$$\tfrac{3}{4}[(19 - 19)^2 + (20 - 19)^2 + (18 - 19)^2 + (19 - 19)^2] = 1.5,$$

and for the second set of data it is

$$\tfrac{3}{4}[(19 - 20)^2 + (23 - 20)^2 + (14 - 20)^2 + (24 - 20)^2] = 46.5.$$

Recalling that the within-samples estimate of σ^2 (or the pooled variance) is, for *both* sets,

$$\tfrac{1}{4}(S_1^2 + S_2^2 + S_3^2 + S_4^2) = \tfrac{28}{12} = 2.33,$$

one notices that whereas the between and within estimates are not much different for the first set of data, the between estimate is *much* larger in the case of the second set. The statistic to be used is the *ratio* of the within to the between estimate:

$$\text{data set 1:} \quad \frac{\frac{6}{4}}{\frac{28}{12}} = \frac{1.5}{2.33} = .643;$$

$$\text{data set 2:} \quad \frac{\frac{186}{4}}{\frac{28}{12}} = \frac{46.5}{2.33} = 19.93.$$

The ratio should be near 1 when there is *no* operator effect and will tend to be rather larger than 1 when there *is* an operator effect. That is, the ordinary variation expected under H_0 from one sample mean to another is inflated by the presence of an operator effect.

To decide how large a ratio should be considered significant—grounds for *rejecting* the null hypothesis of no operator effect—it is necessary to know the sampling distribution of the test ratio. This has been derived, computed, and tabulated, but, more precisely it is readily available only for a slight modification of the above test ratio in which the divisors in the variance estimates are not sample sizes but degrees of freedom. The test ratio, so modified, is usually called F and has a distribution (under H_0) called the *F-distribution*. For the given sets of data, the F-ratio is

$$\text{data set 1:} \quad F = \frac{\frac{6}{3}}{\frac{28}{8}} = .57;$$

$$\text{data set 2:} \quad F = \frac{\frac{186}{3}}{\frac{28}{8}} = 17.7.$$

The degrees of freedom are as follows. In the numerator, the estimate of variance is based on the variance of a sample of k means ($k = 4$ in the example), and the number of degrees of freedom is one less:

$$\text{*numerator degrees of freedom} = k - 1.$$

In the denominator, the estimate is a combination of estimates (the k sample variances) each having $m - 1$ degrees of freedom:

$$\text{*denominator degrees of freedom} = (m - 1) + (m - 1) + \cdots + (m - 1) = n - k,$$

where $n = mk$ is the total number of observations comprising the data.

In the present example, the degrees of freedom are $k - 1 = 3$ in the numerator and $n - k = 12 - 4 = 8$ in the denominator. Table IV of the Appendix gives critical values of the F-ratio corresponding to $\alpha = .05$ and to $\alpha = .01$. From the first page of that table, one finds 4.07 under 3 and across from 8; this means that (under H_0)

$$P(F > 4.07) = .05,$$

so the rule: "Reject H_0 if $F > 4.07$" has an α of .05. For the given data,

$$\text{first data set:} \quad F = .67 < 4.07,$$

$$\text{second data set:} \quad F = 17.7 > 4.07,$$

so the first data set (and the given rule) calls for accepting H_0, but the second set calls for rejecting H_0 and concluding that there is an operator effect.

The analysis of data in problems of this type is usually presented in a table based on sums of squares, and the following identity:

$$\sum_{i=1}^{k} \sum_{j=1}^{m} (X_{ij} - \overline{X})^2 = \sum\sum (X_{ij} - \overline{X}_i)^2 + \sum\sum (\overline{X}_i - \overline{X})^2$$

$$= \sum\sum (X_{ij} - \overline{X}_i)^2 + m \sum_{i=1}^{k} (\overline{X}_i - \overline{X})^2.$$

* These numbers of degrees of freedom are stated as facts. They can be shown mathematically to be correct, but it is not meant to imply that they are obviously so. The notion of "degrees of freedom" is actually a mathematical one, but students usually acquire an intuitive feel for it with continued use.

Rather than derive this algebraically (not a difficult thing to do), we verify it for the problem at hand. Using the second set of data, we obtain

$$\begin{aligned}\sum\sum(X_{ij} - \bar{X})^2 &= (17 - 20)^2 + (20 - 20)^2 + (20 - 20)^2 + (22 - 20)^2 \\ &\quad + (24 - 20)^2 + (23 - 20)^2 + (16 - 20)^2 + (13 - 20)^2 \\ &\quad + (13 - 20)^2 + (21 - 20)^2 + (26 - 20)^2 + (25 - 20)^2 \\ &= 214.\end{aligned}$$

The terms on the right had been calculated previously to be 28 and 3×62, so

$$214 = 28 + 186.$$

The total sum of squares, 214, divided by 12 or by the number of degrees of freedom ($n - 1 = 11$) is what one would use to estimate σ^2 *if* he knew H_O to be true, in which case the 12 observations could be thought of as making up one sample from a single population.

The *analysis of variance*, or *ANOVA*, for the second set of data is set up as shown in Table 11.4. The term "analysis of variance" refers to the decomposition of a sum of squares about the grand mean into components that relate to specific sources of variation. One component is generally an error sum of squares, being related to the inherent random variation measured by σ^2, a term that is not influenced by the presence or absence of factor effects.

Table 11.4

SOURCE	SUM OF SQUARES	DEGREES OF FREEDOM	VARIANCE ESTIMATE
Factor	186	3 $(k - 1)$	$\frac{186}{3}$
Error	28	8 $(n - k)$	$\frac{28}{8}$
Total	214	11 $(n - 1)$	$\frac{214}{11}$

EXAMPLE 11.1

In an agricultural experiment, the yield (in bushels/acre) of spring wheat was determined for each of several treatments, each treatment being a certain mixture of nitrogen, potassium, and potash. Five replicates were obtained for each treatment, with yield as given in Table 11.5. The factor here is the fertilizer used, and the hypothesis to be tested is that there is no factor effect (no treatment differences). The ANOVA table is given in Table 11.6. The test statistic is the ratio of variance

Table 11.5

TREATMENT	YIELD	AVERAGE
1	15.1, 18.5, 16.9, 18.4, 11.8	16.14
2	15.4, 17.9, 18.2, 14.7, 16.6	16.56
3	15.3, 21.3, 13.7, 13.6, 15.5	15.88
4	17.1, 22.3, 20.9, 20.6, 15.6	19.30
5	15.9, 19.0, 20.6, 15.4, 11.4	16.46
		Overall average = 16.87

Table 11.6

SOURCE	SUM OF SQUARES (SS)	DEGREES OF FREEDOM (d.f.)	VARIANCE ESTIMATOR, SS/d.f.
Treatment	38.41	4	9.60
Error	162.68	20	8.134
Total	201.09	24	8.379

estimates:

$$F = \frac{9.60}{8.134} = 1.18.$$

Since the 95th percentile of the F-distribution with (4, 24) degrees of freedom is 2.78, and since $1.18 < 2.78$, one accepts the hypothesis of no treatment differences at the 5 per cent significance level. This does not necessarily mean that fertilizers in general have no value but simply that among the five particular fertilizer combinations used, the average yields are not significantly different (under the given test conditions of soil, weather, etc.).

Problems

11.1. To help in fixing the computational scheme without obscuring it with heavy arithmetic, set up the ANOVA tables and carry out the F-test for equality of means against the alternative hypothesis of the presence of a "factor effect", for each of the following artificial data sets:

(a) Sample with factor at level 1: (10, 12); sample with factor at level 2: (4, 6).

(b) Sample with factor at level 1: (24, 36, 42); sample with factor at level 2: (60, 66, 78).

11.2. When there are only two "levels" of a factor involved, the problem becomes that of testing the equality of means of two normal populations, for which a t-test was given in Chapter 7. Compute the t-statistic for the data of (b) in Problem 11.1 and compare the inequality $t > t_{crit}$ with $F > F_{crit}$. (*Hint*: Square both sides of the t-inequality.)

11.3. Four classes of 20 each took a television course at different times of day. Average, final composite scores were as follows (with standard deviations):

early morning class: $\bar{X}_1 = 76.2 \quad (S_1 = 6.0)$

late morning class: $\bar{X}_2 = 74.6 \quad (S_2 = 6.4)$

early afternoon class: $\bar{X}_3 = 72.5 \quad (S_3 = 5.8)$

late afternoon class: $\bar{X}_4 = 73.0 \quad (S_4 = 6.1)$

Test the hypothesis of no factor effect, where the factor is time of day. [*Note*: To calculate the error sum of squares, you will need the sum of squares for each sample, obtained by multiplying the sample variance by the sample size. Thus, for the first sample,

$$(SS)_1 = 20(6.0)^2 = 720,$$

and so on.

11.4. A recent elementary statistics text gives the following data on gasoline tax per gallon (as of 1970) and asks the student if there is a significant difference in taxes among the three areas:

New England		Mideast		Far West	
Maine	8	New York	7	Washington	9
New Hampshire	7	New Jersey	7	Oregon	7
Vermont	8	Pennsylvania	8	Nevada	6
Massachusetts	6.5	Delaware	7	California	7
Rhode Island	8	Maryland	7	Alaska	8
Connecticut	8	Washington, D.C.	7	Hawaii	5

Can an F-test be properly applied in this situation? (Do you have random samples from normal populations with equal variances? To say that there is a significant difference means that the observed differences are too great to be reasonably explained away as sampling fluctuations; if there is no sampling, is the question of significant difference meaningful?)

11.5. To test the hypothesis that all gasolines are alike, 12 drivers are selected whose cars and driving routines are similar. Each of four major gasoline brands is assigned to 3 drivers, and they report mileages per gallon (after a given number of miles) as follows:

brand A: 12.6, 10.9, 11.3;
brand B: 10.4, 10.1, 11.6;
brand C: 12.2, 11.0, 11.9;
brand D: 12.0, 9.6, 10.8.

Carry out, at the 5 per cent level, the F-test for no factor effect.

11.2 Two-factor Analysis

Even in the supposedly one-factor experiments described in Section 11.1, there were other factors in the background. In growing spring wheat, (Example 11.1) the soil may be a factor, and the replicates were actually obtained by planting at five different locations. The traditional approach of holding all factors constant except the one of interest not only may be awkward or even impossible, but it also restricts the applicability of any conclusion to just those situations where the extra factors are again held fixed at the levels used in the experiment. An approach will now be presented that permits testing the effect of one factor when a second factor is present at various levels.

Suppose, then, that two factors are possibly contributing to a response, factors A and B. Each level of A combined with a level of B defines a treatment, and one or more responses would be obtained for each treatment. However, only the simplest case of one observed response for each treatment is considered here. That observation will be denoted by X_{ij}, corresponding to the ith level of factor A in combination with the jth level of factor B.

The response X_{ij} is assumed to be made up of a deterministic part plus a random error with mean zero:

$$X_{ij} = \mu_{ij} + \varepsilon_{ij}.$$

Thus

$$E(X_{ij}) = \mu_{ij} \quad \text{and} \quad E(\varepsilon_{ij}) = 0.$$

It will be further assumed that the error variance is the same for all treatments:

$$\operatorname{var} \varepsilon_{ij} = \sigma^2.$$

The mean μ_{ij} will be assumed to have a very special structure, to be a *sum* of a contribution associated with factor A and a contribution associated with factor B; for simplicity these contributions will be measured as deviations, so that

$$\mu_{ij} = \mu + \gamma_i + \delta_j.$$

Here, μ is a mean response that would be present even when the factors have no effect, γ_i is a factor A effect with $\sum \gamma_i = 0$, and δ_j is a factor B effect with $\sum \delta_j = 0$. (Thus "no effect" means really that there is no *variation* in average response among the various levels of a factor; the constant part of an effect, if any, is incorporated into the μ.)

Some notation is needed: Let $\bar{X}_{i\cdot}$ denote the average of the responses for level i of factor A, let $\bar{X}_{\cdot j}$ denote the average response at level j of factor B, and let $\bar{X}$ denote the average of all responses:

$$\bar{X}_{i\cdot} = \frac{1}{b}\sum_{j=1}^{b} X_{ij}, \qquad \bar{X}_{\cdot j} = \frac{1}{a}\sum_{i=1}^{a} X_{ij}, \qquad \bar{X} = \frac{1}{ab}\sum_{j=1}^{b}\sum_{i=1}^{a} X_{ij}.$$

Note that $\bar{X}$ is the average of the $\bar{X}_{i\cdot}$'s, as well as the average of the $\bar{X}_{\cdot j}$'s. A numerical example may help fix the notation; Table 11.7 presents artificial

Table 11.7

	LEVELS OF FACTOR B				
LEVELS OF FACTOR A	1	2	3	$\bar{X}_{i\cdot}$	$\bar{X}_{i\cdot} - \bar{X}$
1	3	5	4	4	−7
2	11	10	12	11	0
3	16	21	17	18	7
$\bar{X}_{\cdot j}$	10	12	11	$\bar{X} = 11$	
$\bar{X}_{\cdot j} - \bar{X}$	1	1	0		

data for the case of three levels of each factor ($a = b = 3$), so constructed as to exhibit an obvious factor A effect.

Now, it can be shown that natural estimates of the various components of μ_{ij} are corresponding means or mean differences:

$$\hat{\mu} = \bar{X},$$

$$\hat{\gamma}_i = \bar{X}_{i\cdot} - \bar{X},$$

$$\hat{\delta}_j = \bar{X}_{\cdot j} - \bar{X}.$$

So, although the actual decomposition of, say, $X_{31} = 16$ is

$$16 = \mu + \gamma_3 + \delta_1 + \varepsilon_{31},$$

an approximate decomposition is given by

$$16 \doteq \bar{X} + (\bar{X}_{3\cdot} - \bar{X}) + (\bar{X}_{\cdot 1} - \bar{X}) + \varepsilon_{31},$$

or

$$16 \doteq 11 + 7 + (-1) + \varepsilon_{31}.$$

The random component is then approximately

$$16 - 11 - 7 - (-1) = -1.$$

This and the others like it are used to provide an estimate of error variance (σ^2) that would be independent of factor effects. In general, these estimates of ε_{ij} are

$$X_{ij} - \bar{X} - (\bar{X}_{i\cdot} - \bar{X}) - (\bar{X}_{\cdot j} - \bar{X}) = X_{ij} + \bar{X} - \bar{X}_{i\cdot} - \bar{X}_{\cdot j}.$$

[For $i = 3$, $j = 1$, $16 - 11 - 7 - (-1) = 16 + 11 - 18 - 10$.] The sum of squared deviations is

$$\text{SSE} \equiv \sum\sum (X_{ij} + \bar{X} - \bar{X}_{i\cdot} - \bar{X}_{\cdot j})^2.$$

The deviations are computed in detail in Table 11.8 with the result

$$\text{SSE} = 0^2 + 0^2 + 0^2 + 1^2 + 2^2 + 1^2 + 1^2 + 2^2 + 1^2 = 12.$$

Table 11.8

3 + 11 − 4 − 10	5 + 11 − 4 − 12	4 + 11 − 4 − 11
11 + 11 − 11 − 10	10 + 11 − 11 − 12	12 + 11 − 11 − 11
16 + 11 − 18 − 10	21 + 11 − 18 − 12	17 + 11 − 18 − 11

The presence of a factor A effect is evident in the wide variation among the $\bar{X}_{i\cdot}$ (4, 11, 18). There would be *some* variation here even with no factor A effect, but not this much. The average of the squared deviations about the mean would be close to σ^2/b if there were no factor A effect, so that an estimate of σ^2 is given by

$$\frac{b}{a-1}\sum_{i=1}^{a}(\bar{X}_{i\cdot} - \bar{X})^2 = \frac{1}{a-1}\sum_{i=1}^{a}\sum_{j=1}^{b}(\bar{X}_{i\cdot} - \bar{X})^2$$

$$= \frac{3}{2}[(-7)^2 + 0^2 + 7^2] = 147.$$

Similarly, an estimate of σ^2 is available if there is no factor B effect:

$$\frac{a}{b-1}\sum_{j=1}^{b}(\bar{X}_{\cdot j} - \bar{X})^2 = \frac{1}{b-1}\sum_{i=1}^{a}\sum_{j=1}^{b}(\bar{X}_{\cdot j} - \bar{X})^2$$

$$= \frac{3}{2}[(-1)^2 + 0^2 + 1^2] = 3.$$

Again there is a relationship among the various sums of squared deviations:

$$\sum\sum (X_{ij} - \bar{X})^2 = \sum\sum (\bar{X}_{i\cdot} - \bar{X})^2 + \sum\sum (\bar{X}_{\cdot j} - \bar{X})^2$$
$$+ \sum\sum (X_{ij} + \bar{X} - \bar{X}_{i\cdot} - \bar{X}_{\cdot j})^2$$

or, symbolically,

$$\text{SST} = \text{SSA} + \text{SSB} + \text{SSE}.$$

The degrees of freedoms and their interrelation are then as follows:

$$\text{d.f.(T)} = \text{d.f.(A)} + \text{d.f.(B)} + \text{d.f.(E)}$$

or

$$ab - 1 = (a - 1) + (b - 1) + (a - 1)(b - 1).$$

In the 3 by 3 example under discussion, the sums-of-squares relationship is

$$312 = 294 + 6 + 12,$$

and the degrees of freedom,

$$9 - 1 = 2 + 2 + 2 \times 2.$$

The ANOVA table is, in general, as shown in Table 11.9, and in particular, for the artificial data above, as shown in Table 11.10.

Table 11.9

SOURCE	SUM OF SQUARES	DEGREES OF FREEDOM	VARIANCE ESTIMATE
Factor A	SSA	$a - 1$	SSA/$(a - 1)$
Factor B	SSB	$b - 1$	SSB/$(b - 1)$
Error	SSE	$(a - 1)(b - 1)$	SSE/$(a - 1)(b - 1)$
Total	SST	$ab - 1$	

Table 11.10

SOURCE	SUM OF SQUARES	DEGREES OF FREEDOM	VARIANCE ESTIMATE
A	294	2	147
B	6	2	3
Error	12	4	3
Total	312	8	

The test statistic for factor A, which tends to be larger than it ought to be when there *is* a factor A effect, is the ratio of the factor A variance estimate to the error variance estimate:

$$F_A \equiv \frac{\frac{1}{a-1}\sum\sum(\overline{X}_{i.} - \overline{X})^2}{\frac{1}{(a-1)(b-1)}\sum\sum(X_{ij} + \overline{X} - \overline{X}_{i.} - \overline{X}_{.j})^2}$$

Under the null hypothesis of no factor A effect, and if the observation X_{ij} is *normally* distributed, F_A has an F-distribution with $(a-1, (a-1)(b-1))$ degrees of freedom, *whether or not there is a factor B effect*. For the given data,

$$F_A = \tfrac{147}{3} = 49,$$

far in excess of the 5 per cent rejection limit of 6.94, the 95th percentile of $F(2, 4)$. The null hypothesis would be rejected.

(It may be noted in passing that $F_B = 1$, so if factor B had been of interest, one would accept the hypothesis of no factor B effect.)

Example 11.2

A wear-testing machine consists of four weighted brushes under which samples of fabric are affixed to measure their resistance to abrasion. The loss of weight of the material (after a given number of cycles) is used as a measure of this resistance. Four fabric types were tested in each of the four brush positions, with results as shown in Table 11.11.

Table 11.11

	Brush position (factor B)				$\overline{X}_{i.}$
Fabric (factor A)	1.93	2.38	2.20	2.25	2.19
	2.55	2.72	2.75	2.70	2.68
	2.40	2.68	2.31	2.28	2.4175
	2.33	2.40	2.28	2.25	2.315
$\overline{X}_{.j}$	2.3025	2.545	2.385	2.37	$2.40 = \overline{X}$

To test for a fabric difference (in the presence of a possible difference among brush positions), one uses the test ratio

$$F_A = \frac{\text{SSA}/3}{\text{SSE}/9},$$

where $3 = a - 1, 9 = (a - 1)(b - 1)$, and

$$\text{SSA} = \sum\sum (\overline{X}_{i\cdot} - \overline{X})^2 = .5201,$$

$$\text{SSE} = \sum\sum (X_{ij} - \overline{X}_{i\cdot} - \overline{X}_{\cdot j} + \overline{X})^2 = .1171.$$

With these values, $F_A = 13.32$, which exceeds the 5 per cent rejection limit of 3.86 [the 95th percentile of $F(3, 9)$]. The null hypothesis of no fabric difference is rejected at the 5 per cent significance level.

Problems

11.6. To help fix the method without being confused by heavy computation, construct an ANOVA table and corresponding F-tests for the following simple artificial data for a two-factor problem:

	Level of factor A	
Level of factor B	1	2
1	14	32
2	22	28

11.7. In testing the hypothesis that all gasolines are alike (see Problem 11.5), suppose that one is not sure about the possible effect of the make of automobile and that for each gasoline the three drivers drive, respectively, makes C, P, and R, with results (as before) as follows:

	Make		
Brand	C	P	R
A	12.6	10.9	11.3
B	10.4	10.1	11.6
C	12.2	11.0	11.9
D	12.0	9.6	10.8

(a) Carry out the F-test for no difference among gasoline brands.

(b) Carry out the F-test for no automobile effect.

11.8. Suppose that plant growth (the response) is affected by seed brand and by fertilizer, with no random error. In particular, suppose that seed brand 1 contributes 4 units to growth, and seed 2 contributes 6; and that fertilizer 1 contributes 2 and fertilizer 2 contributes 8. If these combine additively, then the responses are as in the following table, for each combination of seed and fertilizer:

	SEED	
FERTILIZER	1	2
1	6	8
2	12	14

(a) Determine μ, γ_i, and δ_j so that the response to seed i used with fertilizer j is $\mu + \gamma_i + \delta_j$, where $\sum \gamma_i = \sum \delta_j = 0$.
(b) If the responses in the table were actual data, what would be the error variance, that is, (SSE)/d.f.(E)? What would the F-ratios be? (*Note:* If there is no random-error component in the response, then *any* observed difference in the means would be significant.)
(c) If, instead, one had $\mu = 10$, $\gamma_1 = \gamma_2 = 0$, $\delta_1 = -3$, and $\delta_2 = 3$, what would be the F-ratio for the factor corresponding to γ?

Appendix: Tables

Table I Values of the standard normal distribution function.

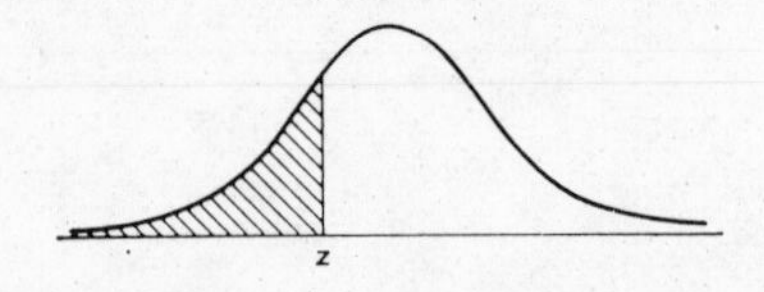

z	0	1	2	3	4	5	6	7	8	9
−3.0	.0013	.0010	.0007	.0005	.0003	.0002	.0002	.0001	.0001	.0000
−2.9	.0019	.0018	.0017	.0017	.0016	.0016	.0015	.0015	.0014	.0014
−2.8	.0026	.0025	.0024	.0023	.0023	.0022	.0021	.0021	.0020	.0019
−2.7	.0035	.0034	.0033	.0032	.0031	.0030	.0029	.0028	.0027	.0026
−2.6	.0047	.0045	.0044	.0043	.0041	.0040	.0039	.0038	.0037	.0036
−2.5	.0062	.0060	.0059	.0057	.0055	.0054	.0052	.0051	.0049	.0048
−2.4	.0082	.0080	.0078	.0075	.0073	.0071	.0069	.0068	.0066	.0064
−2.3	.0107	.0104	.0102	.0099	.0096	.0094	.0091	.0089	.0087	.0084
−2.2	.0139	.0136	.0132	.0129	.0126	.0122	.0119	.0116	.0113	.0110
−2.1	.0179	.0174	.0170	.0166	.0162	.0158	.0154	.0150	.0146	.0143
−2.0	.0228	.0222	.0217	.0212	.0207	.0202	.0197	.0192	.0188	.0183
−1.9	.0287	.0281	.0274	.0268	.0262	.0256	.0250	.0244	.0238	.0233
−1.8	.0359	.0352	.0344	.0336	.0329	.0322	.0314	.0307	.0300	.0294
−1.7	.0446	.0436	.0427	.0418	.0409	.0401	.0392	.0384	.0375	.0367
−1.6	.0548	.0537	.0526	.0516	.0505	.0495	.0485	.0475	.0465	.0455
−1.5	.0668	.0655	.0643	.0630	.0618	.0606	.0594	.0582	.0570	.0559
−1.4	.0808	.0793	.0778	.0764	.0749	.0735	.0722	.0708	.0694	.0681
−1.3	.0968	.0951	.0934	.0918	.0901	.0885	.0869	.0853	.0838	.0823
−1.2	.1151	.1131	.1112	.1093	.1075	.1056	.1038	.1020	.1003	.0985
−1.1	.1357	.1335	.1314	.1292	.1271	.1251	.1230	.1210	.1190	.1170
−1.0	.1587	.1562	.1539	.1515	.1492	.1469	.1446	.1423	.1401	.1379
− .9	.1841	.1814	.1788	.1762	.1736	.1711	.1685	.1660	.1635	.1611
− .8	.2119	.2090	.2061	.2033	.2005	.1977	.1949	.1922	.1894	.1867
− .7	.2420	.2389	.2358	.2327	.2297	.2266	.2236	.2206	.2177	.2148
− .6	.2743	.2709	.2676	.2643	.2611	.2578	.2546	.2514	.2483	.2451
− .5	.3085	.3050	.3015	.2981	.2946	.2912	.2877	.2843	.2810	.2776
− .4	.3446	.3409	.3372	.3336	.3300	.3264	.3228	.3192	.3156	.3121
− .3	.3821	.3783	.3745	.3707	.3669	.3632	.3594	.3557	.3520	.3483
− .2	.4207	.4168	.4129	.4090	.4052	.4013	.3974	.3936	.3897	.3859
− .1	.4602	.4562	.4522	.4483	.4443	.4404	.4364	.4325	.4286	.4247
− 0	.5000	.4960	.4920	.4880	.4840	.4801	.4761	.4721	.4681	.4641

Table I *(cont.)*

z	0	1	2	3	4	5	6	7	8	9
.0	.5000	.5040	.5080	.5120	.5160	.5199	.5239	.5279	.5319	.5359
.1	.5398	.5438	.5478	.5517	.5557	.5596	.5636	.5675	.5714	.5753
.2	.5793	.5832	.5871	.5910	.5948	.5987	.6026	.6064	.6103	.6141
.3	.6179	.6217	.6255	.6293	.6331	.6368	.6406	.6443	.6480	.6517
.4	.6554	.6591	.6628	.6664	.6700	.6736	.6772	.6808	.6844	.6879
.5	.6915	.6950	.6985	.7019	.7054	.7088	.7123	.7157	.7190	.7224
.6	.7257	.7291	.7324	.7357	.7389	.7422	.7454	.7486	.7517	.7549
.7	.7580	.7611	.7642	.7673	.7703	.7734	.7764	.7794	.7823	.7852
.8	.7881	.7910	.7939	.7967	.7995	.8023	.8051	.8078	.8106	.8133
.9	.8159	.8186	.8212	.8238	.8264	.8289	.8315	.8340	.8365	.8389
1.0	.8413	.8438	.8461	.8485	.8508	.8531	.8554	.8577	.8599	.8621
1.1	.8643	.8665	.8686	.8708	.8729	.8749	.8770	.8790	.8810	.8830
1.2	.8849	.8869	.8888	.8907	.8925	.8944	.8962	.8980	.8997	.9015
1.3	.9032	.9049	.9066	.9082	.9099	.9115	.9131	.9147	.9162	.9177
1.4	.9192	.9207	.9222	.9236	.9251	.9265	.9278	.9292	.9306	.9319
1.5	.9332	.9345	.9357	.9370	.9382	.9394	.9406	.9418	.9430	.9441
1.6	.9452	.9463	.9474	.9484	.9495	.9505	.9515	.9525	.9535	.9545
1.7	.9554	.9564	.9573	.9582	.9591	.9599	.9608	.9616	.9625	.9633
1.8	.9641	.9648	.9656	.9664	.9671	.9678	.9686	.9693	.9700	.9706
1.9	.9713	.9719	.9726	.9732	.9738	.9744	.9750	.9756	.9762	.9767
2.0	.9772	.9778	.9783	.9788	.9793	.9798	.9803	.9808	.9812	.9817
2.1	.9821	.9826	.9830	.9834	.9838	.9842	.9846	.9850	.9854	.9857
2.2	.9861	.9864	.9868	.9871	.9874	.9878	.9881	.9884	.9887	.9890
2.3	.9893	.9896	.9898	.9901	.9904	.9906	.9909	.9911	.9913	.9916
2.4	.9918	.9920	.9922	.9925	.9927	.9929	.9931	.9932	.9934	.9936
2.5	.9938	.9940	.9941	.9943	.9945	.9946	.9948	.9949	.9951	.9952
2.6	.9953	.9955	.9956	.9957	.9959	.9960	.9961	.9962	.9963	.9964
2.7	.9965	.9966	.9967	.9968	.9969	.9970	.9971	.9972	.9973	.9974
2.8	.9974	.9975	.9976	.9977	.9977	.9978	.9979	.9979	.9980	.9981
2.9	.9981	.9982	.9982	.9983	.9984	.9984	.9985	.9985	.9986	.9986
3.	.9987	.9990	.9993	.9995	.9997	.9998	.9998	.9999	.9999	1.0000

1. If a normal variable X is not standard, its values must be standardized: $Z = (X - \mu/\sigma)$, i.e., $P(X \leq x) = \Phi[(x - \mu)/\sigma]$.
2. For $z \geq 4$, $\Phi(z) = 1$ to four decimal places; for $z \leq -4$, $\Phi(z) = 0$ to four decimal places.
3. Entries opposite 3 are for 3.0, 3.1, 3.2, etc.

Table II Percentiles of the chi-square distribution.

Degrees of Freedom	p					
	.70	.80	.90	.95	.975	.99
1	1.07	1.64	2.71	3.84	5.02	6.63
2	2.41	3.22	4.61	5.99	7.38	9.21
3	3.66	4.64	6.25	7.81	9.35	11.3
4	4.88	5.99	7.78	9.49	11.1	13.3
5	6.06	7.29	9.24	11.1	12.8	15.1
6	7.23	8.56	10.6	12.6	14.4	16.8
7	8.38	9.80	12.0	14.1	16.0	18.5
8	9.52	11.0	13.4	15.5	17.5	20.1
9	10.7	12.2	14.7	16.9	19.0	21.7
10	11.8	13.4	16.0	18.3	20.5	23.2
11	12.9	14.6	17.3	19.7	21.9	24.7
12	14.0	15.8	18.5	21.0	23.3	26.2
13	15.1	17.0	19.8	22.4	24.7	27.7
14	16.2	18.2	21.1	23.7	26.1	29.1
15	17.3	19.3	22.3	25.0	27.5	30.6
16	18.4	20.5	23.5	26.3	28.8	32.0
17	19.5	21.6	24.8	27.6	30.2	33.4
18	20.6	22.8	26.0	28.9	31.5	34.8
19	21.7	23.9	27.2	30.1	32.9	36.2
20	22.8	25.0	28.4	31.4	34.2	37.6
21	23.9	26.2	29.6	32.7	35.5	38.9
22	24.9	27.3	30.8	33.9	36.8	40.3
23	26.0	28.4	32.0	35.2	38.1	41.6
24	27.1	29.6	33.2	36.4	39.4	43.0
25	28.2	30.7	34.4	37.7	40.6	44.3
26	29.2	31.8	35.6	38.9	41.9	45.6
27	30.3	32.9	36.7	40.1	43.2	47.0
28	31.4	34.0	37.9	41.3	44.5	48.3
29	32.5	35.1	39.1	42.6	45.7	49.6
30	33.5	36.2	40.3	43.8	47.0	50.9
40	44.2	47.3	51.8	55.8	59.3	63.7
50	54.7	58.2	63.2	67.5	71.4	76.2
60	65.2	69.0	74.4	79.1	83.3	88.4

Table III Percentiles of the *t*-distribution.

Degrees of Freedom	p					
	.8	.9	.95	.975	.99	.995
1	1.38	3.08	6.31	12.7	31.8	63.7
2	1.06	1.89	2.92	4.30	6.96	9.92
3	.978	1.64	2.35	3.18	4.54	5.84
4	.941	1.53	2.13	2.78	3.75	4.60
5	.920	1.48	2.01	2.57	3.36	4.03
6	.906	1.44	1.94	2.45	3.14	3.71
7	.896	1.42	1.90	2.36	3.00	3.50
8	.889	1.40	1.86	2.31	2.90	3.36
9	.883	1.38	1.83	2.26	2.82	3.25
10	.879	1.37	1.81	2.23	2.76	3.17
11	.876	1.36	1.80	2.20	2.72	3.11
12	.873	1.36	1.78	2.18	2.68	3.06
13	.870	1.35	1.77	2.16	2.65	3.01
14	.868	1.34	1.76	2.14	2.62	2.98
15	.866	1.34	1.75	2.13	2.60	2.95
16	.865	1.34	1.75	2.12	2.58	2.92
17	.863	1.33	1.74	2.11	2.57	2.90
18	.862	1.33	1.73	2.10	2.55	2.88
19	.861	1.33	1.73	2.09	2.54	2.86
20	.860	1.32	1.72	2.09	2.53	2.84
21	.859	1.32	1.72	2.08	2.52	2.83
22	.858	1.32	1.72	2.07	2.51	2.82
23	.858	1.32	1.71	2.07	2.50	2.81
24	.857	1.32	1.71	2.06	2.49	2.80
25	.856	1.32	1.71	2.06	2.48	2.79
26	.856	1.32	1.71	2.06	2.48	2.78
27	.855	1.31	1.70	2.05	2.47	2.77
28	.855	1.31	1.70	2.05	2.47	2.76
29	.854	1.31	1.70	2.04	2.46	2.76
30	.854	1.31	1.70	2.04	2.46	2.75
∞	.842	1.28	1.64	1.96	2.33	2.58

Note: The area to the right of the table entry is $1 - p$. The distribution is symmetric. For example, for 10 degrees of freedom,

$$P(-1.37 < t < 1.37) = .9 - .1 = .8.$$

Table IVa Ninety-fifth percentiles of the *F*-distribution.

Denominator Degrees of Freedom	Numerator Degrees of Freedom												
	1	2	3	4	5	6	8	10	12	15	20	24	30
1	161	200	216	225	230	234	239	242	244	246	248	249	250
2	18.5	19.0	19.2	19.2	19.3	19.3	19.4	19.4	19.4	19.4	19.4	19.5	19.5
3	10.1	9.55	9.28	9.12	9.01	8.94	8.85	8.79	8.74	8.70	8.66	8.64	8.62
4	7.71	6.94	6.59	6.39	6.26	6.16	6.04	5.96	5.91	5.86	5.80	5.77	5.75
5	6.61	5.79	5.41	5.19	5.05	4.95	4.82	4.74	4.68	4.62	4.56	4.53	4.50
6	5.99	5.14	4.76	4.53	4.39	4.28	4.15	4.06	4.00	3.94	3.87	3.84	3.81
7	5.59	4.74	4.35	4.12	3.97	3.87	3.73	3.64	3.57	3.51	3.44	3.41	3.38
8	5.32	4.46	4.07	3.84	3.69	3.58	3.44	3.35	3.28	3.22	3.15	3.12	3.08
9	5.12	4.26	3.86	3.63	3.48	3.37	3.23	3.14	3.07	3.01	2.94	2.90	2.86
10	4.96	4.10	3.71	3.48	3.33	3.22	3.07	2.98	2.91	2.85	2.77	2.74	2.70
11	4.84	3.98	3.59	3.36	3.20	3.09	2.95	2.85	2.79	2.72	2.65	2.61	2.57
12	4.75	3.89	3.49	3.26	3.11	3.00	2.85	2.75	2.69	2.62	2.54	2.51	2.47
13	4.67	3.81	3.41	3.18	3.03	2.92	2.77	2.67	2.60	2.53	2.46	2.42	2.38
14	4.60	3.74	3.34	3.11	2.96	2.85	2.70	2.60	2.53	2.46	2.39	2.35	2.31
15	4.54	3.68	3.29	3.06	2.90	2.79	2.64	2.54	2.48	2.40	2.33	2.29	2.25
16	4.49	3.63	3.24	3.01	2.85	2.74	2.59	2.49	2.42	2.35	2.28	2.24	2.19
17	4.45	3.59	3.20	2.96	2.81	2.70	2.55	2.45	2.38	2.31	2.23	2.19	2.15
18	4.41	3.55	3.16	2.93	2.77	2.66	2.51	2.41	2.34	2.27	2.19	2.15	2.11
19	4.38	3.52	3.13	2.90	2.74	2.63	2.48	2.38	2.31	2.23	2.16	2.11	2.07
20	4.35	3.49	3.10	2.87	2.71	2.60	2.45	2.35	2.28	2.20	2.12	2.08	2.04
21	4.32	3.47	3.07	2.84	2.68	2.57	2.42	2.32	2.25	2.18	2.10	2.05	2.01
22	4.30	3.44	3.05	2.82	2.66	2.55	2.40	2.30	2.23	2.15	2.07	2.03	1.98
23	4.28	3.42	3.03	2.80	2.64	2.53	2.37	2.27	2.20	2.13	2.05	2.01	1.96
24	4.26	3.40	3.01	2.78	2.62	2.51	2.36	2.25	2.18	2.11	2.03	1.98	1.94
25	4.24	3.39	2.99	2.76	2.60	2.49	2.34	2.24	2.16	2.09	2.01	1.96	1.92
30	4.17	3.32	2.92	2.69	2.53	2.42	2.27	2.16	2.09	2.01	1.93	1.89	1.84
40	4.08	3.23	2.84	2.61	2.45	2.34	2.18	2.08	2.00	1.92	1.84	1.79	1.74
60	4.00	3.15	2.76	2.53	2.37	2.25	2.10	1.99	1.92	1.84	1.75	1.70	1.65

Table IVb Ninety-ninth percentiles of the *F*-distribution.

Denominator Degrees of Freedom	Numerator Degrees of Freedom												
	1	2	3	4	5	6	8	10	12	15	20	24	30
1	4050	5000	5400	5620	5760	5860	5980	6060	6110	6160	6210	6235	6260
2	98.5	99.0	99.2	99.2	99.3	99.3	99.4	99.4	99.4	99.4	99.4	99.5	99.5
3	34.1	30.8	29.5	28.7	28.2	27.9	27.5	27.3	27.1	26.9	26.7	26.6	26.5
4	21.2	18.0	16.7	16.0	15.5	15.2	14.8	14.5	14.4	14.2	14.0	13.9	13.8
5	16.3	13.3	12 1.	11.4	11.0	10.7	10.3	10.1	9.89	9.72	9.55	9.47	9.38
6	13.7	10.9	9.78	9.15	8.75	8.47	8.10	7.87	7.72	7.56	7.40	7.31	7.23
7	12.2	9.55	8.45	7.85	7.46	7.19	6.84	6.62	6.47	6.31	6.16	6.07	5.99
8	11.3	8.65	7.59	7.01	6.63	6.37	6.03	5.81	5.67	5.52	5.36	5.28	5.20
9	10.6	8.02	6.99	6.42	6.06	5.80	5.47	5.26	5.11	4.96	4.81	4.73	4.65
10	10.0	7.56	6.55	5.99	5.64	5.39	5.06	4.85	4.71	4.56	4.41	4.33	4.25
11	9.65	7.21	6.22	5.67	5.32	5.07	4.74	4.54	4.40	4.25	4.10	4.02	3.94
12	9.33	6.93	5.95	5.41	5.06	4.82	4.50	4.30	4.16	4.01	3.86	3.78	3.70
13	9.07	6.70	5.74	5.21	4.86	4.62	4.30	4.10	3.96	3.82	3.66	3.59	3.51
14	8.86	6.51	5.56	5.04	4.69	4.46	4.14	3.94	3.80	3.66	3.51	3.43	3.35
15	8.68	6.36	5.42	4.89	4.56	4.32	4.00	3.80	3.67	3.52	3.37	3.29	3.21
16	8.53	6.23	5.29	4.77	4.44	4.20	3.89	3.69	3.55	3.41	3.26	3.18	3.10
17	8.40	6.11	5.18	4.67	4.34	4.10	3.79	3.59	3.46	3.31	3.16	3.08	3.00
18	8.29	6.01	5.09	4.58	4.25	4.01	3.71	3.51	3.37	3.23	3.08	3.00	2.92
19	8.18	5.93	5.01	4.50	4.17	3.94	3.63	3.43	3.30	3.15	3.00	2.92	2.84
20	8.10	5.85	4.94	4.43	4.10	3.87	3.56	3.37	3.23	3.09	2.94	2.86	2.78
21	8.02	5.78	4.87	4.37	4.04	3.81	3.51	3.31	3.17	3.03	2.88	2.80	2.72
22	7.95	5.72	4.82	4.31	3.99	3.76	3.45	3.26	3.12	2.98	2.83	2.75	2.67
23	7.88	5.66	4.76	4.26	3.94	3.71	3.41	3.21	3.07	2.93	2.78	2.70	2.62
24	7.82	5.61	4.72	4.22	3.90	3.67	3.36	3.17	3.03	2.89	2.74	2.66	2.58
25	7.77	5.57	4.68	4.18	3.86	3.63	3.32	3.13	2.99	2.85	2.70	2.62	2.54
30	7.56	5.39	4.51	4.02	3.70	3.47	3.17	2.98	2.84	2.70	2.55	2.47	2.39
40	7.31	5.18	4.31	3.83	3.51	3.29	2.99	2.80	2.66	2.52	2.37	2.29	2.20
60	7.08	4.98	4.13	3.65	3.34	3.12	2.82	2.63	2.50	2.35	2.20	2.12	2.03

Table V Cumulative distribution of the rank-sum statistic.

c \ m, n	m=3, n=3	m=3, n=4	m=3, n=5	m=3, n=6	m=3, n=7	m=3, n=8	m=3, n=9	m=3, n=10
6	.050	.028	.015	.012	.008	.006		
7	.100	.057	.036	.024	.017	.012	.009	.007
8			.071	.048	.033	.024	.018	.014
9				.083	.058	.042	.032	.025
10					.092	.067	.050	.039
11						.097	.073	.056
12								.080
Max	15	18	21	24	27	30	33	36

c \ m, n	m=4, n=4	m=4, n=5	m=4, n=6	m=4, n=7	m=4, n=8	m=4, n=9	m=4, n=10
10	.014	.008					
11	.029	.016	.010	.006			
12	.057	.032	.019	.012	.008		
13	.100	.056	.033	.021	.014	.010	.007
14		.095	.057	.036	.024	.017	.012
15			.086	.055	.036	.025	.018
16				.082	.055	.038	.027
17					.077	.053	.038
18						.074	.053
19						.099	.071
20							.094
Max	26	30	34	38	42	46	50

Table V *(cont.)*

m	5	5	5	5	5	5
c \ n	5	6	7	8	9	10
15	.004					
16	.008					
17	.016	.009				
18	.028	.015	.009			
19	.048	.026	.015	.009		
20	.075	.041	.024	.015	.009	
21		.063	.037	.023	.014	.010
22		.089	.053	.033	.021	.014
23			.074	.047	.030	.020
24				.064	.041	.028
25				.085	.056	.038
26					.073	.050
27					.095	.065
28						.082
Max	40	45	50	55	60	65

m	6	6	6	6	6
c \ n	6	7	8	9	10
21					
⋮					
24	.008				
25	.013	.007			
26	.021	.011			
27	.032	.017	.010		
28	.047	.026	.015	.009	
29	.066	.037	.021	.013	.008
30	.090	.051	.030	.018	.011
31		.069	.041	.025	.016
32		.090	.054	.033	.021
33			.071	.044	.028
34			.091	.057	.036
35				.072	.047
36				.091	.059
37					.074
38					.090
Max	57	63	69	75	81

m	7	7	7	7
c \ n	7	8	9	10
28				
⋮				
34	.009			
35	.013	.007		
36	.019	.010		
37	.027	.014	.008	
38	.036	.020	.011	
39	.049	.027	.016	.009
40	.064	.036	.021	.012
41	.082	.047	.027	.017
42		.060	.036	.022
43		.076	.045	.028
44		.095	.057	.035
45			.071	.044
46			.087	.054
47				.067
48				.081
49				.097
Max	77	84	91	98

m	8	8	8
c \ n	8	9	10
36			
⋮			
46	.010		
47	.014	.008	
48	.019	.010	
49	.025	.014	.008
50	.032	.018	.010
51	.041	.023	.013
52	.052	.030	.017
53	.065	.037	.022
54	.080	.046	.027
55	.097	.057	.034
56		.069	.042
57		.084	.051
58			.061
59			.073
60			.086
Max	100	108	116

m	9	9
c \ n	9	10
45		
⋮		
59	.009	
60	.012	
61	.016	.009
62	.020	.011
63	.025	.014
64	.031	.017
65	.039	.022
66	.047	.027
67	.057	.033
68	.068	.039
69	.081	.047
70	.095	.056
71		.067
72		.078
73		.091
Max	126	135

m	10
c \ n	10
55	
⋮	
74	.009
75	.012
76	.014
77	.018
78	.022
79	.026
80	.032
81	.038
82	.045
83	.053
84	.062
85	.072
36	.083
87	.095
Max	155

Table VI Wilcoxon signed-rank statistic.

c \ n	3	4	5	6	7	8	9	10	11	12	13	14	15
0	.125	.062	.031	.016	.008								
1	.250	.125	.062	.031	.016	.008							
2		.188	.094	.047	.023	.012							
3			.156	.078	.039	.020	.010						
4				.109	.055	.027	.014	.007					
5				.156	.078	.039	.020	.010					
6					.109	.055	.027	.014					
7					.148	.074	.037	.019	.009				
8					.188	.098	.049	.024	.012				
9						.125	.064	.032	.016				
10						.156	.082	.042	.021	.010			
11							.102	.053	.027	.013			
12							.125	.065	.034	.017			
13							.150	.080	.042	.021	.009		
14								.097	.051	.026	.011		
15									.062	.032	.013	.008	
16									.074	.039	.016	.010	
17									.087	.046	.020	.012	
18										.055	.024	.015	
19										.065	.029	.018	.009
20										.076	.034	.021	.011
21										.088	.040	.025	.013
22											.047	.029	.015
23											.055	.034	.018
24											.064	.039	.021
25											.073	.045	.024
26											.084	.052	.028
27											.095	.059	.032
28												.068	.036
29												.077	.042
30												.086	.047
31												.097	.053
32													.060
33													.068
34													.076
35													.084
36													.094

Table entry is

$$P(S \leqslant c) = P\left(S \geqslant \frac{n(n+1)}{2} - c\right).$$

Example: For $n = 9$, a two-sided test at $\varphi = .10$ would allow up to .05 in each tail. Since

$$P(S \leqslant 8) = P\left(S \geqslant \frac{9 \cdot 10}{2} - 8\right) = .049,$$

test is to reject null hypothesis if $S \leqslant 5$ *or* $S \geqslant 37$.

Table VII Acceptance limits for the Kolmogorov–Smirnov test of goodness of fit.

Sample size (n)	Significance level .20	.15	.10	.05	.01
1	.900	.925	.950	.975	.995
2	.684	.726	.776	.842	.929
3	.565	.597	.642	.708	.829
4	.494	.525	.564	.624	.734
5	.446	.474	.510	.563	.669
6	.410	.436	.470	.521	.618
7	.381	.405	.438	.486	.577
8	.358	.381	.411	.457	.543
9	.339	.360	.388	.432	.514
10	.322	.342	.368	.409	.486
11	.307	.326	.352	.391	.468
12	.295	.313	.338	.375	.450
13	.284	.302	.325	.361	.433
14	.274	.292	.314	.349	.418
15	.266	.283	.304	.338	.404
16	.258	.274	.295	.328	.391
17	.250	.266	.286	.318	.380
18	.244	.259	.278	.309	.370
19	.237	.252	.272	.301	.361
20	.231	.246	.264	.294	.352
25	.21	.22	.24	.264	.32
30	.19	.20	.22	.242	.29
35	.18	.19	.21	.23	.27
40				.21	.25
50				.19	.23
60				.17	.21
70				.16	.19
80				.15	.18
90				.14	
100				.14	
Asymptotic Formula:	$\frac{1.07}{\sqrt{n}}$	$\frac{1.14}{\sqrt{n}}$	$\frac{1.22}{\sqrt{n}}$	$\frac{1.36}{\sqrt{n}}$	$\frac{1.63}{\sqrt{n}}$

Note: Reject the hypothetical distribution $F(x)$ if $D_n = \max|F_n(x) - F(x)|$ exceeds the tabulated value.

Table VIII Random digits.

27417	37466	02634	88669	40048	08214	26969	99765	88081	55897
38954	57301	55195	61827	68762	49737	94572	83779	00261	31017
63288	83157	82610	54988	86451	01137	68099	22368	78097	22814
03857	55306	43080	46580	80276	49049	09335	43444	42364	11343
23156	47373	82235	80873	70787	39294	72004	60456	61679	70083
26704	62413	47551	70601	35963	70579	79965	48012	29519	25935
71074	36017	12410	98131	37185	66769	40324	75192	15397	99213
63660	21039	90850	75335	44590	64661	27258	59402	87138	62719
00047	33405	84663	86436	69541	56266	00371	41292	39250	46203
48663	92549	45095	74678	26127	28716	66992	10129	14894	63870
88932	18086	33261	79176	96024	80772	69613	18780	34177	27814
39907	14677	92605	93176	92130	62843	66016	30816	97506	61924
58611	06636	94454	17717	34219	54262	05093	43003	01160	84026
61100	95696	57413	69688	56774	02459	57104	74179	31156	76371
32729	38187	16717	73386	76436	55586	65076	84409	36479	28616
03841	92173	70560	84500	33663	04739	69540	51932	01083	47499
35824	86387	49058	36372	12809	45085	66857	05203	96070	80848
56840	39483	18589	58205	40236	36508	12602	93550	71181	27029
93743	42148	63617	11277	79357	31829	59084	22450	31138	00554
16987	19109	84955	30511	05353	67065	98314	83411	32312	23167
72616	29968	18777	70421	59577	32625	49622	10155	75311	56733
67179	43481	09120	11762	00807	67566	84662	16132	13115	43016
65272	25753	22378	66210	89639	37482	57799	59979	77914	45624
90947	79279	08307	60124	99022	87802	05172	03349	32094	35209
41359	26049	74696	92393	04126	46791	48695	90355	67020	74380
67113	58433	43737	66740	88458	75580	80956	54845	40171	28785
36829	24167	73952	81638	62795	68101	00638	95272	63829	52414
06834	10105	67375	91668	11808	47950	86592	40986	28899	54464
81568	33762	80294	45631	64198	11616	11088	43348	78775	34610
06555	53079	79795	78115	87180	32465	95939	65525	86907	03813
51900	59234	08783	21595	00828	08366	02279	05456	51756	19950
32034	71495	28149	67612	09808	78818	07209	58358	60334	82616
11004	84122	08109	00850	75281	66311	31738	44799	74394	53821
33112	34843	75082	46597	36828	57707	51505	73661	72515	15422
34676	32859	69593	36231	00570	24333	55450	80738	49414	43909
14983	70033	26583	34833	27117	89707	80893	10229	91852	16566
10784	20917	54419	83281	48890	06119	43045	25756	42774	66388
20768	57585	17185	90688	64196	03288	43199	07695	98581	13933
57102	00082	65327	60863	20655	36014	18507	72598	96950	17637
58402	63311	37942	08571	00674	46677	16815	59005	28515	23363
89943	73042	56283	19089	59422	10212	31107	24937	23260	45055
72267	52610	57849	70679	32670	25072	85341	53254	36650	03058
45910	99654	12651	89473	17900	17025	13468	59460	07529	83832
91848	28451	02209	85536	22184	44048	35594	24703	18057	85665
76184	18280	36249	88903	17360	66496	62484	69830	26077	43829
78521	49138	38056	27285	33355	27016	26673	81124	58531	14123
28941	31372	90410	68318	64855	01861	04188	09743	40754	52288
42202	09402	14968	38596	80569	50644	05102	79146	91983	30543
23623	50510	09313	10294	42681	47540	79372	22936	66247	34149
61725	92558	17887	60673	92238	47362	15660	56379	97327	29574

Table VIII *(cont.)*

64696	64857	21227	26529	86917	75227	00379	51347	19317	38751
14778	47072	28921	51157	29835	60520	03322	73253	76555	35470
88599	16344	73364	22136	27112	19880	97150	97586	78186	07049
11173	16375	35703	84318	24765	37612	54473	88901	11870	04658
25639	68250	67113	37342	62149	90550	66985	73552	14655	86303
76586	72679	50277	44637	58456	58444	15593	41960	80033	94522
38963	00333	52997	32740	74618	88240	38472	70342	32607	42636
25055	21127	03933	73150	50756	24782	68802	73198	82400	17965
44550	96727	19618	57481	32117	59123	74554	96793	31437	62533
32524	72839	26851	99508	62882	54222	78774	48676	77987	99287
62312	00436	93683	82640	16574	74898	30747	51488	29014	96976
34959	55375	12768	93504	07721	13185	91279	93883	52600	86255
48808	24055	31273	58525	07459	69621	72651	81549	06622	25949
99263	65504	55424	97824	50196	24808	72790	90379	18791	88840
74193	81704	38887	57696	09837	13943	98203	58230	41690	15244
42166	93853	84877	57226	01611	40996	88573	61439	54266	67287
95072	53717	20956	85391	05413	16148	58056	02084	29997	23313
59205	96521	36571	90396	17995	31695	89782	72796	64047	85185
45294	70745	24670	26970	98656	68841	16147	96907	08206	57986
67121	23593	20371	89207	41445	20901	77089	33034	19442	85147
93156	02905	19331	42521	54049	22163	20727	93675	01481	04241
41991	87340	80089	32669	74570	96618	11450	45008	24258	93639
41328	11691	20048	41412	76754	78657	12527	00908	76834	18684
37650	42004	25808	80230	64248	24405	22169	48208	95346	29483
95713	88495	87179	08110	18757	65360	98039	09189	26794	47030
95494	90102	28920	61704	37409	38143	89423	89682	84349	29711
00921	61836	50205	57705	78102	86497	94303	22528	22515	40284
92133	72643	38514	79149	62997	20155	59817	04988	77411	68208
23435	97711	72860	02540	01850	98534	05521	22880	11353	93646
91874	60900	66412	67418	65952	26868	00291	00254	97800	50716
94345	74264	28008	66245	32475	70778	47621	70857	82295	60910
46880	48498	62966	35014	64634	62680	89650	66581	41967	60990
91123	46834	57183	79812	91659	92179	78303	24820	70060	71939
85193	76899	57655	62112	68312	83951	58521	51289	49593	26309
00533	89871	09070	24179	04385	23494	80977	72858	43910	97935
64021	76784	12020	26744	57333	43183	75611	63812	83324	12357
30669	23470	76111	33931	43533	57741	80007	03286	43448	82536
33098	59137	80547	90035	37723	84263	94869	74185	92526	06158
49407	16937	62631	61317	58848	55490	45814	41796	76252	31381
58337	32968	92312	89022	99631	60542	11719	63625	64250	96715
19786	48413	50995	74094	35694	02446	66344	15064	57937	71182
21936	23604	78773	91313	95296	57662	66345	78412	11924	28151
77329	68797	08760	30258	90455	89034	84140	56487	86956	32811
98365	75054	59850	10278	30822	12800	95205	41824	17526	62545
87856	02002	80939	36659	60715	69028	54285	47988	05594	47856
34970	68054	78295	91884	33005	74483	81017	56018	38611	77007
37342	79395	45799	06677	32994	86569	09749	91351	04804	72093
94011	69622	77007	35632	34899	48432	05174	23107	13975	72834
20658	12069	64678	03276	44060	07834	74443	16776	18405	85948
17761	03731	49910	87154	68575	19853	74848	79913	94123	82248

Table VIII *(cont.)*

56315	48471	97116	74033	32008	86001	03248	76110	72718	87132
34953	86422	23665	75111	63236	76659	71158	95212	94502	70177
88971	22665	98584	86442	17821	91569	43124	66751	58851	91739
53192	05633	29317	47183	37796	88954	47766	33111	38572	62958
99446	60028	68716	92684	39974	10901	77409	55607	09911	13040
82056	45547	91033	49035	09785	68576	92199	52904	38024	87580
49165	65216	82029	35348	43155	23153	69395	64751	68952	33093
29071	82641	33462	14437	89436	30693	94470	95231	38709	58722
86632	03222	28400	56629	15483	93928	92874	06406	86368	51669
45535	71075	23630	13690	14164	60048	46283	68186	03924	72366
40297	61499	01088	25579	22276	14286	89072	71552	92425	78081
31836	24195	51192	77951	57526	52512	61985	10803	22403	15584
72856	21486	51966	56651	83545	56127	79406	25725	53523	78965
22696	29613	66349	33652	62644	88183	28390	57001	65043	54793
71201	01109	65830	40179	85274	86680	44140	11162	11256	21508
47393	23959	15886	76954	37026	72761	87742	68516	19603	45566
25789	94892	40389	50933	30931	96622	16467	23167	88175	48009
71756	73447	92393	80828	94026	78580	52428	28854	54944	38339
60776	10179	98467	64236	86330	47519	80125	58744	58987	37225
18958	67633	77016	01029	99852	97885	13320	41088	69639	91068
13878	58155	84322	84585	30305	37061	89544	16284	46535	18359
57846	49565	45012	39134	05992	44637	90722	55304	03500	89870
93260	31490	61899	45584	47354	16749	47013	03209	27689	22375
54654	79263	40641	55449	84016	70636	83461	30766	92430	30388
87232	73290	40975	14888	31536	43748	51794	38323	32221	72215
44894	68764	53268	93672	68114	82216	20509	96036	61433	81030
99724	03300	81720	04570	81920	95765	48996	52595	75041	87815
76537	29132	05188	13037	09674	22928	31725	41970	62336	32985
17602	23614	44685	87928	29490	95765	91632	81614	62804	97778
40909	65008	84889	50960	69458	54896	30927	53587	79362	00151
03961	84458	43665	85930	04432	03371	43509	51674	16737	27874
75960	12175	97765	48126	04967	26963	01845	25203	95088	71121
21664	64912	07742	27406	69062	36750	87567	61571	26136	80931
10838	15046	51879	84148	45365	55493	69049	55321	91419	34435
44595	05138	55603	86748	55072	52161	08704	63120	85513	34352
00313	11597	97935	28087	02744	99393	50251	90612	05612	48577
57274	96193	76790	18872	48431	08239	77645	05879	02083	00365
40335	67876	77116	17367	95913	10037	03749	69417	31246	62927
76807	45782	78149	96586	30339	14177	23771	38644	53888	37454
12896	33857	35792	31155	58360	04882	35664	23493	89052	25993
32752	09135	66931	02604	86074	76522	37122	23947	12282	70862
77108	28260	92821	76967	88135	69181	67626	61356	36371	37294
52331	25839	63251	17042	22928	64422	39920	10214	45110	37191
92028	35866	47434	89214	92350	94021	47730	01010	22477	06430
36214	52053	09480	63176	83975	75825	03156	75740	25778	89976
20553	56800	46655	37398	03915	15547	26606	66728	16559	48430
06654	12662	25118	91240	03538	98651	68993	06059	40925	09397
01762	56544	83380	78381	71287	22296	70074	19975	17416	88572
79137	93701	23596	66243	58888	48651	76652	72789	22213	33738
36085	52201	29287	05932	03702	02356	88024	42089	13634	42559

Table VIII *(cont.)*

91025	20023	48231	68548	55984	86517	34481	95029	24106	09473
44462	79220	65948	86719	85361	91214	34761	83781	47378	11865
94176	70087	17707	39399	30031	87081	35535	33838	72953	74668
14345	81566	41288	38747	08528	80192	61073	48280	23029	06428
89218	62413	26284	28472	01234	74150	25607	46877	40261	98071
97806	65247	01446	87034	16555	12313	21250	89709	59692	99028
77490	86442	62291	59432	24736	50676	12111	95769	89103	44210
07861	91463	25521	06947	25055	73817	38530	92817	20253	55010
49777	49044	67185	21957	11325	90180	22260	72257	94950	87650
43180	95627	64133	77491	03042	71819	92675	42349	75394	98363
78982	23679	88405	59312	23557	40579	90120	68105	32126	43548
81044	66607	11437	95268	00784	60634	15308	02703	14846	90420
66642	59344	91580	36595	84881	51900	30200	83877	73492	97497
74307	64584	81773	24288	51174	44524	43291	90217	18253	93041
70398	25969	92129	95162	07597	73101	60619	20890	74837	95842
92316	91427	66145	06046	36457	83456	89833	07042	85150	23111
20253	14725	97526	82143	70661	48486	00854	66077	68436	92760
20664	39842	38005	38524	99514	88260	65096	60563	77854	69968
68093	70687	44832	50617	71541	29688	17144	00519	68816	94860
96629	56782	28700	06371	00339	87889	88922	47162	17820	14477
12651	83453	96170	50250	70077	39365	70962	37580	74478	08439
73397	34279	32537	68087	42429	31858	72181	42924	68572	72958
97160	79676	73607	67181	43297	97819	59015	99001	18719	20450
42734	17179	87541	67978	25428	04748	35073	43836	50130	37535
75096	15214	89002	72104	48776	62395	01792	01636	16137	16471
70562	78929	48915	08477	23662	65558	45764	24481	55813	56516
06156	62809	23538	01594	24533	57683	32906	19653	81596	99896
93792	53867	89411	50002	90257	52650	84247	88020	60885	97282
56212	58345	80525	80935	21627	23253	06842	35675	60608	30474
73950	78904	11174	80205	25978	28030	57452	14598	82953	63549
06652	85160	72750	64135	83803	53705	10239	34909	88778	54496
68313	85139	52373	61256	06248	37181	42320	69902	82575	26271
45775	79924	78650	35082	16769	13005	70816	29367	11976	66461
05268	64854	50911	56559	93577	63687	07149	61594	57957	29899
98762	52557	17929	91529	98156	96274	17494	99522	01619	97158
53152	96281	75843	25398	79632	11605	00568	41929	88467	43781
51065	49702	80829	90993	95861	30821	04384	34471	08058	34116
35739	75644	14806	30958	07020	22769	05327	83236	60715	65290
91462	66633	67653	76616	91606	45656	80283	66238	89725	55515
49574	15895	94915	85244	69126	12074	92439	37480	37055	54689
10548	21337	32138	19504	24612	95804	23067	80080	82533	50594
25613	61197	85956	58472	26860	25679	63898	26867	75262	57524
37675	75091	86208	79357	58019	81946	92127	24737	09901	35328
34731	13855	65661	43277	03249	66595	63174	13783	17010	63051
52846	48413	57468	02269	27884	42707	92801	43504	15329	11263
63973	70211	43736	77713	40956	26551	82015	18954	13403	97307
07766	82553	56537	64718	74719	75472	23141	24740	71736	89378
25099	26608	10067	56397	67216	92306	36097	61454	98043	32054
55200	49064	58726	46031	74686	32075	80289	09397	34991	79353
57120	66591	31886	70879	43179	51108	31652	83041	02114	10388

Table VIII *(cont.)*

08149	14776	83594	25074	00184	83366	52998	42104	81540	11126
69393	16793	26625	03626	93300	21854	71725	18831	52906	60561
07127	28219	15917	42137	10713	95124	57960	05683	43756	05025
67448	61562	40266	23938	90957	52162	01177	90182	08454	96561
86214	53821	81970	23142	42933	93478	30378	18495	48373	97689
21568	19342	07821	29941	84094	38074	22429	67170	41001	57467
80376	95821	97763	91328	25063	02004	71503	82970	61669	28335
04265	23100	73964	45774	15753	31831	59817	20633	92268	82581
52382	67432	94394	67879	11303	41356	96669	46031	80327	98777
41948	99708	55353	81160	43200	48243	37044	46103	65464	21671
09201	35481	83003	23514	02844	78075	65420	67535	32447	34575
12379	36696	08556	03309	21587	07065	87588	07954	46624	39729
96328	82959	31874	87005	08728	91636	18986	98453	18663	10773
33004	64495	76596	87233	49788	45582	95749	84642	89299	92760
56796	12936	76308	03529	33361	43442	11350	59165	70198	88570
20887	43157	74092	19950	81312	82117	80948	49051	78442	26003
58773	50675	68623	36991	65293	62509	22069	33326	33164	24959
17542	12554	64286	76034	90353	04714	13614	11349	92165	99212
95002	80153	31722	66744	63397	41530	81338	67986	51478	98720
57807	77433	65367	64861	36395	45595	17574	48568	73159	36830
78394	97930	72476	81628	66801	06838	53938	02116	31290	35484
36010	00874	61554	03930	76813	04110	80237	56992	58885	82395
35736	49271	60789	09339	34868	09009	72530	18735	64332	03697
25323	56652	55557	19128	99012	68240	74399	63812	57213	17417
85278	21251	40588	02478	56420	99299	28412	83989	89082	83173
96114	58859	90474	81029	52698	06727	13753	62728	72986	64711
67430	57097	39476	63242	22233	20613	25136	09443	83084	22925
06964	90193	70344	46617	88997	39555	76652	34255	82586	04720
89983	69718	33004	19795	83779	90326	90185	18240	11673	70968
60718	80714	47399	25051	18095	28726	08346	98810	02209	58208
74024	90794	99100	24031	69005	12944	67732	22424	29221	79654
37033	17764	59482	12790	95015	27273	65531	20187	60790	57358
49246	59630	43635	02831	95923	14749	02529	02694	79631	97723
52978	20248	07296	53257	55027	21790	75655	87355	23715	05433
31425	39865	60729	70449	43054	85834	73832	52815	37898	25696
04402	26377	19057	86671	58288	31367	95663	70948	82881	42306
45389	43993	28279	14144	61256	47718	19017	24024	51852	85436
74187	62120	63159	75454	29647	59915	74184	47261	46941	61024
65624	31299	63494	85822	54612	25054	27475	89887	07347	55895
58428	17582	18339	13867	98191	21556	34049	42139	29373	33628
50626	40047	41078	65809	93332	05759	36533	76154	53981	96183
35706	97649	32802	26107	01466	83932	75217	07618	63642	74751
87633	10424	93235	65113	11385	23673	48544	98040	54496	92004
83126	63377	81018	51621	98097	95515	96879	16966	06685	40393
70090	05750	43225	63728	66615	68807	22251	60175	22681	73598
17382	39493	85125	52023	41381	72824	76163	52644	55398	81201
22214	57511	99807	64671	50244	12345	99347	21606	31123	72781
51139	50509	66346	75458	80228	47405	37597	12683	74525	71035
18864	71938	84707	81152	27322	80417	10993	52930	68340	51329
40108	66030	03600	77583	27036	94382	10674	09531	11786	90382

Table VIII *(cont.)*

73671	14024	60023	27499	77781	24882	08974	74437	78355	19616
98039	61503	19311	42424	75633	08699	05320	65979	59272	69347
37420	39895	56851	15194	85118	53344	30789	89912	69025	11113
07820	76675	60154	94957	21601	09421	61672	02135	05527	52573
58206	19959	83117	97179	13240	92073	28913	37428	96289	22961
14767	21201	30758	57813	54987	64030	82181	10730	75526	84880
91601	01194	26441	99002	90422	45993	79062	75742	03485	54294
05558	21504	67952	29221	49093	73514	40909	70154	93646	74071
91505	09644	12571	97018	29374	11290	86691	59185	14719	07528
57053	44385	50859	94569	45434	37209	13442	68724	15254	36454
66139	73745	82406	10660	84421	63294	35070	73689	24441	45612
72864	63608	65226	63528	49942	84330	27956	91816	09151	25254
25424	37935	31892	58685	26860	56667	22789	50107	60776	17419
58674	11080	51591	77955	45826	55351	25787	73941	25611	00800
84321	96870	71402	06902	00995	78820	66583	97800	78826	11639
64972	67978	01222	88485	24406	80484	22358	85475	71193	40522
38865	69193	46418	63155	20652	86224	09813	94505	75860	68271
83009	56449	00918	03235	19922	83015	34803	18871	72020	85579
64441	21178	72477	74735	90513	56930	52080	34665	22518	38361
40238	52052	57850	98925	57288	71574	39686	43244	02347	52801
06881	08717	46865	02900	29101	33790	24663	97946	85044	34154
16882	68979	65538	75557	56835	41527	62953	90333	70644	71044
82230	59159	32395	45840	47019	13841	44252	77755	24251	92641
19399	96711	32655	98876	80671	01892	65816	04113	68097	34291
07473	43637	63744	75681	96694	22726	67924	62888	71086	59191
86863	07776	63947	76456	64151	14424	99408	16009	24622	88450
55336	36436	41560	78385	87324	01971	92084	55574	50412	82134
36556	97347	96321	78021	17917	63101	25700	76237	68951	83865
91719	90927	71748	89341	68813	08203	66350	46571	97435	74176
46589	87112	69367	95174	30836	70207	23072	54573	52047	59532
74295	19272	40915	11323	62989	91132	66175	94913	63995	44339
62791	81122	43983	50972	43526	99654	00391	40383	19063	90971
87735	39583	18076	35850	38627	46920	03515	42168	28577	99145
54994	76693	32676	19986	70840	71433	87069	97147	93529	79573
34122	31420	70693	27459	81844	39691	31009	06028	12158	44719
08708	31868	84881	75305	16013	81716	84552	94197	65003	83178
88554	13754	11479	69967	82898	40559	46732	88057	31619	18214
70684	99215	47585	08179	19829	32573	27691	06526	65733	28646
66921	57655	07006	28685	43478	10923	94747	33908	76555	64451
38739	00438	40832	72446	10384	19560	74714	24908	43386	79331
02352	48080	43082	64395	96847	59609	20708	77350	58902	34449
31506	06020	65139	83443	77321	78833	60514	48225	74935	15465
17002	38457	60093	02787	10692	62747	69620	63843	29874	85075
48112	87933	68998	69677	67736	81815	75607	39412	83589	23887
39409	24122	36322	26662	82055	42066	16098	75558	54200	28798
32124	82210	62497	28758	32018	31359	46385	12782	98226	15402
34683	61856	84154	19815	65588	31747	68104	67329	18281	16581
74807	53942	77091	38514	34909	23288	81185	89212	75185	47550
85857	15545	54846	95054	91653	97762	97477	87331	39302	38425
45848	65584	75923	48978	01515	25680	32399	24056	23706	46324

Table VIII *(cont.)*

62954	02943	65587	93648	71459	79980	28401	13928	59801	18097
34771	92872	87561	70710	61725	51733	69911	24290	94542	85912
38156	77341	98413	37243	11140	65973	99577	50568	04874	23744
89297	47718	91704	84418	01356	54443	31515	82462	46019	82663
45927	57240	45107	56166	57669	08681	41413	28347	78919	80032
40570	07999	27709	22292	25318	36416	00968	41343	11441	97557
36258	85551	30780	86239	98738	00842	97112	37565	69096	16905
08890	32086	67167	28407	65670	74144	51482	66902	95495	65661
18862	92089	03870	05961	24762	94088	65169	63298	71837	66225
11509	37950	46818	66864	93752	31830	21265	04910	76226	73786
04436	56894	38059	46507	76027	99982	23464	50154	96990	03045
41844	69445	27560	60189	99491	15326	89247	43452	89145	28649
50441	15736	19542	99346	18367	98128	50356	30649	44313	86569
60577	16095	12464	03583	15893	86673	48034	28918	85183	92458
56618	84121	43262	58878	24560	68557	27560	35890	34463	71345
95429	95393	69931	31454	35457	32114	35408	46677	98879	59742
44733	71660	72100	53648	80716	31601	68699	92500	89609	20388
75982	38721	09036	70405	45720	84486	19044	08000	52797	87054
89593	56259	10542	09095	16192	82202	78721	25394	67440	61072
07407	39827	30592	24023	39305	04874	05378	35151	47829	37308
45595	03178	93880	03528	13480	46995	03827	50746	77007	68836
64224	62101	83374	49742	42754	84698	84332	18130	58356	58151
99801	88887	04296	63808	67534	42045	92632	05993	71937	50576
80977	62350	23849	17477	93234	78006	57630	28011	46253	61683
61000	84473	47488	27908	81003	63324	23938	54787	89310	37145
83403	08404	00973	06210	27249	21667	31618	15522	50619	12102
47782	70170	21444	06784	81758	74735	87330	63034	93470	85035
09410	79959	13061	99963	82403	75261	74390	12908	44603	73536
81719	81141	20378	61717	41156	37491	62277	75353	42642	09929
17282	86203	07730	54032	14819	88150	95263	43141	17839	77082
01652	32499	01986	42199	92277	30871	77971	64075	02461	75530
78766	97737	70189	49201	07477	20852	84061	28354	84006	92466
53504	03566	41086	70463	39285	78563	75542	89512	66436	17710
33928	22958	18041	13192	90017	32455	09310	83995	89774	69982
81214	31882	50246	18248	92838	75015	23886	88938	71607	47623
80515	88953	17652	64693	13376	13525	42778	09456	60431	79066
81117	28474	03825	05564	85545	40990	27144	66674	84105	12371
93301	29964	14781	00471	64049	62815	80048	94571	49872	91080
21876	60343	24767	40523	68568	33519	43852	17202	97087	84913
36101	24555	24691	55314	80656	87694	92042	10967	80227	78435
29654	05268	26621	51286	54016	92704	06935	92320	09244	98334
67156	93151	11560	76932	71539	99676	53773	22484	61198	83201
48812	07692	96934	83919	55687	40257	29991	51477	37996	87887
73864	86418	90413	19727	98760	07064	93709	24719	04341	36512
59359	68814	74673	44694	59545	09790	22429	43257	30728	38266
28918	42344	18538	50848	17113	43212	13437	66905	09155	83648
45271	38945	95420	60714	54523	41284	81939	14482	21121	08328
33829	86076	09452	40702	14006	30104	51181	06850	98173	38577
44997	26849	79901	76981	68225	01494	38864	71460	89319	60154
82237	86534	27153	51543	64074	41350	05753	00264	59186	67117

Table VIII *(cont.)*

61055	19392	95601	05134	18046	41780	64121	93550	39649	33440
60083	63789	90211	45898	82541	81508	40606	59318	45772	72930
41270	56184	10155	98837	22742	43211	51521	37401	48405	90061
35304	20811	84872	64215	13990	68726	58713	63220	70508	68032
44396	67803	02874	75098	34373	95812	22576	12725	59859	13371
34916	52029	11401	85583	83645	81751	16868	31536	09987	68722
93036	78065	18937	86687	74076	89317	09757	64711	46683	09876
25434	35776	97060	11204	74403	57422	45725	51088	22203	85570
79208	76636	58954	96841	27180	09644	64053	47642	26335	08783
15081	95371	05623	48156	55084	03440	25307	64497	84675	52612
08757	17635	31688	57856	73401	05251	12412	22926	54124	25266
07815	99845	12505	76385	15821	89560	07213	72199	82704	03431
67400	15167	92450	30436	91284	71474	07319	59811	54467	92567
97476	48344	49665	34800	82696	82291	10244	29821	00031	23567
63030	82195	49452	07037	03792	12562	58638	27133	94889	12372
94350	85989	51844	33801	11816	02663	68008	02883	94027	26394
64118	25484	90015	12945	68039	67110	37833	04542	06136	70344
76229	45749	28589	90287	75976	01676	72153	12585	82403	71043
79378	11934	72136	27507	98146	58896	54624	30386	98667	90016
32912	65278	11023	47545	47951	13411	48233	32799	60395	47383
77548	07273	89956	17488	70738	63990	28461	01303	78785	45222
93777	73240	73668	39353	33011	38333	87541	37552	33467	65245
81895	43691	96938	00724	56356	17257	82394	01103	77312	73088
17333	51756	35901	08608	71551	98335	57443	66045	01170	95216
75578	40499	96692	29183	26439	14270	96884	74260	74574	56516
16295	07137	40349	77523	03855	32436	74834	73273	86439	10320
79184	08083	42323	17978	41346	24633	32571	39487	17751	17431
30872	53665	24217	38904	22479	96079	27256	44947	28350	42300
12814	19353	98187	95392	97575	75803	65002	25052	39324	07507
78618	69938	55813	69077	72005	08206	67849	98574	45854	98822
94267	92158	02377	94677	37946	15240	93618	78506	58548	19846
08925	21577	63013	99028	41524	56018	14442	63606	83649	73110
69171	38625	04619	14563	42181	68863	80121	31294	05679	10391
24446	26696	87113	89227	84813	17508	08895	14220	11682	77046
63750	54601	21686	24373	33100	29039	49456	37504	82249	57252
26572	17339	64805	10579	16136	58305	73264	89988	30841	42318
98108	99447	14133	22402	33969	83233	83719	34379	03128	86872
50861	64413	83292	36025	20098	88603	44523	27205	54285	52264
50843	12994	63844	49487	08136	16682	25951	30401	90741	45528
98743	06068	66632	64101	52378	71232	09629	18732	36568	27904
09632	28848	11055	50443	49486	01388	19730	32161	15167	41926
28516	97385	27275	94299	38571	11238	34219	19949	76664	59880
47252	00624	80635	73592	49631	49945	94378	87716	89944	52491
83095	36850	73787	22163	13226	31233	40905	18738	73609	62624
52763	88253	32168	12097	01578	95421	43522	47895	52620	22791
94520	82996	26493	54232	15162	20562	28570	25919	03136	56601
33428	81309	16910	74161	10050	89018	76604	30681	73618	45262
26843	48344	58005	50447	51427	49247	95442	53365	15717	16078
37035	92880	93900	61357	84552	29436	94784	23916	80554	64522
37521	11597	77500	57284	53226	37389	39783	82583	87850	75927

Table IX Normal random numbers.

(Mean 0, standard deviation 1)

.154-	.442	.830	1.585-	1.086-	.544-	1.562	.644	.445-	.115-
.047	.999-	1.968	.829	.500	.866-	.354	.122	.051-	1.472
1.420	.262-	2.206	.903	.232-	.562	.580-	.008	2.664-	.265
1.063-	.416	.042	1.277-	.226	.693-	.022	1.353-	.408-	.229
1.936-	.723	.828	.295-	1.975	.836-	.942	1.009-	1.725-	.911-
1.442-	.269-	1.131-	.060-	.484-	.483	1.512-	.808-	.884	.400
.968	1.625-	.604	.504-	.068	1.520	.350-	.514-	.467	1.181-
1.781-	.341	.050-	.936	1.053	.524	.746	1.214-	.713	.767
.584-	.647-	.536	1.103	.176	.928	1.411-	.700	.288-	1.071
.108	1.618	.580-	.475-	.048-	.169	.849	2.463-	1.069	1.002
.825	.614-	1.538	.478	.158	.498-	1.770-	2.178-	.990-	1.450
.922-	.789-	.030-	.863	.253	1.712-	.580-	.408-	.472	1.959-
.156	.177-	.943	.324-	.751	.323-	1.391	1.620-	.488-	.516
.681	1.020	1.792-	.847	.391	.826-	.021	.073-	.614	.413
.137-	.868	.569-	.135-	.002-	.405	.163-	.193-	.422	.472-
1.198-	.675-	.107-	.372-	.802-	.716-	.097-	.415-	1.818	1.422-
.603-	1.598	.066-	.781-	.603	2.154	.648	2.356-	.463	.073-
.394	2.094-	.972	.192-	.247-	.406-	1.004	.222-	.032-	.970
.120-	1.907	.513-	.485	.373-	1.620	.628	1.108-	.268-	.801
.770	.869-	.354-	1.963	.280	1.793-	.564	.269-	1.506	.863-
.110	1.452-	1.221	1.442-	.982-	.147-	.874-	2.111-	.029	.467-
1.370-	.166	.398	.302	.028-	1.245-	.210	1.969	.709-	1.119-
.333-	.277-	.462	.140	.910	.877	.229	.389-	.052	2.026-
.573-	.159-	.418-	.039	.161	.465	.968-	.780	.566-	1.113-
.964-	.355	.219-	1.671	.680	.681-	.424	.984-	.698	.454
.435-	.392	.517	1.056	1.213-	1.293	.014-	.440-	1.107	.131
.629-	1.205	1.494	1.239	1.225	1.041-	.480-	1.176	1.585-	.582-
.344-	.723	1.071-	1.489	.584	.577-	.251-	1.278	.282	.396-
.435	1.541	1.502	.049	1.333-	.242	1.908	1.041	.692-	.231-
1.319-	1.352-	1.907-	.164	.372-	.233-	1.476	1.451	.379-	.850
1.454-	.566-	.684	.011	.529-	.172	.804	.217	.050-	.711
2.288-	.247	.030-	.034	.236-	1.205	1.788	1.093	1.320	.909
1.005-	.430	2.943	.327	1.175	.455-	.347-	1.503	1.816	.110
.973-	1.239-	.034	.028-	.308-	1.814	.399	.288-	.256-	1.752-
1.807-	1.679	.188-	.005-	.012	1.748-	.101	1.515-	.892-	.727-
.626	.387-	.384-	1.046	2.039	.135-	.961	.196-	.195	.407-
.051	.167-	2.701	.552-	.333	.375	1.133	.158-	1.163	1.847-
.351	.924	.177	1.307-	1.500-	.153	.089-	.418-	1.458	1.027
.967-	.112	2.056-	1.230-	.284-	1.539	1.024	.470	.704-	.610
2.204-	1.527-	.334-	.389-	.017-	.269	.690-	.770-	.282	.775-
.006-	.586-	.302	.562	1.570-	.506	1.003	1.440	.909-	1.241-
.234-	1.003	.900	.259	1.847	1.910	.476	.177	.408-	.176
.364	1.959	1.403	.879	.321	1.552	2.414-	1.633	.778-	.505-
.690-	.517	.222-	.019	1.876-	1.069-	.392-	.430	.479-	.061-
.406-	.489-	.486-	1.599	.776	.170-	.023-	.687-	1.271-	1.011
.259	.180-	1.518-	.325	.287	.036	.095	.395-	.929-	.312
.339-	.570-	.972-	.427-	.384-	1.411-	2.808	.735	.080	.219-
1.568	.425-	.182-	.796-	.125	.158-	.216	.065-	1.154-	.206-
.348-	1.580-	1.644	1.123-	.045	.390-	.234	.100	1.135	.547-
1.000-	.019	1.671	1.866-	1.333	.751-	.423	1.633	.606	1.479-

Table IX *(cont.)*

1.644	2.770-	1.174-	.275	.740	.368	.261	2.177-	.278-	1.071-
.594	.552	.228-	.562	1.305-	.425-	1.266-	1.209-	1.303-	.697
.673	1.228-	.426	.587	.990	.877	.954	.416	.681-	.913-
.636-	.711-	.457-	.253-	.893	1.411-	.066-	.251	1.551	.656-
1.586-	1.384-	.481	.763	.463	.228-	1.062	.617-	2.035-	1.241
.450	.126	.217-	.475	.933-	.209-	.497-	1.031	1.154	1.142
1.456	1.557	.385-	.078	1.864-	.587-	.872	.423-	.661	.230-
.907-	.721	1.349	.048	.261-	2.031-	.255-	.010	.753	.175
.868	1.318-	1.365	.146-	1.882-	1.164-	1.199-	.406	.463	.406
.675	1.188-	1.066	.498-	1.186	.904	.128	1.316-	1.012-	.930
.255	.214	.858	.815-	.498-	.512	.674	.239-	.281-	.023
.516	.574	.567-	.198	.643	.642	2.570	.028	.016	.245-
1.412	.411-	.162-	1.122	.028-	.447-	.272	.003-	1.408-	1.201
.372-	.321-	.341	1.783	1.399-	.068	.430	.066	.554	2.163
.678-	.472-	.774	.813-	.281-	.547-	.802-	.012-	.570-	1.627-
.495	.340-	.266	.667	.205	.787	1.425-	.217-	.646-	2.256
1.825	.264	1.850	.237-	.812-	1.192	.536	1.438	.663	.356-
.432	2.290-	.458	.499-	1.498	.650-	1.560-	.030	.555	1.905
.272-	.147-	.204-	.404	.248-	1.087-	.347-	1.041-	1.252	.392
.188-	.613-	.780-	.147	.219	.028-	.566	.396	1.442	3.074-
.779	.076	.749	.889	1.379-	1.806	1.729	2.029-	1.001-	.435-
.488-	1.553-	1.939-	.278-	1.507	.121	1.773	.149-	1.204	.570
.736	.796-	1.218-	.444-	.213	.682	.518-	1.433	.171-	.241
.476-	.124	.996-	1.243	1.797-	.175	1.408-	1.625	1.029	1.176-
.735-	.939-	.581	.757-	.925	1.093-	.188	2.089	1.090	.977
1.663	.402-	1.089-	.205-	.185	.284	1.036-	.176-	.540	.070-
.519	2.586	.110	.190-	.082	.121	.549	.323-	.400	1.947
.762-	.428-	1.047-	1.061-	.320-	1.130-	.010-	2.039-	1.802	.546
1.394	1.687	1.698-	.477	1.137-	1.230	.399-	2.187-	.359	.933-
.892-	.116-	.088	.272-	.612-	.323-	.399-	2.119-	.902	1.067
1.915	.912	.404	.818-	.700	1.202	.067-	.111-	1.110	.183
.622-	1.466	.586-	.427	.707	.392-	.417-	.923	2.540-	.885-
.570	.166-	.195	.012	.232	1.404-	1.212	.218-	.679	.667-
.079-	.434	1.392-	1.059	.188-	.117	.397-	2.563	.097-	1.550
.495	1.336	.800	.915-	.721-	1.479	.963-	.455	.589	.223-
.938	.084-	2.066	.587-	.542	.622	.002-	.184	.735-	.259-
.618-	.653	.784-	.604-	1.793	.959-	.737	.181	1.687	.971-
.434	.143-	.191-	1.115-	.357-	.329	1.199-	.710-	1.275-	.948-
.617-	.292	.989	1.192-	1.436-	1.174-	.258	.382	.568	.176-
.976	1.681-	.077-	1.606	.269-	.349-	.851-	.322	1.343-	.080
.515	.077	.407-	.291	.127-	1.597	1.146-	1.264-	1.053-	.166
1.395	.955	.238	.342-	.128-	1.532	.276-	1.114-	1.295-	2.809
1.623	.718-	.504-	.616	.347-	.455	1.131	.514	.125-	.148
1.515-	.447	.993-	.422	.740	2.455	.717-	.850	2.521-	.032-
.303-	1.927	.041	.348-	.601-	2.569	.859	.877-	.555-	.681
.186-	.937	1.343-	.358-	.241-	.886-	.679	.548	.379-	.229
.695	1.022	1.188-	.375	.398-	1.368	1.786	.108-	.284	.224
.219	1.445-	1.033	.070	1.076	.329	1.567	.208	.229-	.630-
.015-	.583-	.466	1.379	1.430-	1.059	1.840-	.246-	1.047	.333
2.019-	1.284	.713-	.642	1.824	.868-	1.590-	.920	1.555	.039

Table IX *(cont.)*

1.545	2.142-	.522-	.260-	1.219-	.338	1.499-	.372	.755	.266-
.729	.085	.193	.058-	2.456	.003	.522-	.713-	1.950-	.006-
1.703-	1.555-	.738	.722-	.580	.200-	.701	.746-	.474-	.356-
1.187	1.448	.334-	.358-	.435	.539	1.283-	1.032	.927-	1.642
.782-	1.088-	2.384-	.026-	1.300	.429-	.381	.042-	.361	.642
.797	.458	.454-	.632-	.159	1.577	.295-	.345-	2.082	1.291
.947	.237	.394-	.020	.562	.243-	.703	.895-	.013-	.696-
2.716	.088	2.029	.584-	.370-	.170-	1.170	.704	1.735-	1.086
1.399-	.829-	.437	.663-	.120-	.539	.351	1.715-	1.991	.327
.017-	.354	1.547	.532-	.171-	.949-	1.507	.114-	.963-	.373-
.851-	.676	.802-	.222	1.545-	1.193-	1.163	.330	.569	.525
1.384-	.247	.522-	1.320	.858-	1.749-	.561	.220-	.926	1.085-
1.521-	.160-	.220-	.082-	.340-	.022-	2.451	1.383	.914-	.040
.904-	.794-	.430-	.510-	1.143-	.381	1.431-	.914	.253	1.583
1.956	.789	.780	1.848	.579-	.001-	1.248	.428	.256	.712
.254-	.973-	.453	.478	.775	.219	.831	.322-	2.338-	1.162-
.583-	.916	.123	.231	.794	.317-	.469	.898-	.452	.453
.109-	2.324	.001-	.354-	1.709-	.213-	.528	2.625-	1.585	1.475-
1.104-	.039-	.166-	.826-	1.799-	.183-	.632	1.673	.024-	.584
.587-	.152-	.660-	.829	.805-	.609-	.532-	.033-	.037-	1.397-
.228	.597-	1.325	.015	.177-	1.966-	.513	.293	.830-	.434
.138	.856	.468-	.519-	1.046	1.373-	.765-	.226-	1.032	.366-
.695-	.365-	.169-	1.067-	.246	.305-	.327	.571-	1.013-	1.092-
.398-	1.247-	1.026-	.413-	1.296-	.195	.022-	.284	.471	1.636-
.011-	1.128-	.288-	.066-	1.156	1.867-	.652	.862-	.447	1.191
2.376	.184	.260-	.247-	.673-	.947	1.184-	.126	2.627	1.029
1.657-	.779	.182-	.966	.297	.109-	.124	1.741	.473-	.396
.015	.682	.868-	.351-	.086	1.090	.226	1.173-	.108	.029-
.516-	.526	1.870-	.279	1.192-	.742-	.989-	.304-	.440-	1.434-
.575-	.389-	.027-	.148-	.595-	1.387	1.398	.054	.850	1.393-
.956-	.160-	.730	.023	.968-	.348	1.101-	.278-	.816-	1.803
1.645-	.734	2.447	.687	1.624	.391-	.130	1.567	1.042-	.846
.656-	.116	1.034-	1.284	1.446-	1.679	.302	.387-	.148	.570
.342	1.472-	.412-	.581	1.215	.787	1.593	2.945-	1.025	3.010-
1.360-	2.107	.427-	1.352	.539-	1.100	.882-	1.347-	.190	1.009
.001	.373	.399	.319	.750-	1.790	.666	1.674	1.034	1.018
.555-	.077	.836	.710	1.946-	.985	.695-	.287	.246-	.545
.762-	1.184-	1.279	1.568	.213	.167	.402-	2.180-	.012-	.024-
2.195	.074	.726-	.743	.844-	1.185	.455	.639	.658	.691
.914-	.553-	.263-	.529	.671	2.515-	1.445-	1.760	1.030-	.809
1.715-	.254	1.486-	1.295	.134	1.358-	.149-	1.173-	.485	.952-
.436	.313-	.117	.042	.027-	.203	.642-	1.694-	.409	.241
.530-	.708-	.056-	1.722	.328	.831	1.013-	.042	.021-	1.940-
.633-	.695	.080	.706-	.182-	.221-	1.271-	.285	1.588-	1.373
.117-	.389-	.886-	.409-	.501	1.589	.181	1.212	.253	.351
.650-	1.327-	.077-	.212	1.385	.272-	.101	.204	.960-	.602
.403-	.803-	.233	.850-	1.797-	.677-	.765-	1.332	.359	.126
.406	.666	.138-	1.018	1.484	1.010	.850	.258-	1.105	1.908
.228-	1.138-	.934-	.084	.751-	.480-	.287	.692-	1.328	.225-
1.002	.018	1.401	.612-	.344	.198	.566	.479	1.168	.780-

Table IX *(cont.)*

2.689	.408-	1.144-	.234-	1.212	.600	.846-	.485	1.775	2.397-
.287	1.442-	.747	.223	2.070-	.624-	.828	1.147	.141-	.841
1.136	.706	.221-	1.333	.575	.195-	1.386-	.497	1.338-	.986
1.482-	.447-	1.145-	.675-	1.619	.677	2.443	.588-	.497	1.297
1.107-	.503	1.831	1.497	1.533	1.367-	.731-	1.259	1.534	.276
.110-	.356-	.298	.105	.069-	.683	1.485	.368	2.496	2.025
.614	1.007-	.198-	1.510-	.794-	1.119-	.279-	.380-	.262	.191
1.385-	.869	.483	1.207-	.150	.596-	.818	.671-	.192	1.265-
.636-	.610	.853-	2.817	.331-	.959-	.603	.823-	1.900	1.559
.492	1.721-	.561-	1.172	.334-	.721-	1.729-	1.328-	.829	.298-
.698-	.985	1.780	2.934-	.298-	.446-	2.239	.585	.602-	.410-
.055	1.740-	1.338	.741-	.131-	1.071-	2.042	.326	1.409	2.189-
.487	.024	.755-	.158	.066	.480-	1.134	1.081	.549-	.083-
.087-	.163-	.240-	1.032-	.071-	1.197	1.609-	.520-	.427-	1.236
.599	.299-	1.346	1.099-	.696	1.550-	.295-	.544-	1.129-	.532
.157	.019	.112-	1.665	2.587-	2.037	.610-	.335	.433	1.021-
.405	.918	.483-	1.311	.051-	.709	.802	.545-	3.066	.677
.717	.445-	.212	1.463	.040-	.522	2.481-	1.069	1.124	.763-
.459-	.074	.355	.393-	1.198	.761	1.343-	.730-	.915-	.279-
.955	2.253	1.188	2.261-	.147	1.615-	.074	.196	.981-	1.087-
.092-	.669-	2.217-	1.110-	1.085-	.386	.843	.364-	.011-	.775-
.453	.223	.365	.523	.039-	.262-	1.517	.419-	.563	.153-
.777-	.441-	.913-	.314	.276	1.424-	.804	.824	.210	.627-
.174-	.024-	.102-	.629	.576	1.196-	.198	.149	.737	1.224
.122	.192-	.826	1.677	.669	.906-	1.366	.614	.813-	.815
1.027	.511-	.032-	1.117	.702-	.773-	1.259-	.333	1.920-	.667
.684-	.714	.715-	.754-	.715-	1.385-	.626-	1.440	.860	1.114
.185-	1.151-	.648-	.506	1.982-	1.030	.642	.671	1.196-	2.169-
.002	.726	.143	.703-	.117-	.053-	1.074	.508	.447	1.314-
.227-	.261	2.677	.852-	.551-	2.044	1.130	.860	.488	.218-
.736-	1.046	.060-	1.737-	.188	.515	.637	1.945	1.504-	.148-
.093-	.093	.321-	.810	.535-	1.897-	1.031	.138-	.644	.214
.418-	.022-	.581-	.192	.093	.801-	.520	.056	.875-	.774-
.343-	.743	1.041	1.299	.828	.340-	.003	1.084-	.248-	.563-
.698	.998-	.909-	.568-	2.846-	1.132-	1.210	1.521-	.869	.154
.268	.769	1.563	.143-	.276-	.052-	.462	.916	1.556-	.789
.209	.215	.960-	1.161	.277-	.861-	.745-	.362-	.086-	.032
2.246	.514-	.407	.851-	.139-	.017	1.831-	.339	.282-	1.161
.275-	.181	.208	1.691-	.504-	1.134-	1.317-	.288	1.749-	.789
1.395-	.358-	1.356	.268	.733-	.525-	.030	1.952	.053	1.143-
.435	.172-	.593	.486-	.372	1.887	1.233	.118	1.103-	.422-
1.386-	.941-	.947	1.598	.345-	.367	1.802	.543-	.905	.696-
1.127-	.030	.488-	.984	.567	1.632	.590	.914-	.615-	2.176
.147-	1.567	.344	1.530	.506-	.919-	.313	.663-	.163	.011-
.353	.782-	.915-	1.210-	1.279-	.718	.024-	1.220-	.559-	1.371-
.274-	1.364	.249	.736	1.351	.750	.467-	1.220	.918	.912-
.254-	1.156-	.809	1.196	2.020	2.101	1.732	1.001	.623-	1.012-
.925-	.726	1.160	.520-	1.475-	1.352	1.735	.020-	1.244	.482
1.719-	1.263-	.357	1.114	1.437-	.866-	.268	.805	.219-	1.488
1.820-	.351	2.012	.297-	.321-	.450-	.689-	1.942-	.348-	1.060

Table X Squares and square roots.

n	n^2	$\sqrt{n}$	$\sqrt{10n}$	n	n^2	$\sqrt{n}$	$\sqrt{10n}$
1.0	1.00	1.000	3.162	5.5	30.25	2.345	7.416
1.1	1.21	1.049	3.317	5.6	31.36	2.366	7.483
1.2	1.44	1.095	3.464	5.7	32.49	2.387	7.550
1.3	1.69	1.140	3.606	5.8	33.64	2.408	7.616
1.4	1.96	1.183	3.742	5.9	34.81	2.429	7.681
1.5	2.25	1.225	3.873	6.0	36.00	2.449	7.746
1.6	2.56	1.265	4.000	6.1	37.21	2.470	7.810
1.7	2.89	1.304	4.123	6.2	38.44	2.490	7.874
1.8	3.24	1.342	4.243	6.3	39.69	2.510	7.937
1.9	3.61	1.378	4.359	6.4	40.96	2.530	8.000
2.0	4.00	1.414	4.472	6.5	42.25	2.550	8.062
2.1	4.41	1.449	4.583	6.6	43.56	2.569	8.124
2.2	4.84	1.483	4.690	6.7	44.89	2.588	8.185
2.3	5.29	1.517	4.796	6.8	46.24	2.608	8.246
2.4	5.76	1.549	4.899	6.9	47.61	2.627	8.307
2.5	6.25	1.581	5.000	7.0	49.00	2.646	8.367
2.6	6.76	1.612	5.099	7.1	50.41	2.665	8.426
2.7	7.29	1.643	5.196	7.2	51.84	2.683	8.485
2.8	7.84	1.673	5.292	7.3	53.29	2.702	8.544
2.9	8.41	1.703	5.385	7.4	54.76	2.720	8.602
3.0	9.00	1.732	5.477	7.5	56.25	2.739	8.660
3.1	9.61	1.761	5.568	7.6	57.76	2.757	8.718
3.2	10.24	1.789	5.657	7.7	59.29	2.775	8.775
3.3	10.89	1.817	5.745	7.8	60.84	2.793	8.832
3.4	11.56	1.844	5.831	7.9	62.41	2.811	8.888
3.5	12.25	1.871	5.916	8.0	64.00	2.828	8.944
3.6	12.96	1.897	6.000	8.1	65.61	2.846	9.000
3.7	13.69	1.924	6.083	8.2	67.24	2.864	9.055
3.8	14.44	1.949	6.164	8.3	68.89	2.881	9.110
3.9	15.21	1.975	6.245	8.4	70.56	2.898	9.165
4.0	16.00	2.000	6.325	8.5	72.25	2.915	9.220
4.1	16.81	2.025	6.403	8.6	73.96	2.933	9.274
4.2	17.64	2.049	6.481	8.7	75.69	2.950	9.327
4.3	18.49	2.074	6.557	8.8	77.44	2.966	9.381
4.4	19.36	2.098	6.633	8.9	79.21	2.983	9.434
4.5	20.25	2.121	6.708	9.0	81.00	3.000	9.487
4.6	21.16	2.145	6.782	9.1	82.81	3.017	9.539
4.7	22.09	2.168	6.856	9.2	84.64	3.033	9.592
4.8	23.04	2.191	6.928	9.3	86.49	3.050	9.644
4.9	24.01	2.214	7.000	9.4	88.36	3.066	9.695
5.0	25.00	2.236	7.071	9.5	90.25	3.082	9.747
5.1	26.01	2.258	7.141	9.6	92.16	3.098	9.798
5.2	27.04	2.280	7.211	9.7	94.09	3.114	9.849
5.3	28.09	2.302	7.280	9.8	96.04	3.130	9.899
5.4	29.16	2.324	7.348	9.9	98.01	3.146	9.950

Answers

Chapter 2

2.1.

x_i	11	12	13	14	15
f_i	3	3	10	8	1

2.2.

x_i	8	9	13	14	15	16	17	18	19	20	21	22
f_i	1	1	4	3	2	4	4	13	4	2	1	1

2.3. (a) 25

2.4. (a) Age is a continuous variable.
(b) Class intervals $n - \frac{1}{2}$ to $n + \frac{1}{2}$, where n is an integer

2.6. Categories ordered but not necessarily of the same width

2.7. The first interval might be taken as 10–14, and the last as 40–49 (unless there are births at ages under 10 or over 50). The height of the last bar would need to be adjusted.

Chapter 3

3.1. (a) The set of nonnegative integers (0, 1, 2, . . .)
(b) The 52 possible cards
(c) The list of voters used

3.2. Elements in the sample space: ace, face card, anything else

3.3. (a) $\frac{11}{20}$ (b) $\frac{7}{20}$ (c) $\frac{4}{20}$ (d) $\frac{15}{20}$

3.4. Set of real numbers; set of integers (1, 2, 3, . . .)

3.5. (a) .53 (b) .28 (c) .64 (d) .17 [*Note:* (c) = (a) + (b) − (d)]
(e) .14 (f) .14 = .08 + .04 + .02

3.6. $\frac{30}{155}, \frac{11}{155}$

3.8. $\frac{1}{19}, \frac{9}{19}, \frac{9}{19}$

3.9. (a) $\frac{1}{6}, \frac{1}{6}$ (b) $\frac{2}{3}$

3.10. (a) $\frac{2}{3}, \frac{1}{4}$ (b) $\frac{11}{12}$ (c) $\frac{7}{12}$

3.11. (a) $\frac{1}{52}$ (b) $\frac{1}{4}, \frac{1}{13}, \frac{3}{13}, \frac{11}{26}$

3.12. (a) $\frac{78}{1326}$ (b) $\frac{32}{663}$ (c) $\frac{40}{663}$

3.13. (a) $p + q + r + s + t$ (b) $c + r + w$
(c) $a + b + d + e$ (d) $1 - e - t - y$

3.15. (a) .2 (b) .8 (c) .4

3.16. (a) $1 - q, t + u, u + q$ (b) $q + r + s + t + u = 1$

3.17. (a) .977 (b) .023 (c) .023 (d) .046 (e) .910

3.18. (a) $P(|\text{difference}| < .1) = .19$ (b) .25

3.20. (a) .975 (b) .025 (c) .050 (d) .90

3.21.

x_i	0	1	2	3	4
f_i	2	2	1	4	1

3.22. (a) .84 (b) .84 (c) .68 (d) .05

3.23. (a) $C + D + E$ (b) $A + B + C$ (c) C (e) $A + E$

3.28. (a) $\frac{1}{8}, \frac{3}{8}, \frac{3}{8}, \frac{1}{8}$

2pm
Set
Band

Sophomore Summer - Mobile Powerwash (apr - oct)
Junior Summer - Stadium, Landscaping
Senior Year - Farrell's (oct-aug)
Freshman + Soph Year - Pop's (aug - may)
2nd Soph - Tiffany Corp. Wis Packing Co, Sentry
Soph Summer - Paul Martin's (may →)

200

240 → Indy

390 → Columbus

500 Colo. → SRC

K-8 Grace Lutheran
9-12 Fall's North
F - Marquette, UW Waukesha, UW Madison
1/2 soph - Madison

Chapter 4

4.1. (a) 12 (b) 11 (c) 12
4.2. 4.418
4.3. 1.32
4.4. 172.5
4.5. (b) 3.13 (c) 82.3
4.6. (a) 4.125 (b) $\frac{1}{5}$
4.7. (a) 7.093 (b) 13.52
4.8. (a) $\frac{28}{9}$, 3 (b) 3.65 (with n), 3.87 (with $n-1$) (c) 12
4.9. $S = .389$ ($\overline{X} = \$2.756$)
4.10. \$1.42, \$2.59
4.11. 1.81
4.12. $\overline{X} = 13.52$, $S_x = 1.841$
4.13. $\overline{X} = 13.04$, $S = 1.038$
4.14. 5.45, 1.66
4.15. $\overline{X} = 16.65$, $S = 2.88$
4.16. $\mu = 4.5$, $\sigma = 2.87$
4.17. $\mu = \frac{10}{13}$, $\sigma = 1.31$
4.18. (a) \$4.74 (b) \$.395
4.19. $\mu = 7$, $\sigma^2 = \frac{35}{6}$ (mean of sum = sum of means; variance of sum = sum of variances)
4.20. $\mu = \frac{83}{24}$ ($\overline{X}$ will depend on your sample)
4.21. $\mu = 5$, $\sigma = 10/\sqrt{12}$
4.22. (b) .023 (c) .159 (d) 712.7 hr
4.23. .994
4.24. (a) 95.5 (b) 433.5
4.25. (a) \$12,000 (b) 70 per cent (c) \$10,000
4.26. (a) $\mu_X > \mu_Y$, $\sigma_X = \sigma_Y$
(b) $\mu_X = \mu_Y$, $\sigma_X > \sigma_Y$
(c) $\mu_X > \mu_Y$; $\sigma_X = \sigma_Y$

Chapter 5

5.3. .0062, .106
5.4. .110
5.5. .036

Chapter 7

7.1. 3.083, .294
7.2. 122.6, 4.77

7.3. (a) 16.52, 3.37 (b) 16.88, 4.03
7.4. 100 or more
7.5. (a) 7.093, .177; .970, .125 (b) 1.15, .175
7.6. $24 \pm .052$, $24 \pm .082$
7.7. (10.15, 11.61)
7.8. (a) Narrower (b) Wider
7.9. (a) 1600 (b) 67.3 per cent
7.10. $\alpha = \frac{1}{2}$, $\beta = 0$; yes, accept H_O no matter what.
7.11. (a) $\alpha = .0026$, $\beta = .014$
(b) Reject for 18 or more, $\beta = .004$
(c) $\alpha = .0012$ (rejecting for 21 or more) ($\beta = .11$)
7.12. (a) 13 (b) 5 per cent (c) Two-sided
7.13. (a) $Z = -4$; reject H_O at usual levels of 5 or 1 per cent.
(b) $63 \pm .98$
7.14. (a) $Z = -2 < -1.96$; reject H_O at $\alpha = .05$.
(b) .023 (c) .84, (.036)
7.15. $Z = -3.2 < -1.645$; reject H_O in a one-sided test.
7.16. $Z = -1.67 < -1.645$; barely significant at 5 per cent level
7.17. $Z = 2.5 > 2.33$; reject.
7.18. (a) Measurement should be normal (or approximately so).
(b) $\alpha = .21$ (c) Reject if $|\bar{X} - 20| > .31$. (d) .397
7.19. $T = 1.59 < 1.782$ accept $\mu = 45$.
7.20. (113.62, 131.55)
7.21. (6.726, 7.460), 360 (approx)
7.22. $T = 1.85 < 2.064$; accept $\mu = 30$.
7.23. (34.03, 41.97)
7.25. $Z = 2.15 < 2.33$; not significant at 1 per cent
7.26. $Z = 7$, very strong evidence that the average male salary exceeds the average female salary. (In a real-life case, sampling would not likely be needed, since university records would provide the "population" average.)
7.27. $Z = 1.5$: not very significant
7.28. $Z = 2.57$ (= 99.5 percentile); if test is one-sided, reject H_c at $\alpha = .01$
7.29. $T \doteq .8$; not significant at ordinary levels
7.30. $T \doteq .94 < 1.30$; not significant at $\alpha = .10$
7.31. (a) $R_B = 45 < 52$; reject H_O.
(b) $T = 2.78 > 1.76$; reject H_O.
7.32. (a) $T = 2.26 > 1.73$; reject H_O at 5 per cent level in one-sided test.
(b) $R_T = 78.5 < 83$; reject H_O at $\alpha = .05$ (one-sided).
7.33. (a) $T = .90$ (4 degrees of freedom); accept H_O (assume differences normal).

(b) $R = 4 > 2$ = critical value (18.8 per cent; two-sided)

7.34. (a) $T = 2.47 > 1.533$; reject H_O.
(b) $R = 1$; reject H_O.

7.35. (a) Need variance information of some sort
(b) Z at least as big as $3.4 = .5/(1.37/\sqrt{88})$; reject H_O.

7.36. (a) $T = .7316$; accept H_O (at $\alpha = .20$, two-sided).
(b) $R = 12.5 > 8$; accept H_O (at $\alpha = .20$, two-sided).

7.39. $D_n = .21 < .325$; accept at $\alpha = .10$.

7.40. $D_n = .092 < .22$; accept at $\alpha = .10$.

Chapter 8

8.1. .0013

8.2. .11

8.3. .45, .025

8.4. (.78, .82)

8.5. (a) (.09, .21) (b) (.131, .169)

8.6. Standard error about .062

8.7. Yes, the 95 per cent confidence interval includes .50: $.49 \pm .026$.

8.8. (a) $Z = 2.53, 1.645$; reject H_O at 5 per cent level.
(b) $Z = .798$; not significant

8.9. $Z = 3.0$; significant, even at $\alpha = .01$

8.10. $Z = 2.16 > 1.96$; reject H_O at $\alpha = .05$.

8.11. Population differences could be 0 (the sample difference of .03 is only 1.34 standard deviations away from 0).

8.12. $Z = 1.45$, not significant at 5 per cent

8.15. $\chi^2 = .64 < 9.49$; accept at $\alpha = .05$.

8.16. Z^2 in (b) $= \chi^2$ in (a)

8.17. (a) $\chi^2 = 15.4$, (b) $T \doteq 4$

8.18. (χ^2-value is square of Z-value obtained earlier.)

8.19. $\chi^2 = 49$; reject

Chapter 9

9.1. $\chi^2 = 44.3 > 3.84$; reject independence at $\alpha = .05$.

9.2. $\chi^2 = 13.4 > 9.49$; they are related ($\alpha = .05$). (From the nature of the data it would seem to have been collected as a two-sample problem of comparing populations.)

9.3. $\chi^2 = 48.8 > 9.49$; reject independence at $\alpha = .05$.

9.4. $\chi^2 = 46.2 > 12.6$; reject independence at $\alpha = .05$.

9.5. (a) $\frac{5}{13}$ (b) $\frac{5}{13}$ (c) $\frac{1}{2}$ (d) $\frac{1}{2}$ (The event "honor" and the event "heart" are independent, as are also the event "black" and the event "honor.")

9.7. $r = .80$

9.8. $r = .67$

9.9. Heavier linemen tend to have higher numbers.

9.10. (a) Persons of ability tend to acquire education.
(b) The larger the village, the more places for stork nests and the more babies.

9.11. $r = .49$

9.15. mse = 1.87, using regression line (and = 9.025 using μ_Y); ideal reduction factor .36

Chapter 10

10.1. (b) $3x + 14y = 16$ ($a = \frac{8}{7}$ and $b = \frac{3}{14}$)
(c), (d) $\frac{25}{42}$

10.2. (a) $y = 1.15 + .13x$
(b) 3.1 (but there may be an effect of diminishing returns that would invalidate the assumption of linearity)

10.3. (a) Slope = 2.28 days/decade
(b) Slope = .1 day/decade. (It would be foolish to draw a conclusion about changing climate.)

10.4. $y = 56.6 + 19.8x$

10.5. 17.4 to 22.2

10.6. 26.9 (rms error = 1.3)

10.7. -1.64

10.8. $T = .95$; accept $\beta = 0$ at ordinary levels.

10.9. (a) .36 (b) Predict $Y = 3.15$ (rms error = .29).

Chapter 11

11.1. (a)

	SS	d.f.
Factor	36	1
Error	4	2
Total	40	3

$$F = \frac{36/1}{4/2} = 18$$

(b)

	SS	d.f.
Factor	1734	1
Error	336	4

$$F = \frac{1734/1}{336/4} = 20.6$$

11.2. $T = 4.54 > 2.78 = t_{\text{crit}}$
$F = 20.6 = (4.54)^2 > (2.78)^2 = 7.7 = F_{\text{crit}}$

11.3.

	SS	d.f.
Factor	168.6	3
Error	2956	76

$$F = \frac{168.6/3}{2956/76} = 1.44 < 2.73$$

(Accept H_O at the 5 per cent level.)

11.4. F-test is not really proper.

11.5.

	SS	d.f.
Factor	2.46	3
Error	6.50	8

$F = 1.1$ (accept H_O).

11.6.

	SS	d.f.	F
A	144	1	$\frac{144}{36} = 4$
B	4	1	$\frac{4}{36} = .111$
Error	36	1	—
Total	184	3	

11.7.

	SS	d.f.	F
Brand	2.46	3	2.10
Make	4.16	2	5.33
Error	2.34	6	—

Accept no brand difference (2.1 < 4.76).
Reject no auto-make difference (5.33 > 5.14).

11.8. (a) $\mu = 10, \gamma_1 = -1, \gamma_2 = 1, \delta_1 = -3, \delta_2 = 3$
(b) Ratio is infinite.
(c) Ratio is undefined (0/0).

Index